ECOLOGY AND DESIGN
IN LANDSCAPE

ECOLOGY AND DESIGN IN LANDSCAPE

THE 24TH SYMPOSIUM OF
THE BRITISH ECOLOGICAL SOCIETY
MANCHESTER 1983

EDITED BY

A. D. BRADSHAW
Department of Botany,
University of Liverpool

D. A. GOODE
Greater London Ecology Unit,
County Hall, London

E. H. P. THORP
18 Leegate Road, Heaton Moor,
Stockport SK4 3NH

BLACKWELL SCIENTIFIC PUBLICATIONS
OXFORD LONDON EDINBURGH
BOSTON PALO ALTO MELBOURNE

© 1986 by
Blackwell Scientific Publications
Editorial offices:
Osney Mead, Oxford, OX2 0EL
8 John Street, London, WC1N 2ES
23 Ainslie Place, Edinburgh, EH3 6AJ
52 Beacon Street, Boston
  Massachusetts 02108, USA
667 Lytton Avenue, Palo Alto
  California 94301, USA
107 Barry Street, Carlton
  Victoria 3053, Australia

First published 1986

Printed at
The Alden Press, Oxford
and bound by
Green Street Bindery, Oxford

DISTRIBUTORS

USA and Canada
  Blackwell Scientific Publications Inc
  PO Box 50009, Palo Alto
  California 94303

Australia
  Blackwell Scientific Publications
  (Australia) Pty Ltd
  107 Barry Street,
  Carlton, Victoria 3053

British Library
Cataloguing in Publication Data

British Ecological Society. *Symposium*
  (24th: 1983: Manchester)
  Ecology and design in landscape: the 24th
  symposium of the British Ecological
  Society.
  1. Landscape architecture   2. Ecology
  I. Title   II. Bradshaw, A.D.   III. Goode,
  D.A.
  IV. Thorp, E.
  712   SB469.3

ISBN 0-632-01408-3

Library of Congress
Cataloging-in-Publication Data

British Ecological Society. *Symposium (24th:*
  1983: Manchester, Greater Manchester)
  Ecology and design in landscape.

  Bibliography: p.
  Includes index.
  1. Landscape protection—Congresses.   2.
  Landscape protection—Great Britain—
  Congresses.   3. Landscape architecture—
  Congresses.   4. Landscape architecture—
  Great Britain—Congresses.   5. Urban
  ecology (Biology)—Congresses.   6. Urban
  ecology (Biology)—Great Britain—
  Congresses.   7. Ecology—Congresses.   8.
  Ecology—Great Britain—Congresses.   I.
  Bradshaw, A. D. (Anthony David)   II.
  Goode, D. A.   III. Thorp, E.   IV. Title.
  QH75.A1B75   1983   712   86-13673.

ISBN 0-632-01408-3

CONTENTS

V · REPAIR AND RENEWAL

PREFACE

Landscape design stems from people, who in their turn are influenced by their own immediate needs and their cultural background. Over the last 600 years, ideals and approaches to landscape have changed with the needs and understanding of landowners and designers. We have moved from a simple early approach, where the characteristics and uses of the individual plants were paramount and designs were simple and formal, to one in which plants lost their individuality in favour of grand effects to please the romantic aspirations of the great landowners of the eighteenth and nineteenth centuries. Since then we have been overwhelmed by the wealth of species brought in from all over the world and stimulated into gardenesque. Recently we have been led into constructivism as a result of the embraces of the modern architectural movement.

Today, we have a new landscape movement, stemming from science, where the approach is based on ecology. Its major protagonists reject the aestheticism of former schools and point to its failure to provide landscapes, particularly in cities, which urban populations can understand. They argue that, as a result of increasing industrialization, many more people wish to live in and experience a more natural world. They also suggest that a landscape based on nature will be more resilient and require less maintenance than one that is artificial. So the new design approach involves wild plants almost exclusively, in naturalistic arrangements, with the planting technique involving ecological principles as far as possible.

At the same time, ecologists have become very involved in nature conservation and landscape renewal, recording what is left of our natural and semi-natural vegetation and working out ways, based on hard scientific understanding, in which our landscape can be preserved, enhanced and renewed for future generations. There are startling achievements. Ecologists have also stimulated our interest in wild plants because of their concern for the future of our wildlife heritage.

All this has, curiously, led to a confusion. The new design approach, which seems to represent the ecological viewpoint, has been called 'ecological planting'. But in fact it does not by any means represent all that is, or can be, ecology *in* landscape design. Nor does it represent all that could be applied from the newly developed science of ecology *to* landscape design. The care for and use of wild plants is a naturalistic approach in design; an ecological approach embraces much more. At the same time, parts of an ecological

ix

approach to design have been integral aspects of good landscape design ever since gardeners began to appreciate that different species have different preferences, and began to realize that if wild species are harnessed to the service of mankind the products are durable and resilient.

This collection of essays seeks to explore the width of the relationship between ecology and design in landscape, and to set this relationship into the more mundane context of everyday management. They were presented as papers to a joint meeting of the British Ecological Society and the Landscape Institute at Manchester Polytechnic in September 1983. The meeting attracted over 200 people, most of them young professionals, suggesting that the time for ecology to be harnessed for the service of landscape design has arrived. While these essays are not intended to form a complete treatise, they will, we hope, give the reader a better understanding of the important contributions that design and ecology can give to each other in the achievement of landscapes, which will bring joy and delight not only to us but also to those many who will succeed us.

A. D. BRADSHAW
D. A. GOODE
E. H. P. THORP

I
APPROACHES TO LANDSCAPE

INTRODUCTION

The important first step is to understand the different approaches to landscape. In these four papers, representing design, ecology, management and history, it is immediately possible to see both the separate contributions of each approach and how much they overlap. Equally, it is possible to see from the different contributions—which have not been tailored to give the sort of unified approach found in a single author book—that there are very different understandings.

There are indeed four quite different approaches. The *designer* says he has the crucial responsibility of appreciating all the requirements which will be put upon his creation, and find ways to satisfy them effectively. He, or she, is expected, and aims, to contribute both imagination and technical competence. The *ecologist* maintains that he understands particularly the requirements of plants and appreciates the ways in which nature, without the assistance of man, achieves effective and often remarkable solutions at no cost. He, therefore, hopes to contribute fundamental understanding and effective techniques. The *manager* has to ensure that the product works in practice. He is therefore very concerned to involve the contribution of all who might have something to offer, in the final solution. Evolution of the landscape, over the centuries, is unravelled for us by the *historian*; it is a complex story, for many different contributions have been involved in its development.

To some extent, each contributor implies that his discipline has the whole understanding and solution already, and perhaps gently suggests that other contributions are less important. The reader must be left to judge how true this is. But it is interesting to speculate what would happen if each were denied the speciality of the others, or what would happen if only one of the four approaches prevailed.

In practice, effective and attractive landscapes are only achieved and maintained by contributions from all these approaches. If any one component, any one understanding, is left out, it is unlikely that the resulting landscape will be satisfactory. The two major streams of thought—design and management; ecology and history—are surely complementary and not exclusive. The truth of this will become apparent as we move on to succeeding chapters, in which specific approaches are expounded and justified.

1. LANDSCAPE DESIGN:
USER REQUIREMENTS

J. B. CLOUSTON
St Cuthbert's House, Framwellgate Peth, Durham DH1 5SU

SUMMARY

Urban landscape is in a crisis. Our older parks, generally planted according to ecological principles, were a bounty conferred on deprived populations. Today these parks are themselves deprived, while inner city programmes spend millions of pounds on urban non-descript, when what is needed is areas which are dynamic and attractive, have mystery and excitement, and offer many facilities. So arguments about a 'new ecological' approach are unprofitable. Ecology should be used to guarantee the outcome of any kind of landscape design. In parallel, city-wide landscape master plans, incorporating management, are essential, backed up by appropriate research. Without clear design, management and financial objectives, the changes which are so necessary cannot be brought about.

INTRODUCTION

Early protests against man's exploitation of his environment were mainly aesthetic. Victorian writing abounds with descriptions of poor housing, London fogs, the dark satanic mills of industry, exploitation of child labour. There are contemporary descriptions of the visual effects upon the landscape of canal and railway buildings. This written protest against ugliness and depravity in Victorian towns, and the changes brought to the rural landscape echoes the spirit of the poet John Clare, who years earlier wrote wistfully of the disappearing landscape of the day and its replacement by the hedged fields of enclosure. It is this landscape which our generation in turn regrets losing, largely because of depletion of established habitats within the hedgerows, rather than for purely aesthetic reasons.

By the middle of the nineteenth century it was clear that the consequences of industrialization had produced towns which were not only aesthetically distasteful but were hazardous to health. In response to this knowledge, Parliament enacted a series of Public Health Acts requiring urban authorities to provide clean water, discharge sewage and provide parks and open spaces. It is perhaps ironic that the fine Victorian parks built from 1850 onwards should be the product of legislation enacted primarily to introduce piped water and to build sewers. These sewers, incidentally, were the means of polluting many thousands of miles of hitherto clean streams and rivers.

Parks were built by Victorian reformers out of a concern for the health and recreation of the working class. The design and form of these parks was, to a large extent, inspired by earlier parkland estates created for the pleasure and leisure of a wealthy minority, by such landscape designers as Capability Brown, Repton, their contemporaries and amateur imitators.

Concern for the health and well-being of the masses brought a great increase in the number of parks in London and in the major industrial cities of the north, designed by Paxton, Loudon and others. Prior to the 1850s, city parks were rare and confined in London to a handful of Royal Parks and to the older meeting 'greens' such as 'Glasgow Green'. It is interesting to note, however, that the vast majority of major urban parks now forming the public estate of cities such as Liverpool, Glasgow and Sheffield, were not designed as public parks at all, but were in fact private leisure parks donated by their owners to the City or purchased by the City for public use. If we look at the parks of Glasgow we find:

Glasgow Green: acquired in 1662, not as a park but as a place to graze cattle, dry salmon nets and bleach linen. The Green became a public park by degrees.

Kelvin Park: was designed by Sir Joseph Paxton following its aquisition in 1852 as a public park.

Of the remaining major public parks, Tolcross Park, Overtoun Park, Linn Park, Pollock Park, Russ Hall Park were all private park estates given to or acquired by Glasgow City Corporation.

During the inter-war years few major parks were designed and built. Perhaps this was because city authorities relied heavily on the purchase of existing private estates, engulfed by rapid suburban growth, to meet the needs of its population, and because authorities acknowledged the fact that low-density suburban housing, both public and private, benefited from relatively large garden plots. Since the Second World War even fewer major urban parks have been built, due initially perhaps to austerity, but more likely to the fact that massive growth in private car ownership has given working people access to coast, countryside, country and national parks and stately homes, making the provision of new city parks a low priority.

The majority of our parks and open spaces were not, therefore, designed for public use, and most certainly not for the use of people living and taking their recreation in the later decades of this century. We now find ourselves in a somewhat unusual position.

1   We now have a stable, and in some areas a declining, population.

2 Our major cities are losing population, e.g. both Liverpool and Glasgow are decanting population at a rate of 15–20 000 per annum.

3 Many major urban centres have more houses than households.

4 Government is grant-aiding the development of new open space and the reclamation of derelict industrial land to open space use, within and around major urban centres.

5 The ability of local authorities to maintain existing parks and open spaces is under severe pressure.

6 Newly created open spaces are seen by many local authorities as a burden they could well do without.

7 New town corporations are concerned that the landscape they have created will not be managed well by district councils taking over the assets created by the new towns.

All of this is, of course, a change from the heady days of the 1960s and early 1970s, when planners, landscape architects and park managers were clammering for more parks and more open space. Now it is appropriate to ask the purpose of more parks, and to rethink the function of existing open space. We may now be in a position to agree between us two factors:

(a) that we have inherited more town landscape, in terms of acres planted, than we are likely to construct before the end of this century and probably during the first half of the twenty-first century;

(b) that any new urban landscape we may build is likely to be constructed in inner city areas or in the rural–urban fringe, i.e. in environments already much modified by the activities of man.

Out of this we can probably agree that in future the landscape profession is likely to become more heavily involved in the restructuring of established urban landscape than in the design and construction of entirely new urban landscape. I say this simply because I doubt very much whether society can afford to build more while the landscape we have remains obsolete in terms of its design and management.

In a nutshell, our urban landscape is in crisis. Trees, which form such a vital element, are passing maturity. Few authorities replant at all, and those that do rarely use forest species likely to achieve the scale of those trees they are to replace. Grass areas ill-treated by both management and public show severe signs of stress. The planting matrix in between is often of inappropriate design and its maintenance neglected. Facilities and equipment are worn out and obsolete.

APPROACHES

I make no apology for giving a long introduction outlining the present situation of our urban parks and open space; nor for my lack of faith in an ecological approach answering our current problems in landscape design. The idea of working with nature rather than against it is by no means new. If we examine a typical Brown or Repton park—a park of the kind ultimately incorporated into the public parks of many cities—we find the basic matrix of species planted in fact included native species. Contemporary records show that planting methods, planting sizes and indeed planting mixes differ little from those now proposed by advocates of 'ecological' planting. Many Victorian town park designers copied planting forms from earlier times. Consider, however, the conditions within which they were attempting to create parks: the atmosphere of industrial towns was choked with smoke; soils were damaged by the fallout of soot and sulphur. All of this restricted the palette of plants available to the designer. By trial and error the designers found a number of both native and exotic species capable of surviving in an environment polluted by the atmospheric filth of coal burning. We should, in my view, praise our forebears for their achievements, rather than heap lofty criticism upon them. References to horticultural literature at the time illustrate the conditions faced by designers of the day. Nowadays, because of the success of Clean Air Acts, there are fewer problems of air and soil pollution, whereas native species were in many cases ill-matched to the conditions prevailing in towns a century or more ago.

Towards the end of the nineteenth century, William Robinson was busy promoting ideas not dissimilar to those now advanced by enthusiasts in current landscape publications. He wrote about the use of native plants, the introduction of plants from similar climatic zones in northern Europe, and about their use in wild gardens, woodland gardens and heath gardens. We may have dressed up William Robinson's message with contemporary jargon, but there is little that is really new in modern proposals. Just as Brown and Repton used woodland, copse, covert, lake and meadow in their designs, and Robinson and Jeckyll later worked with native and exotic plants in woodland, heath and aquatic habitats in their landscapes, so today ecologically-minded designers aim to create landscapes with a diversity of natural habitats and plant communities. Certain writers are promoting old ideas and well tried techniques as new concepts, but remember many of the great gardens established in Britain during this century, e.g. Bodnant, Inverue, Logan and the Younger Botanic Garden. All have used ecological principles and have worked with nature to create conditions suitable for a wide range of species and habitats. Using such principles, we are able to establish semi-natural landscapes in our cities. But, in my opinion, whether or not to plant so-called natural landscape is not the most pressing problem faced by society, nor by the profession.

In the UK, a new interest is dawning in urban parks for the greening of cities. This is an interest prompted by general increases in (i) leisure time (often the enforced leisure of unemployment, short time and shift working); (ii) leisure activities; and (iii) the price of fuel, high costs tending to reduce visits to the countryside, coast, stately homes and national or country parks.

Nineteenth century parks were considered to be a bounty by those responsible for the political decisions to provide, finance and build them; a bounty conferred on the deprived populations of often overcrowded, unsanitary and ugly cities. Today we must turn this thought around. Our open spaces, for reasons I will discuss later, are deprived places: deprived of money, of creative management and, sadly, deprived of appreciation and love by the public at large. In truth, it is the people who must now confer the bounty of appreciation on parks and, in doing so, make them successful once again. By withholding appreciation they withhold use, and abandon parks to rejection and failure.

Parks are volatile places: they tend to be extremely popular or extremely unpopular. Much of their popularity derives from their design, i.e. their total design in terms of: (i) their relation to their context; (ii) the facilities they offer; (iii) their mystery; and (iv) their intricacy. As land uses surrounding our parks change, so must parks themselves be adapted; they should be dynamic. To a large extent, we have inherited acres of Victorian museum.

Current Government policy is enabling authorities to provide more and more open space, but few new parks. Inner city partnership programmes in, e.g. Glasgow, Liverpool, Tyneside spend millions of pounds each year creating what I can only describe as 'urban non-descript': open space which no one really needs and which few public authorities are able to afford to maintain. It is a paradox that at a time when revenue budgets are being cut, capital budgets are being increased. Open space is being created without the benefit of a City Landscape Management Plan.

REASONS

For years the professions, the planners, the park managers and now the ecologists have asked for the creation of more open space. But what we should really ask ourselves is more open space for what? To create bleak vacuums between buildings? For rapists, drug addicts, hooligans, vandals and drop-outs? Those in politics, both national and local, and those in the professions should realize that the public do not use open space just because it is there, nor because designers, planners and managers intend that they should; and parks deserted by ordinary people may then become areas for deviants. The best policed parks are those used to the full, but people will only use them if:

    (a) they are correctly sited;

(b) they are easily reached;

(c) they have a diversity of edge development;

(d) they are dynamic and filled with attractive elements;

(e) they provide mystery and intricacy.

Parks are creatures of their surroundings more than of their content. Sadly, too many of our parks are bland and predictable when they should be a place of mystery. They need spaces of differing character defined by landform and planting, with tempting paths between. Sometimes formality is appropriate, but it must be combined with an enriching variety of materials and style. Parks should also provide facilities which everyone can enjoy, e.g. (i) play facilities for children and adults; (ii) restaurants/cafes/bars; (iii) theatres; (iv) exhibition areas; (v) transport modes; (vi) sports areas; (vii) cycle ways; (viii) trim trails; (ix) pets' corners. A wider range of facilities attracts a wider range of people, and thus increases the likelihood of continued use. Many of our parks rely too heavily on horticultural excellence to draw in the public, whereas it is the careful use of hard materials, furniture and planting which reinforces the concept and character of the place. The difficulty and, therefore, the cost of maintaining parks relies much on the detail of their design. Outmoded, outworn and inappropriate Victoriana costs precious time, valuable funds and kills the spirit of the staff.

It is accepted practice to allocate major budgets for the renewal or refurbishment of certain elements of urban fabric: sewers and other public utilities are replaced when they are old; schools and hospitals are closed when they fail to meet contemporary standards of provision; housing is refurbished to raise it to acceptable modern standards. Government funds are allocated for all these purposes. If we look at a Victorian park or a mid-twenties recreation ground using the same criteria as those applied to Victorian housing or a twenties hospital, there is little doubt that funds should be made available for their improvement. But the money is not available because park managers, the landscape profession, and industry have no concerted lobby to convince government and elected members of the need for change.

In my opinion, it is nonsense to spend time discussing the 'pros' and 'cons' of a new ecological approach to landscape design, when this approach has always been present in the minds of the great designers. Perhaps the promulgators of the new ecological landscapes are confusing two parallel issues: those of technique and philosophy. Although unprecedented recent growth in the scientific understanding of natural processes has enabled new ecologically-based techniques to be employed successfully in the realization of designs, this has not instituted a new design philosophy. However, ecological theory and its techniques of survey and analysis, if properly applied, should speed and guarantee the outcome of any kind of landscape design more effectively than the former intuitive methods.

THE WAY FORWARD

The major challenge of reintroducing nature to our cities, particularly to deprived working class residential areas, lies in persuading central and local government to commission city-wide landscape master plans, incorporating landscape management plans. Undoubtedly, we need to redistribute open spaces, perhaps selling off little-used parkland or park fringes to raise revenue for the creation of space in deprived areas, or for the restructuring of older parks.

Questions of whether or not to reallocate space within housing areas, bringing more land into private use, should be raised and answers found. Germany and Holland have successfully adopted schemes of this kind, reducing the burden of maintenance on the public sector. Tenants and private owners also benefit from the additional private landscape areas, with all the rich diversity of planting which is traditional in private gardens. The preparation of such plans would involve the ecologist and the environmental educationalist.

Early in the 1970s we were commissioned to prepare a landscape master plan for Catterick Garrison, a military town in North Yorkshire of some 22 000 population, covering a land area of 2500 acres. Many service families live in married quarters. Other soldiers were quartered in barrack lines. Officers and Warrant Officers occupied married quarters or were housed in officers' and W.O. messes. Civilian families also lived within the Garrison area. Catterick Garrison is therefore, in many ways, like many other small towns. The prime difference is that a high proportion of its population is transient.

Our work at Catterick mainly concerned the reclamation of former hutted encampment areas and the rationalization of land use outside married quarters areas, barrack lines, administrative areas, workshop areas, the military hospital and established formal recreation areas. During the first year of our commission we prepared a master plan report outlining a landscape programme for the decade ahead and a budget covering capital spending for that period. The report was prepared in full consultation with the PSA landscape maintenance branch and with all other agencies having responsibility for grounds maintenance and forestry within the Garrison. Our recommendations were aimed at:

1 providing a landscape structure within the Garrison, the object of which was to dispel the image associated with Catterick as a town of hutted barrack areas, mud and duckboards;

2 providing a landscape structure within which the military, their families and the civilian population of the camp could enjoy informal recreational pursuits;

3   bringing as much reclaimed land as possible into use, as woodland, grazing land or natural landscape areas;
4   minimizing the cost of maintaining the total system of open spaces within the Garrison.

Many of these objectives can be identified as being similar to those faced by landscape managers throughout the country, and in particular by those in our older towns and cities.

During the past decade over £1m has been spent on landscape improvement works at Catterick. An essential part of the programme has been the involvement of school age children, particularly in planting naturalized areas of spring bulbs. The overall result of our work can be judged by the following factors.

1   The Garrison has become more popular as a posting for soldiers and their families than in years past. The prime objective of a plan, i.e. to change the image of the military town, has been achieved.
2   Maintenance tasks have been reduced considerably despite the increased size of the landscape estate, because the landscape is designed to simplify maintenance tasks and therefore to reduce costs.
3   Vandalism is almost unknown and is mainly confined to damage through 'horseplay' by soldiers returning to barrack lines after evening entertainment.

As a practice we are concerned with the preparation of a number of landscape master plans for new towns in Hong Kong, Kuwait, Iran and Saudi Arabia. In all of these commissions an essential part of the master plan has been a section dealing with:

(a) the administrative organization required to establish and maintain the new landscape of the town;
(b) management policies to be adopted in order to maintain newly established and maturing landscapes;
(c) Programmes devised to achieve minimum costs in landscape maintenance.

Sadly, we seldom take time to question why we adopt particular regimes of maintenance to the landscape under our care. It is only when a new town is created, or perhaps when a commission is of the kind I described for Catterick Garrison, that we are able to return to first principles.

CONCLUSIONS

City-wide landscape plans may take a decade or more to implement, yet without clear design, management and financial objectives, how can fundamental changes in terms of design, maintenance programmes and budgets be brought about? We see all around us worn-out urban parkland, in some cases incapable of taking user pressure, without the benefit of recent capital expenditure to upgrade design, facilities, functions or management. Sterile open space surrounds much of our current housing stock. Why not plan instead a visually interesting landscape, a useful landscape maintained largely by the community?

Our city highways are often devoid of landscape interest and are costly to maintain, because fashion dictates that we mow the grass at frequent intervals. When travelling through our towns, not only our northern towns, we are seldom out of sight of urban wasteland—land unused and useless. Why not replace these desert areas with low cost maintenance urban forest? Within almost every town and city we can find examples of ecologically stable and balanced landscape which demand few resources in terms of labour or money in their maintenance.

Highly manicured landscape maintained at high cost can seldom be equated with public popularity. Often highly maintained Victorian parks are shunned in favour of urban wilderness areas such as commons, overgrown cemeteries or river banks. In my adopted home town of Durham, more people use the river banks below the Cathedral, the open river banks next to the old racecourse, and naturally regenerated sand quarries in Flass Vale than use the more formal facilities at Whorton Park; the park was infinitely more expensive to create and is now considerably more expensive to maintain than any of the other areas. I would not for one moment say that we do not require highly maintained landscapes in certain areas within our towns. Instead, I advocate variety, diversity and quality, measured against different standards to those in current use.

As a first step towards achieving these changes, I would suggest that our institutes, with the Institute of Leisure and Amenity Management, lobby the ministers concerned with allocating budgets to the various authorities within our responsibility, to demand immediate research funding for studies ascertaining the kind of landscape best suited to the needs and economics of today. We need to create new urban landscapes containing a diverse arrangement of well maintained floral display and man-made wilderness, similar in approach to those created by master landscape designers in the past.

FURTHER READING

Billings, W.D. (1972). *Plants, Man and the Ecosystem*. Macmillan, New York.
Bonnet, P. & Gill, D. (1973). *Nature in the Urban Landscape*. York Press, Baltimore.
Colvin, B. (1948). *Land and Landscape*. Murray, London.
French, J. (1973). *Urban Green, City Parks of the Western World*. Kendall/Hunt, Iowa.
Hackett, B. (1979). *Planting Design*. Spon, London.
Hackett, B. (1980). *Landscape Conservation*. Packard, London.

Hadfield, M. (1969). *A History of British Gardening*. Spring, London.
Laurie, I. (Ed) (1979). *Nature in the City*. Wiley, Chichester.
McHarg, I. (1969). *Design with Nature*. Natural History Press, New York.
Nicholson, E. (1969). *Environmental Revolution*. Hodder & Stoughton, London.
Westmacott, R. (1974). *New Agricultural Landscapes*. Countryside Commission, Cheltenham.

2. ECOLOGICAL PRINCIPLES IN LANDSCAPE

A. D. BRADSHAW
Department of Botany, University of Liverpool

SUMMARY

Ecology is the study of living organisms in relation to their environments. As such it should be the handmaiden of design. It allows us to understand (i) the basic requirements of plants and (ii) the ways in which plants interact when put into communities. But plant communities are themselves part of, and are influenced by, their physical surroundings, forming interacting ecosystems, which develop with time by the acquisition and cycling of nutrients. Because environments and species are diverse so are the ecosystems which result.

Ecologically-based design must take account of all these fundamental processes or there will be poor growth and even failure. They should be incorporated into the actual creation of landscapes, whether in *laissez-faire*, *positive construction*, *manipulation of development* or *restoration*. What ultimately survives in nature is that which is best fitted to its environment. So if we attempt to model landscape designs on natural systems and use natural processes to achieve desired end-points, we are more likely to produce resilient and self-sustaining solutions.

INTRODUCTION

The word 'ecology' was coined by Haeckel in 1850 from *oikos* (house) and *logos* (study), to identify the study of living organisms in relation to their environments. In the last 130 years the subject has expanded from its small beginnings to cover all aspects of the natural environment and the relationships of organisms within it, to each other and to their surroundings. There are at least fifteen major scientific journals entirely devoted to ecology, innumerable degree courses, an infinity of textbooks and a political movement. It is true that primitive farmers, or the eighteenth century landscape architects, were ecologists in the sense that they could only be successful if they appreciated that plants must be fitted to their environment, but the breadth and depth of ecology is now so considerable that it is difficult to see how any aspect of landscape design can be tackled without the fundamental understanding that has come from ecology. It is the aim of this essay to explore these ecological principles and put them into their landscape context. Modern ecology is not a design gimmick; it provides insight into how plants grow and can be encouraged to grow. As such it should be the servant of design, but not the design itself.

BASIC PLANT REQUIREMENTS

The needs of plants for satisfactory growth are simple: (i) carbon dioxide; (ii) light; (iii) water; (iv) mineral nutrients. Of these, we can, for practical purposes, forget carbon dioxide since, as an omnipresent component of the atmosphere, its supplies are assured. (Although in enclosed, glasshouse environments, we now realize that it can be depleted and methods have been developed to supplement it.)

In most situations the supply of light energy is adequate and commonly above that required to saturate the photosynthetic system. But in winter and in shaded habitats, light supply becomes a limiting factor for plant growth. The relationship is linear down to the compensation point, where photosynthesis equals respiration and no growth occurs (Fig. 2.1). Plants can tolerate levels below compensation point for limited periods of time, but ultimately

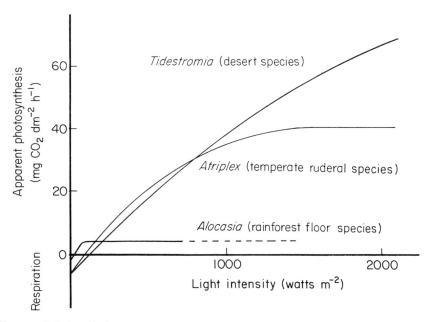

FIG. 2.1. Relationship between photosynthesis and respiration at different light intensities, showing the compensation point below which growth will not occur (from Salisbury & Ross 1978).

they will deteriorate and die. The crucial point, however, is that different species have quite different compensation points, due to both lowered respiration rates and more efficient photosynthesis at low light intensities. These shade species are therefore able to grow successfully at very low light intensities (Boardman 1977). But they may only achieve this after a period of acclimatization at low light intensities, during which they develop leaves with the appropriate structure adapted to very low light intensity—a point of concern to landscape architects concerned with interiors (Manaker 1981) (Fig. 2.2).

Everybody realizes that plants require a supply of water or they will die from loss of turgor and consequent collapse of their tissues. However, what is often not realized is (i) the degree to which growth is reduced when water is in short supply and the plants' water potential (the negative water pressure in the plant) is high, and (ii) the differences in the response of apparently rather similar species to water shortage or excess (Fig. 2.3). The amount of water lost from vegetation has been well studied and quantified, particularly by Penman (1963). As a result standard recommendations are available for agricultural crops.

For individual plants, such as a standard tree, it may be easier to

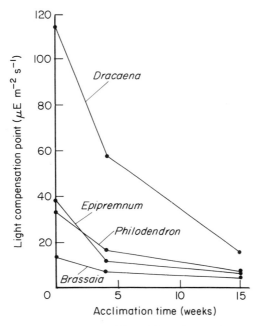

FIG. 2.2. Changes in compensation point of four foliage plant species during acclimatization to low light (from Fonteno & McWilliams 1978).

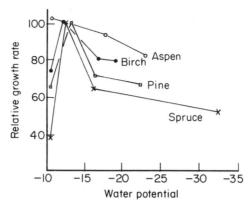

FIG. 2.3. Relationship between relative growth rate and soil moisture stress for four common temperate tree species (from Jarvis & Jarvis 1963).

understand the amount of water required, if full growth is to be maintained, by a brief calculation (Table 2.1). The crucial point is that this water has to be stored and made available by the soil. The amount which can be stored in the vicinity of a newly planted tree with its typical limited root system, is very small. It is exacerbated by the poor soil structure so common in new sites. This can not only halve the water storage (available water capacity) of a soil but

TABLE 2.1. Water requirements of a newly planted standard lime tree

Mean no. of branches	17
Mean no. of leaves per branch	46
Mean area of leaves	22 cm^2
∴ leaf area 17 × 46 × 22	17204 cm^2

Trees transpire about 1 litre of water m^{-2} day^{-1}
∴ this tree needs about 2 litres of water day^{-1}

Its roots will occupy not more than 40 × 40 × 40 cm,
∴ total soil volume $c.$ 6 litres
Available water capacity $c.$ 20%
∴ this soil volume would itself hold the water required for only a single day

More is therefore needed from:
    (i) the soil store by conduction
    (ii) rain
    (iii) artificial watering

also reduce substantially its ability to transfer water from other parts of the soil (hydraulic conductivity). It seems that these fundamentals are not appreciated, otherwise such high percentages of newly planted trees would not die every year or show reduced growth. Yet guidelines are available (Dutton & Bradshaw 1982).

The need of plants for plant nutrients is obvious and is well described (e.g. Russell 1973), yet the required amounts are hidden by the fact that normal soils constantly provide them. We can find out what plants really need by a simple calculation based on their overall growth. The amounts are quite substantial (Table 2.2), and the outstanding need is for nitrogen. If a soil is very poor or even skeletal it is normal to provide the nutrients required by means of fertilizer. Growth can be sustained in this manner, but very commonly there is poor growth, which is nearly always due to deficiency of nitrogen (Bloomfield *et al.* 1982). This is partly due to the considerable plant requirement for nitrogen; but it also arises because, in contrast to other nutrients, nitrogen is only stored in soils in organic matter and released by its slow decomposition. As a result, there has to be a very large and permanent store of nitrogen in the soil, equal to 10–20 times the annual plant requirement (Marrs *et al.* 1983). This can only be built into the soil economically by the normal ecosystem processes, which will be discussed later, although other nutrients can be efficiently added as fertilizers. Nitrogen deficiency is almost inevitable in newly established plants, whether grassland or trees (Bradshaw 1981), and yet it is rarely recognized as a serious long-term problem.

Other nutrients are likely to be deficient, particularly in subsoils. But they can be deficient even in specially important topsoils (Bloomfield *et al.* 1981). The problem is to know when these deficiencies are occurring. Although they

TABLE 2.2. Amount of nutrients required annually by ecosystems with different levels of total production (Bradshaw 1983)

| | Asumed nutrient content | Production level for each ecosystem type (kg ha^{-1} y^{-1}) | | | |
		Tundra + desert	Poorly productive temperate	Productive temperate + poorly productive tropical	Tropical
Total production		1000	5000	10000	20000
N	2·0	20	100	200	400
K	1·1	11	55	110	220
Mg	0·51	5·1	26	51	102
Ca	0·26	2·6	13	26	52
P	0·18	1·8	9·0	18	36

will be indicated by a standard soil analysis, the most critical information comes from after-care experiments involving the application of individual nutrients in factorial combinations. These may show startling interactions between nutrients, so that satisfactory growth is only obtained when two nutrients, e.g. nitrogen and phosphorus, are applied together (Berg 1973). These can usually be applied without difficulty by use of fertilizers. But the crucial point is that, if the deficiencies are not recognized at the outset, plant growth can be so poor that the whole of a landscape design can be jeopardized.

In natural conditions, different species occupy different habitats, because they have different ecological preferences. One way in which preferences may differ is in nutrient requirement. On the whole, fast-growing species require high nutrient levels and will not persist in low nutrient conditions. These preferences can be specific for individual nutrients (Etherington 1975). In planting schemes, species can therefore be chosen in relation to the nutrient levels occurring. But the connection between vigorous growth and nutrient requirement can pose problems, because it is not really possible to obtain vigorous growth under conditions of low nutrients, whatever species are chosen (Grime & Hunt 1975).

If vigorous growth is required the necessary nutrients will have to be added. Equally, where there are problems of extremely poor conditions, choice of species with low nutrient demands may not be sufficient; even these species will have some nutrient demands for growth, which may not be met by the environment.

Plants do not stand like concrete lamp posts inert in the ground. They are in a continual state of flux, requiring a continual supply of materials from the ground, as well as from the air, for maintenance and growth. If these needs are not recognized, problems are inevitable.

PLANTS IN COMMUNITIES

Plants rarely grow alone. Indeed the only places where they do, is under cultivation or in some landscape schemes, where other plants are scrupulously eliminated. Even then they are usually growing in the company of other individuals of the same species. So an essential characteristic of plant life is interaction between individuals and between species.

In all situations invasion by other species is inevitable. Evolution has built into all species a capacity by which they can be dispersed and can invade existing habitats and assemblages of species. We are particularly familiar with the plants we call weeds, species which are characterized by specialized powers of dispersal, whose whole biology is attuned to rapid invasion and multiplication. These are not only annual species, e.g. groundsel (*Senecio vulgaris*) and meadow grass (*Poa annua*), and biennial species, e.g. sow thistle (*Sonchus oleraceus*) and buttercup (*Ranunculus acris*) which spread mainly by seed, but also perennials, e.g. couch grass (*Agropyron repens*) and ground elder (*Aegopodium podagraria*) which spread almost entirely by vegetative means.

These are species of open habitats, the sort of places where we would expect invasion. But invasion also occurs into closed habitats, where there is already plant cover. This will nearly always be by larger species. Some, such as japanese knotweed (*Polygonum cuspidatum*), spread vegetatively and invade in an inexorable and devastating manner. Others invade by seeds adapted to allow survival and growth of the early seedling stages in the face of competition from pre-existing species. This sort of invasion by species such as gorse (*Ulex europaeus*), hawthorn (*Crataegus monogyna*) and trees such as sycamore and even oak, is more erratic and unpredictable because it depends on seeds which are heavier and less mobile; and yet in the end invasion is inevitable.

As a result, different sorts of interactions between species are set up. The simplest is that of competition, where the two species make demands on the same resources, e.g. light, water and nutrients, and there is insufficient resource to supply the needs of both. The result is that one species can only grow at the expense of the other, the total growth of the two remaining constant (Fig. 2.4) so that, eventually, the repressed species will be eliminated. It is easy to think that competition only occurs above ground, for light. But many experiments (e.g. Donald 1958) have shown that competition for below-ground resources is usually more important, and can lead to brutal consequences. Grasses may not seem a serious competitor for trees, but below ground they may remove so much water and nutrients that newly planted trees can suffer badly (Bradshaw & Chadwick 1980; Insley & Buckley, Chapter 8). Sometimes it may be possible to relieve competition for nutrients by the

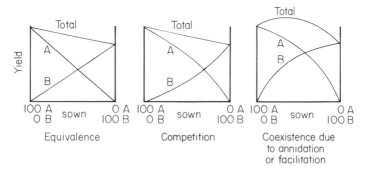

FIG. 2.4. Yields of two components sown at different ratios in a mixture, showing the four most common forms of species interaction.

addition of fertilizer, but this will depend on the weed (Welbank 1963). The
great bulk of evidence, mainly from agriculture, is that invading species cause
substantial reductions in the growth of the species being invaded and must be
removed. Yet, time after time in newly created landscapes they are not, or at
least not until it is too late.

Not all interactions, however, result in the elimination of one species by
another. It is possible either that two species obtain their resources
differently—in ecological terms, annidation or occupation of different
niches—or that two species each provide some resource required by the
other—in ecological terms, facilitation. In these circumstances the presence of
one species does not repress the other and the total growth of the two is more
than each separately (Fig. 2.4). In these cases neither species is eliminated and
both will co-exist in perpetuity. An example of annidation is bluebells in an
oak wood; they do not compete because they grow at different times.
Facilitation is shown between clover and grasses; the clover contributes
nitrogen to the mixture by its ability (via *Rhizobium*) to fix atmospheric
nitrogen.

The significance of these interactions in plant communities is overwhelming. For a full discussion the reader should consult Harper (1977) and Grubb
(1977). But the essential point is that mixtures of species where competition
occurs are unstable: they will change unless something specific is done to limit
the growth of the more successful species. By contrast, mixtures of species
where annidation or facilitation occur can be stable.

The effects of competition may be slow, but they are inevitable and must
not be forgotten. However, stable mixtures of species are achieved in nature;
there is no reason why they cannot be achieved in landscaping schemes.

PLANTS IN ECOSYSTEMS

Plants do not just grow in groups or communities; they grow in particular
places, where they are influenced by the physical environment and other
organisms. The relationship is not simple because the influences can be in both
directions; plants can influence climate as well as be influenced by it. Plants are
certainly influenced by soil as we have already seen, but they can themselves
change soil characteristics by their root growth and the organic matter they
contribute. Animals and man are involved likewise.

As a result, we need to envisage plants as part of an ecosystem of
interacting components (Tansley 1935). Plants are neither independent of one
another nor independent of their surroundings. What plants are present and
will remain in a particular plant community, will depend not only on what
plants are put there, or arrive, in the first place, but also on the interactions of

components in the whole ecosystem (Fig. 2.5). Change in one component may result in change in everything else. Disappearance of rabbits due to myxomatosis on the chalk downs did not only lead to taller growth of existing vegetation, but also to invasion by shrubs, disappearance of attractive herbaceous species, and changes in bird and animal populations. Appearance of man in the wilder parts of the world, e.g. the Himalayas and the Andes, has not just lead to changes in species frequency due to the effects of his grazing animals and, therefore, to vegetation changes, but also to invasions of new species with different ecological tolerances and to drastic soil degradation (Holzner *et al.* 1983).

But ecosystems are not just a series of interacting factors. There are direct physical relationships between the components arising from the movement of materials between them (Fig. 2.6). This is particularly important for any material, such as nitrogen, for which there is a limited supply for the ecosystem as a whole: the amount of plant growth depends on how much the plant can obtain. This depends upon what is present in other parts (or compartments) of the ecosystem and the rapidity of the transfers (or fluxes) from one compartment to the next, as was discussed earlier. The total flux within the ecosystem will depend on the total amount in the ecosystem. This depends on the balance between total inputs and total outputs. An ecosystem will degrade if inputs are reduced or outputs increased, from whatever cause. Everything in an ecosystem has to come from somewhere.

This is why we are concerned with changes to ecosystems brought about by man. When, for instance, a forest is felled and the timber is removed there is a small loss of nutrients in the timber. But ecosystem studies have shown that the most serious effect is an enormous increase in losses of nitrogen due to leaching and erosion, because of the lack of a protecting and conserving vegetation cover (Bormann & Likens 1979). In contrast, when tree lupins

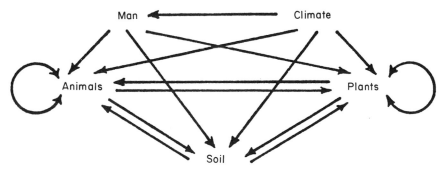

FIG. 2.5. The interactions occurring in an ecosystem.

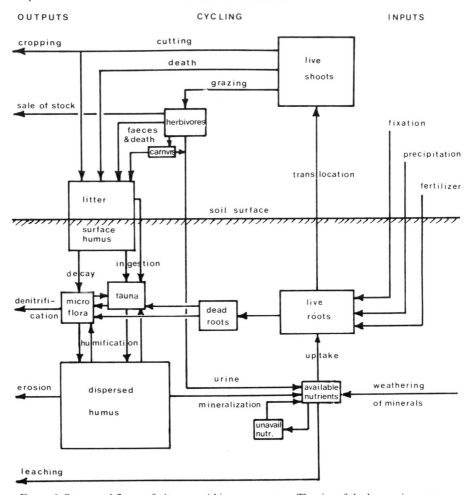

OUTPUTS CYCLING INPUTS

Fig. 2.6. Stores and fluxes of nitrogen within an ecosystem. The size of the boxes gives some indication of the relative sizes of the different stores.

(*Lupinus arboreus*) are sown prior to planting conifers on sandy soils, the nitrogen content of the ecosystem, and as a result the subsequent tree growth, can be enormously increased (Gadgil 1971). Empirically the farming community has discovered these sort of relationships over the past 2000 years. Landscape designers have also learnt by trial and error, but to a much lesser extent. However, we are now able to provide more precision by understanding the underlying processes. If we do not understand ecosystem functioning we cannot expect to build reliable and satisfactory landscapes, any more than an architect can build satisfactory houses without a knowledge of plumbing.

ECOSYSTEM DEVELOPMENT

In an ecosystem nothing is static: the plants and animals themselves grow and die; they may then be replaced by other invading species; their growth will influence climate and soil. In particular, the plants will manufacture organic matter by photosynthesis and accumulate nutrients, drawing them up from deeper layer of the soil and fixing nitrogen from the air. All this material finally arrives on, or in, the surface layers of the soil, where some of it decomposes and the rest is incorporated into the soil by the activities of animals.

At the same time the parental rock material is slowly decomposing, releasing nutrients and providing more particulate soil material. So, in the early stages of ecosystem development on raw soil materials, there is an increase in the depth and fertility of the soil in terms of nutrients available to plants. In most temperate situations this soil development leading to increased biological productivity lasts for about 100 years, e.g. at Glacier Bay (Fig. 2.7) (Crocker & Major 1955). After this, although the soil continues to develop pedologically, in gross structure, fertility often begins to diminish, particularly because of loss of soluble nutrients such as calcium. How soon this occurs depends on rainfall and the underlying rock.

There is a parallel development in the plant and animal parts of the ecosystem as smaller species are replaced by larger ones, and the increased soil

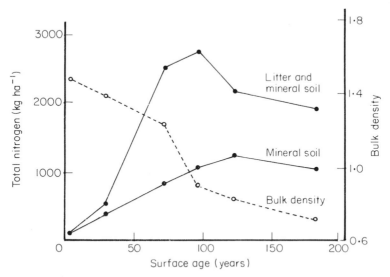

FIG. 2.7. Changes in nitrogen and soil bulk density over a period of 180 years in ecosystems developing naturally over glacial moraines at Glacier Bay, Alaska (from Crocker & Major 1955).

fertility allows these to grow larger. Whether or not the larger species could grow to begin with is a matter of debate, but there is no doubt that they grow better in the more developed soils, as is reported for trees on naturally colonized ironstone wastes in Minnesota (Leismann 1957). The result is, as everyone knows, that in temperate regions the ecosystem develops until it reaches what is called a climax, usually dominated by large trees. The climax is itself complex and will contain many different plants and animals. A good review of the processes is provided by Miles (1979).

In general terms there is a gradual increase in species number and total biomass with time. However, it is possible for the whole process of ecosystem development to be arrested at some stage by some external factor such as fire or grazing. The soil development may continue, and the plant and animal community, although prevented from developing along its normal path, may continue to accumulate species which are specifically adapted to the arrested stage or subclimax. As a result the community is often called a plagioclimax. Some of these communities can be very rich in species and very beautiful, such as old hay meadows and downland grasslands. More than half the sites recognized as being important for nature conservation in Britain are subclimaxes of one sort or another (Ratcliffe 1977).

It is very important to understand this development process. Firstly, we have to realize that the raw materials from which soils are made are not themselves fertile and productive. They only become so as the result of biological as well as weathering processes. But this may not take a long time. The fertility of raw subsoil materials will naturally upgrade rapidly if vegetation is present. This provides a basis for methods of soil improvement (Bradshaw 1983; Roberts & Roberts Chapter 7). Legumes in particular are important (Figs 2.8 and 2.9).

Secondly, we have to remember that all ecosystems will develop until they reach the climax, unless there is a specific arresting factor. Grassland goes to scrub, and scrub, which sometimes seems so permanent, goes to woodland. Apparently stable ecosystems, such as heath or chalk grassland, which owe their maintenance to a specific arresting factor, will not maintain their characteristics nor their species if the arresting factor is removed, e.g. if mown grass roadside verges are left unmown they will sooner or later disappear into scrub or woodland. Similarly, the characteristics of the vegetation will not remain if the type of arresting factor is changed. Mowing is not a substitute for grazing, because grazing can be curiously selective (Duffey et al. 1974). Even apparent climax communities such as woodlands often owe their characteristics to particular management treatments (Peterken 1981).

So the natural processes of ecosystem development can be valuable in landscape design, but they can provide problems for management.

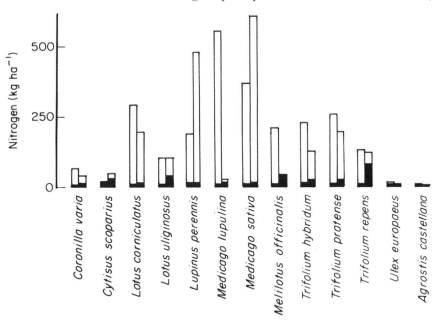

FIG. 2.8. Standing biomass nitrogen content of twelve stands of different legume species □, accompanied by the grass *Agrostes castellana* ■, established on colliery spoil, after 14 months (left columns) and 21 months (right columns) (from Jefferies *et al.* 1981).

DIVERSITY OF ENVIRONMENTS

The processes of ecosystem development and arrest at particular stages mean that in any one environment a series of different ecosystems can occur, each with its own particular community of plants and animals. At the same time, even a small area of land can have considerable variations in environment, due to differences in climate, drainage and soil material. As a result, even a small country like the British Isles has a quite remarkable range of environments, each of which has its own recognizably different plant and animal community. The task of describing them was considerable even in the 1930s (Tansley 1939), and even a modern treatment in almost catalogue form is extensive (Ratcliffe 1977). A full description of European vegetation does not exist.

These parochial accounts emphasize the diversity of ecosystems. But they also emphasize the very specific ecological preferences of the plant species which make them up, whether the 1200 species of Great Britain or the 12000 of Europe. If species are grown in controlled conditions in the absence of competition, they will be found to have particular physiological responses to environmental factors. Usually these mean that the species can grow under a

FIG. 2.9. Newly established urban grassland showing very poor growth except in the patches where dogs have urinated. Some clover patches occur. There is severe nitrogen deficiency because the topsoil used was of poor quality. Clover should have been included in the original seed mix.

fairly wide, although defined, range of conditions. But their actual, ecological, distributions may be much more restricted in natural conditions where they are subject to competition from other species and the effects of a combination of environmental factors (Rorison 1969). Species may even be shifted to one end of their range of physiological response (Fig. 2.10). Excellent examples of the specificity of ecological distributions of species are available for the species of the Sheffield district (Grime & Lloyd 1973), yet even these have considered only certain ecological factors.

The importance of these two principles—environmental diversity and ecological specificity—in landscape is considerable. While there are species of wide ecological tolerance, such as the common oak (*Quercus robur*) and the

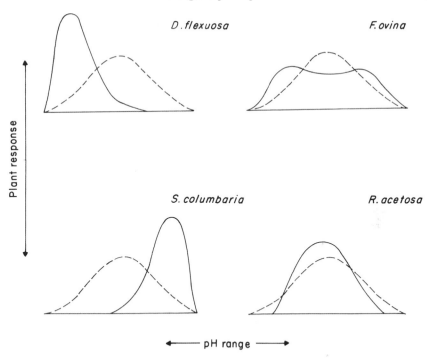

FIG. 2.10. Differences between the theoretical physiological response (– – –) of four contrasting plant species to soil pH and their actual ecological response (——) shown by their distribution in nature (from Rorison 1969).

grass, sheep's fescue (*Festuca ovina*), which can be grown and will persist in a wide range of habitats, most species have much narrower ecological tolerances and can grow and persist only in a narrow range of habitats. The reasons for this range of habitats must be understood if the species is to be grown successfully and a permanent landscape solution provided.

In general terms, trees and other large species have wide ecological preferences and will withstand the competition of invading species. At the other end of the scale, small herbaceous species will be much more sensitive; not only do many of them have precise soil preferences, more than just whether they are calcicoles or calcifuges, but they will also be intolerant of competition. The ecological preferences of species can also be divided on the basis of whether they show stress-tolerant, competitive and ruderal strategies; each group has its own particular characteristics and habitat requirements (Grime 1979).

Beyond this, we must remember that because evolution is a process common to all organisms, effectively all species are composed of different ecological races or ecotypes, i.e. local populations which have evolved specific adaptations to the environments in which they live (Clausen, Keck & Hiesey 1940). In trees this is recognized as differences in provenance (Wright 1962). It allows species to occupy both specialized environments, e.g. metal polluted areas (Bradshaw & McNeilly 1981), and a wide variety or ordinary soils (Snaydon & Bradshaw 1961). On the other hand, the specificity of adaptation of individual populations, when studied under field conditions, can be quite remarkable, e.g. buttercups (Lovett Doust 1981).

Environmental diversity is immense. It is accompanied by a corresponding diversity in ecological preferences among species and even among populations within species. In any landscape design these two have to be matched most carefully, or what seems to thrive initially may fail later.

ECOLOGICALLY-BASED DESIGN

So what is an ecologically-based approach to design? Essentially it must be based on (i) an awareness, and (ii) a use, of the ecological principles which have just been discussed. McHarg (1969) advised us to design *with* nature. If *with* can mean both *in agreement with* and *by means of*, this is surely wise advice. But what will this entail in practice? Certainly it is more than just using a few wild plants in landscaping schemes.

The first and perhaps most crucial element is a proper understanding of the environment of each site. This must include not only the climate but also the soil, even if this is to be brought in. It must also include possible influences and demands from man or animals, and the influences of other plants, especially those which may invade without invitation. The final product cannot be just a particular collection of pleasing plants; it will be a whole ecosystem subject to the same controlling and determining factors as other more natural ecosystems. Indeed, we have to understand 'the genius of the place' and follow in the footsteps of Kent and Pope (Dutton 1937). We can now do this with a much greater background of knowledge than our predecessors.

We can adopt one of a number of different ecologically-based approaches, the first of which is remarkably simple—*laissez faire*. We can allow natural ecosystem development. To let an area 'go wild' is not a silly idea. Many beautiful recent landscapes are due to it. The upper parts of Hampstead Heath are merely an old sand pit which supplied the needs of Georgian London. Many disused quarries, such as Millersdale in the Peak District, are not only havens for attractive wild plants, but are beautiful in their own right. We are now beginning to realize that we possess an immense heritage of wilderness, such as rubbish dumps and gravel pits, often very close to our doorstep (Teagle 1978), which does not require reclamation in the modern 'land reclamation' sense, but merely some careful nurturing (Bradshaw 1979). This approach has a bonus, in that it is one way of carrying out nature conservation. Its possibilities are shown by Goode & Smart (Chapter 14).

The second approach is the opposite, *positive construction*. This can be seen as the traditional approach to landscape design, but the difference is that it uses ecological principles. The soil can be constructed. The materials which are present can be accepted as they are and not replaced or covered up. But they can then be improved as far as is required: perhaps by selective use of fertilizers, etc. to overcome particular limiting problems, but also by the use of plants themselves, particularly legumes which accumulate nitrogen as well as organic matter. This is of course the basis of modern land reclamation (Bradshaw & Chadwick 1980) and is discussed by Roberts & Roberts (Chapter 7). Plants can similarly be used as engineering materials, for soil and slope stabilization (Schiechtl 1980). The climate can also be modified as required by protective planting; splendid extreme examples of what can be done are at Logan and Inverewe (Cowan 1964).

The plant community can itself be constructed to ensure ecological balance and soundness, i.e. by choosing species which are adapted not only to the physical environment, but also to coexist with each other at the outset and subsequently as the ecosystem develops. It is this that has been called ecological planting and is discussed in more detail by Tregay (Chapter 17). However, it is crucial to realize that the mere sowing of a wild flower mixture or the use of particular native species are not themselves examples of ecologically-based design, unless the species have been carefully chosen in relation to the soil and intended. Indeed, an ecological approach may involve alien species.

The third approach lies between the two previous; it is *manipulation of development*. Although ecosystem development is inevitable, we should remember that we can manage it for our own ends if we understand the ecology (Grime 1980). A number of courses of action are open. We can, for instance, inoculate the ecosystem with particular ecologically appropriate species and then leave them to their own devices. One example of this is the technique of tree seeding (Luke & McPherson 1983); another is the inoculation of specifically adapted wild species on industrial waste land: if the species are chosen well the results can be remarkable (Ash 1983).

Ecosystems can be managed by the use of particular mowing, burning or other practices (Green, Chapter 12), heathland being one of the best examples of what can be achieved (Gimingham & de Smidt 1983). But we must actually use these practices; the effects of different management on roadsides are now well understood (Way 1969) but, alas, the Department of the Environment has decreed that no management will be applied. The results will be interesting to watch. We can also manage the soil: this has rarely been done, but the astonishing effects that can be achieved by simple liming or fertilizing are very clear from the first 100 years of the Park Grass experiment at Rothamsted (Brenchley 1958). The diversity and control that can result from different management practices, as well as the lack of control from lack of management, must not be forgotten.

The fourth approach is *restoration*. It is perhaps really a combination of the previous three, applied now to a landscape which has been degraded. Here, almost more than anywhere, an appreciation of the qualities of the environment and the existing ecosystem is essential. Often only one or two ecological problems may exist and have to be treated. This might involve no more than restoration of soil fertility. A good example is the importance of lime to restore the pH of city park grasslands (Vick & Handley 1975). It might otherwise involve the replacement of one species contributing to the structure of the landscape by another more ecologically suited. But it can also be more radical, such as the wholesale removal and replacement of an ecosystem threatened by industrial activity. There are now splendid examples of what can be achieved, such as in pipelines through British moorlands (Putwain, Gilham & Holliday 1983) and Australian mineral sand mining (reviewed in Bradshaw & Chadwick 1980). But none of this can be done without ecological understanding, as can be seen from later contributions such as Chapter 22 by Handley.

CONCLUSIONS

The essential characteristic of an ecological approach is to copy with understanding what nature achieves by blind natural processes, i.e. to maximize the adaptation of the species planted to their environment, to simulate natural conditions in their management, and so to create integrated and harmonious ecosystems. It is an essential paradigm of nature that what ultimately survives is that which is best fitted to its environment. It therefore follows that if we attempt to model landscape designs on natural systems and use natural processes to achieve our desired end-points, we are more likely to finish up with resilient and self-sustaining solutions. Indeed, the nearer our solutions are to natural solutions the less likely they are to be ousted by nature and to require costly management to maintain them.

At the same time, the solutions are likely to be more economical to achieve. *Laissez-faire* requires little input, *construction* requires more. But if the construction is ecologically based and uses ecological processes it should be possible to reduce costly artificial inputs. This is obvious in the use of legumes as a source of nitrogen, but it is also true, for instance, for soil stabilization by plants rather than by chemical means.

It could be argued that these approaches will produce such naturalistic results that they will be indistinguishable from nature. But it is never possible to recreate natural processes exactly: the art of naturalistic landscape design is a subtle use of existing materials to create new entities which improve on nature. At a time when the natural landscapes we enjoy so much are being ravaged by the requirements and pressures of modern civilization, there is a great deal to be said for emulating nature. They can, as at Warrington (Scott, Chapter 9) provide an outstanding setting for an urban development. In this sort of approach, ecological understanding is essential.

Some landscape design has little to do with nature but, even here, plants are a substantial element. Ecological understanding is, therefore, crucial to ensure that the environments produced are optimal and economic, the plants

FIG. 2.11. Tree death because of a simple disregard of ecological principles. The soil generally is poor. Good soil has been used in the tree pits, but grass has been allowed to grow immediately around the trees. Its vigorous growth has meant excessive competition for water, which has actually killed most of the trees after they had begun to grow.

are appropriate, and the designers' aims are achieved. If the principles of ecology are disregarded then the scheme will fail (Fig. 2.11).

The science of ecology has thus provided valuable insight into the relationship between organisms and their environment. This understanding has a crucial part to play in the development of landscape design.

REFERENCES

Ash, H.J. (1983). *The natural colonisation of derelict industrial land.* Ph.D. thesis, University of Liverpool.

Berg, W.A. (1973). Evaluation of P and K soil fertility tests on coal-mine spoils. *Ecology and Reclamation of Devastated Land* (Ed. by R.J. Hutnik & G. Davis) vol. 1, pp. 93–104. Gordon and Breach, New York.

Bloomfield, H.E., Handley, J.F. & Bradshaw, A.D. (1981). Topsoil quality. *Landscape Design*, **135**, 32–34.

Bloomfield, H.E., Handley, J.F. & Bradshaw, A.D. (1982). Nutrient deficiencies and the aftercare of reclaimed derelict land. *Journal of Applied Ecology*, **19**, 151–158.

Boardman, N.K. (1977). Comparative photosynthesis of sun and shade plants. *Annual Review of Plant Physiology*, **28**, 355–377.

Bormann, F.H. & Likens, G.E. (1979). *Patterns and Process in a Forested Ecosystem.* Springer-Verlag, New York.

Bradshaw, A.D. (1979). Derelict land—is the tidying up going too far? *The Planner*, **65**, 85–88.

Bradshaw, A.D. (1981). Growing trees in difficult conditions. *Proceedings of Forestry Commission/Arboricultural Association seminar* pp. 93–106. Forestry Commission, Edinburgh.

Bradshaw, A.D. (1983). The reconstruction of ecosystems. *Journal of Applied Ecology*, **20**, 1–17.

Bradshaw, A.D. & Chadwick, M.J. (1980). *The Restoration of Land.* Blackwell Scientific Publications, Oxford.

Bradshaw, A.D. & McNeilly, T. (1981). *Evolution and Pollution.* Arnold, London.

Brenchley, W.E. (1958). (revised by Warington K.) *The Park Grass Plots at Rothamsted 1856–1949.* Rothamsted Experimental Station, Harpenden.

Clausen, J., Keck, D.D. & Hiesey, W.M. (1940). Experimental studies on the nature of species I. The effect of varied environments on western North American plants. *Carnegie Institute of Washington Publication* **520**.

Cowan, M. (1964). *Inverewe.* Geoffrey Bles, London.

Crocker, R.L. & Major, J. (1955). Soil development in relation to vegetation and surface age at Glacier Bay, Alaska. *Journal of Ecology*, **43**, 427–48.

Donald, C.M. (1958). Competition among crop and pasture plants. *Advances in Agronomy*, **15**, 1–118.

Duffey, E., Morris, M.G., Sheail, J., Ward, L.K., Wells, D.A. & Wells, T.C.E. (1974). *Grassland Ecology and Wildlife Management.* Chapman and Hall, London.

Dutton, R. (1937). *The English Garden.* Batsford, London.

Dutton, R.A. & Bradshaw, A.D. (1982). *Reclamation of Land in Cities.* HMSO, London.

Etherington, J.R. (1975). *Environment and Plant Ecology.* John Wiley, London.

Fonteno, W.C. & McWilliams, E.L. (1978). Light compensation and acclimatization of four tropical foliage plants. *Journal of the American Society for Horticultural Science*, **103**, 52–56.

Gadgil, R.L. (1971). The nutritional role of *Lupinus arboreus* in coastal sand dune forestry. 3. Nitrogen distribution in the ecosystem before tree planting. *Plant and Soil*, **35**, 113–26.

Gimingham, C.H. & de Smidt, J.T. (1983). Heaths as natural and semi natural vegetation. *Man's Impact on Vegetation* (Ed. by W. Holzner, M.J.A. Werger and I. Ikusima), pp. 185–200. Junk, The Hague.

Grime, J.P. (1979). *Plant Strategies and Vegetation Processes.* Wiley, London.

Grime, J.P. (1980). An ecological approach to management. *Amenity Grassland: An Ecological Perspective* (Ed. by I.H. Rorison & R. Hunt), pp. 13–56. Wiley, Chichester.

Grime, J.P. & Hunt, R. (1975). Relative growth rate: its range and adaptive significance in a local flora. *Journal of Ecology*, **63**, 383–422.

Grime, J.P. & Lloyd, P.S. (1973). *An Ecological Atlas of Grassland Plants,* Arnold, London.

Grubb, P.J. (1977). The maintenance of species richness in plant communities: the importance of the regeneration niche. *Biological Reviews*, **52**, 107–145.

Harper, J.L. (1977). *Population Biology of Plants.* Academic Press, London.

Holzner, W., Werger, M.J.K. & Ikusima, I. (1983). *Man's Impact on Vegetation.* Junk, The Hague.

Jarvis, P.G. & Jarvis, M.S. (1963). The water relations of tree seedlings. I Growth and water use in relation to soil water potential. *Physiologica Plantarum*, **16**, 236–253.

Jefferies, R.A., Bradshaw, A.D. & Putwain, P.D. (1981). Growth, nitrogen accumulation and nitrogen transfer by legume species established on mine spoils. *Journal of Applied Ecology*, **18**, 945–956.

Leismann, G.A. (1957). A vegetation and soil chronosequence on the Mesabi Iron Range spoil banks, Minnesota. *Ecological Monographs*, **27**, 221–45.

Lovett Doust, L. (1981). Population dynamics and local specialisation in a clonal perennial (*Ranunculus repens*) II. The dynamics of leaves and a reciprocal transplant-replant experiment. *Journal of Ecology*, **69**, 757–768.

Luke, A.G.R. & McPherson, T.K. (1983). Direct tree seeding: a potential aid to land reclamation in Central Scotland. *Arboricultural Journal*, **7**, 287–299.

McHarg, I.L. (1969). *Design with Nature.* Natural History Press, New York.

Manaker, G.H. (1981). *Interior Plantscapes.* Prentice Hall, New Jersey.

Marrs, R.H., Roberts, R.D., Skeffington, R.A. & Bradshaw, A.D. (1983). Nitrogen and the development of ecosystems. *Nitrogen as an Ecological Factor* (Ed. by J.A. Lee, S. McNeill & I.H. Rorison), pp. 113–136. Blackwell Scientific Publications, Oxford.

Miles, J. (1979). *Vegetation Dynamics.* Chapman and Hall, London.

Penman, H.L. (1963). *Vegetation and Hydrology,* Commonwealth Bureau of Soils, Harpenden.

Peterken, G. (1981). *Woodland Conservation and Management.* Chapman and Hall, London.

Putwain, P.D., Gillham, D.A. & Holliday, R.J. (1983). Restoration of heather moorland and lowland heathland with special references to pipelines. *Environmental Conservation*, **9**, 225–235.

Ratcliffe, D.A. (1977). *A Nature Conservation Review* Vol. I. Cambridge University Press, Cambridge.

Rorison, I.H. (1969). Ecological inferences from laboratory experiments on mineral nutrition. *Ecological Aspects of the Mineral Nutrition of Plants* (Ed. by I.H. Rorison), pp. 155–175. Blackwell Scientific Publications, Oxford.

Russell, E. (1973). *Soil Conditions and Plant Growth* (10th edn). Longmans, London.

Salisbury, F.B. & Ross C.W. (1978). *Plant Physiology* (2nd edn.). Wadsworth, Belmont, California.

Schiechtl, H. (1980). *Bio-engineering for Land Reclamation and Conservation.* University of Alberta Press, Edmonton.

Snaydon, R.W. & Bradshaw, A.D. (1961). Differential responses to calcium within the species *Festuca ovina* L. *New Phytologist*, **60**, 219–234.

Tansley, A.G. (1935). The use and misuse of vegetational terms and concepts. *Ecology*, **16**, 284–307.

Tansley, A.G. (1939). *The British Islands and their Vegetation.* Cambridge University Press, Cambridge.

Teagle, W.G. (1978). *The Endless Village.* Nature Conservancy Council, Shrewsbury.

Vick, C.M. & Handley, J.F. (1975). Lime and fertilizer use in the maintenance of turf in urban parks and open spaces. *Journal of the Institute of Park and Recreation Administration*, **40**, 39–48.

Way, J.M. (Ed.) (1969). *Road Verges: Their Function and Management*. Monk's Wood Experimental Station Symp. Natural Environment Research Council, London.

Welbank, P.J. (1963). A comparison of competitive effects of some common weed species. *Annals of Applied Biology*, **51**, 107–126.

Wright, J.W. (1962). *Genetics of Forest Tree Improvement*. Food and Agriculture Organization of the United Nations, Rome.

3. PROFESSIONAL INTEGRATION IN PLACE OF INEPTITUDE

R. O. COBHAM

Cobham Resource Consultants, 19 Paradise Street,
Oxford OX1 1LF

SUMMARY

In the early 1970s the Landscape Institute decided to widen its membership by forming three branches: architects, scientists and managers. Since then there has been progress in providing society with integrated landscape services. Examples of how managers interpret the integrated approach and are putting it into practice are given. The approach differs considerably according to the nature of the project: (i) large-scale rural or resource projects, (ii) historic landscape restoration, or (iii) urban development work.

However, there are still shortcomings and these stem in part from the inherently different roles and products of the three professions. Means of remedying some of these deficiencies are suggested later in the paper. Firstly, however, a major shortcoming is highlighted; namely, the inadequate data we have on the quantity and qualities of all the features which make up the national landscape. The biggest data deficiencies lie in expenditure. Estimates compiled from a range of sources indicate that the national annual figure for expenditure on all landscapes is likely to exceed £600 million, but by how much is not known. Recommendations are made for undertaking research to compile the necessary data, including support for the establishment of a Centre for a National Conservation Strategy. The paper includes a strong plea for landscape planning to be taken seriously, so that national priorities, policies and plans can be prepared. These are essential in order to provide the three professional interests with a strong basis on which to develop and focus their integrated services.

INTRODUCTION

When members of the Landscape Institute and the British Ecological Society stumble upon each other in the field it is to be hoped that recognition in the supreme style of 'Dr Livingstone, I presume?' is not required. What opening remark might we hope or expect instead of this now classical and almost

37

cheeky demonstration of self-confidence? 'Oh no! you're not working on my patch as well?' or 'Hi! what's another . . . . ist doing in these parts?' or 'Have you discovered the "genius loci"?', 'Does the "golden section" apply?' or 'Is there a management plan?' Naturally we hope that it is something in the vein of the last. Thus, it is very much in a spirit of thought-sharing and with a desire for discovery that this paper has been prepared. Whilst I am conscious of presenting a personal view I hope it is not one which is too biased, isolated or inept.

The invitation to contribute stems from the desire of the symposium organizers to include a landscape manager in setting the scene. I attempt to do so on a large scale. The first two contributions have been from a landscape designer and a landscape scientist, after which some may remark 'and they've said it all'! On the contrary, I hope to demonstrate that the concept of a triumvirate of landscape designers, scientists and managers, conceived by the free thinkers of the Institute in the early 1970s, is working at least partly in practice and is even more relevant to the needs of society today than it was then. I shall also indicate—hence the choice of title—the ways in which I have observed the three disciplines relating to each other and the ways in which the relationships need to improve.

PROFESSIONAL ROLES AND RESPONSIBILITIES

Definitions

At the outset it is important to remind ourselves about the particular roles and skills which the three parties were expected to achieve and contribute respectively.

Landscape architects are trained in the planning and design of all types of outdoor spaces. They use knowledge of the natural elements of the landscape, its materials and components, to create the spatial and aesthetic elements of the new environment. Many practitioners are also qualified in other disciplines such as horticulture, planning or architecture. They may develop projects into contracts and supervise their execution on site. Landscape architecture is classified as a professional occupation group under the International Standard Classification of Occupations (ILO 1968).

Landscape scientists are concerned with the physical and biological principles and processes which underlie the planning, design and management of natural resources. They have the ability to relate their scientific knowledge to the practical problems of landscape work. This can range from small-scale site surveys to the ecological assessment of broad areas for planning or management purposes, as well as preparation of reports on development impact or research into the importance of particular species in given areas. Landscape scientists usually have a scientific background, such as ecology, often with specialist skills such as soil science, hydrology, geomorphology or botany.

Landscape managers use their detailed understanding of plants and the natural environment to advise on the long-term care and development of the changing landscape. This involves them in the financial and physical organization of manpower, machinery and materials. They also have to consider statutory measures such as planning laws and grant aid schemes in order to preserve and enhance the quality of the landscape. Landscape managers usually have a degree in horticulture, forestry or agriculture together with further training in land management or other related disciplines (Landscape Institute 1983).

Nowhere, however, has anyone attempted to portray how this triumvirate is expected to operate. Figures 3.1 and 3.2 attempt to do so, in the interests of providing structure for the discussions which follow. There are obvious similarities between the responsibilities. Equally important, however, are some of the differences, both in function and in degree, which are illustrated later. There are several reasons for these differences because:

(a) ecologists have tended to be concerned primarily with the *conservation* of *existing* features, particularly with the most diverse and therefore often the oldest features; whereas

(b) landscape architects have tended to be as much, if not more, concerned (depending on circumstances) with creating new features as with conserving existing ones.

Whilst the landowner has often come to terms with not using or cultivating land occupied by an existing wildlife-cum-landscape feature, it is quite another matter when he is asked to give up even a small area of potentially productive land (with an asset value of over £5000 per hectare and an annual profit-generating capacity of about £100 per hectare).

However, there are even more fundamental reasons for the differences between the professional responsibilities, associated with the inherent nature of their respective 'products', to which I will refer later.

PROFESSIONAL APPROACHES

General

The title of this paper refers to ineptitude and integration. The former symbolizes the designed and managed landscapes which are either manifestly deficient or do not work, because an isolated or inept approach has been

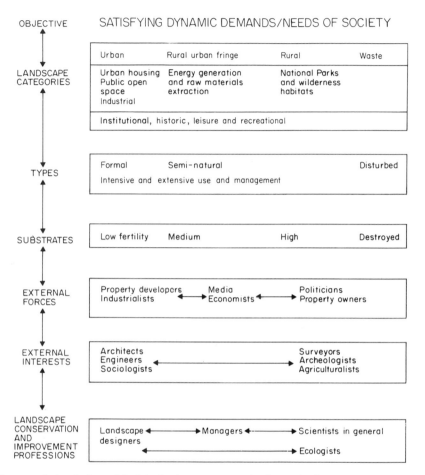

OBJECTIVE | SATISFYING DYNAMIC DEMANDS/NEEDS OF SOCIETY

Urban	Rural urban fringe	Rural	Waste
Urban housing Public open space Industrial	Energy generation and raw materials extraction	National Parks and wilderness habitats	

Institutional, historic, leisure and recreational

Formal	Semi-natural		Disturbed
Intensive and extensive use and management			

| Low fertility | Medium | High | Destroyed |

FIG. 3.1. A simplistic model of the landscape conservation 'ecosystem'.

adopted rather than an integrated or multi-disciplinary one. Ineptness applies where individual professions try to solve problems by operating 'solo'. I do not wish here to dwell on the inept approach, since, unfortunately, examples can be seen all too frequently. There are, of course, straightforward design, ecological or management tasks, which can be undertaken in a single-minded way, but often the problems are more complex.

Whilst an integrated approach is not offered as a panacea, personal experience and observation indicate that the integrated approach is being increasingly adopted. It is for positive reasons that I wish to concentrate on various aspects of this approach.

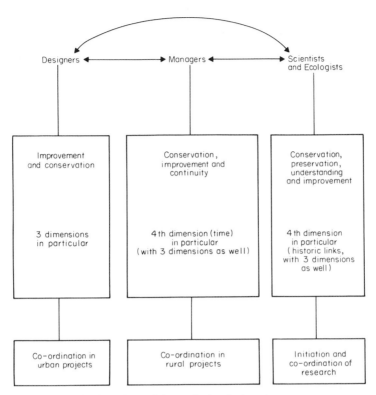

FIG. 3.2. Environmental orientations of the professional triumvirate.

Approaches adopted by managers

Landscape managers are generally unable to operate successfully without inputs from their design and science colleagues, and indeed from a wide range of other specialists. The particular specialists depend, of course, upon the nature of the project. A series of different projects illustrate what can justifiably be termed the 'integrated approach'.

Demonstration farms project. Multiple use and environmental improvement. For this the integrated planning and implementation process is described by Fig. 3.3a, and the professional inputs by Fig. 3.3b.

Blenheim restoration project. Restoration and resource improvement in the context of multiple use. For this the integrated planning and implementation process is described by Fig. 3.4a, and the professional inputs by Fig. 3.4b.

Kuwait and Saudi Arabian projects. Resource development: utilization of treated sewage effluent for commercial and environmental enhancement. The

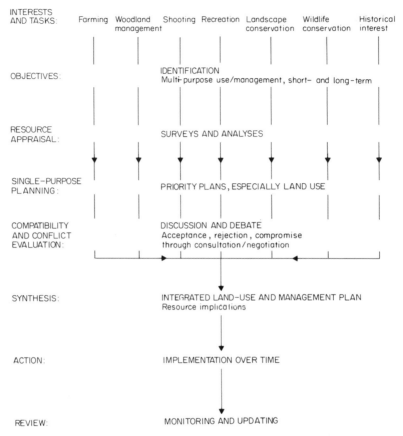

FIG. 3.3. (a) Integrated planning and implementation process for rural landscapes.

processes involved are similar to Fig. 3.3a, but with different interests, i.e. agriculture, forestry, environmental protection, urban beautification, recreation, industry, water conservation.

The approaches adopted differ according to the nature of the project or problem. Major rural or resource development projects are likely to find the land manager fulfilling the lead co-ordination role, if for no other reason than that there is a need for a sound knowledge of both commercial and conservation interests. From experience, landscape architects in particular recognize that decisions on land use and other functional issues need to be taken before they can attend effectively to matters of environmental improvement.

In the case of prestige landscape restoration projects the nature of the

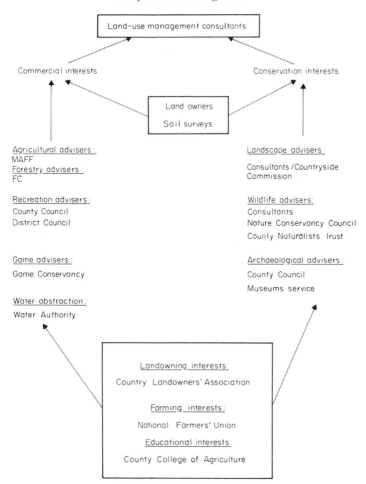

Fig. 3.3. (b) Professional inputs into an integrated plan for a rural landscape.

integration is different, as can be seen from the contribution by Moggridge (Chapter 26) with whom I am working at Blenheim. In such cases, where cultural interests assume an importance at least as great as the functional interests, the disciplines are able to work in parallel throughout.

Where urban environmental improvement projects are involved, the lead co-ordinating role is invariably taken by the landscape architect with important supporting inputs from management, ecology and other disciplines. If these projects are to succeed, the managers and ecologists need to be involved at the conceptual design and cost-planning stages as well as the management planning and continuing maintenance stages.

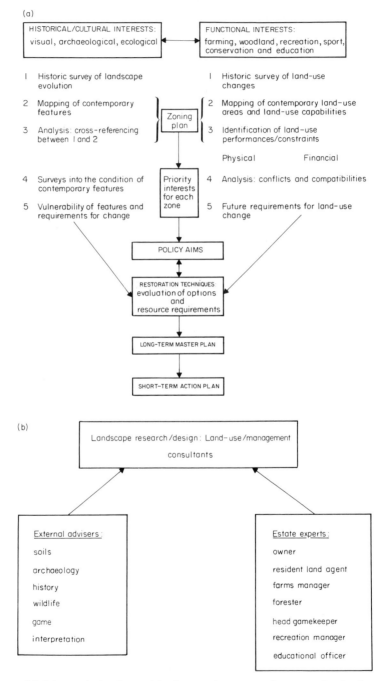

FIG. 3.4. (a) Integrated planning and implementation process for restoration landscapes. (b) Professional inputs into an integrated plan for a restoration landscape.

OBSERVATIONS ON THE INTEGRATED
AND OTHER APPROACHES

Practical achievements

The integrated approaches displayed in the above projects are working, as shown by the results achieved. Readers can judge the success for themselves, either by visiting the sites concerned or by referring to the recently published reports and papers cited.

Demonstration farms project (Cobham *et al.* 1984): ten multi-purpose plans and their implementation/actual improvements achieved.

Private estate projects: Blenheim (Cobham & Hutton 1983; Cobham 1984): corporate plans, involving landscape restoration and improvement.

Kuwait and Saudi Arabian projects (Beynon *et al.*, 1984): resource development.

Roles and performances of designers and scientists

The differences in roles referred to earlier are best illustrated by reference to the results of ten demonstration farm planning-cum-management exercises, in which ten different landscape architects/planners and ten different ecologists (wildlife evaluators) were invited to make single-purpose proposals (SPP) which would achieve the optimization of wildlife or visual landscape values (irrespective of other professional land use/management interests) within the bounds of practicalities. In essence this meant that the farms were required to continue first and foremost as commercial enterprises and could not be converted either to nature reserves or public parks.

It is interesting to observe the fate, at the hands of the final arbiters, of the recommendations made by the various disciplines, particularly the landscape architects and the ecologists (Table 3.1). These results indicate that where a landscape architect is invited to participate in a farm planning exercise he is likely to fare as well as any of the other professional interests. In these situations the figures display no significant differences. However, differences do exist and are largely responsible for the relatively meagre influence which landscape architects have had in rural areas. The difference in influence of ecologists and landscape designers stems primarily from what may crudely be referred to as the inherent differences between the professional products.

(a) The ecologist's products are often more tangible, in the form of annual crops of birds, plants, butterflies and other insects which the landowner or manager can enjoy as a direct result of conserving existing habitats, usually at relatively low cost. Where these habitats

on searching for more effective solutions; solutions which not only embrace, but also go beyond those being developed through the work of Arthur Oldham, Alan Ruff, Chris Baines, Lyndis Cole and others. I have long felt that we need to develop more skill in providing large-scale and robust urban environments: much more urban woodland and urban scrubland, i.e. real adventure areas rather than just playgrounds. In theory, there is surely enough available space, but something has to be done in the property market first. In addition we need to develop new systems of management and personal remuneration which do not provoke the present types of hostilities between union, voluntary and MSC labour.

In short, this type of inventory approach should help us to determine the action priorities for environmental provision and improvement, e.g. how our efforts should be allocated between rural, urban and rural fringe areas. Without such an approach, some major opportunities and decisions will most probably go by default. We will continue in a rather *ad hoc*, reactive manner.

Other recommendations

There are several recommendations which could be made, but at this point it seems appropriate to concentrate on four.

1   Because of the lack of essential resource data, renewed thought should be given to the role of landscape planning as an integral discipline. At present there is no overall custodian of environmental needs and resources; should this role be fulfilled in large measure by planners?

2   To improve the future success of an integrated approach, it is essential that each of the three professional interests should positively seek ways in which they can help the others. For instance:

   (a) landscape designers need considerable help in sorting out (i) the nature of the service required by rural landowners, and (ii) how best to arouse the interests of land-owning organizations;
   (b) all three parties should be asked to indicate the time-scale over which any landscape will require to be renewed;
   (c) all parties should promulgate interest in the preparation of and commitment to management plans where significant landscape projects are involved. No self-respecting shipping, vehicle or washing machine manufacturer would offer its products for sale without an instruction manual. There are many good reasons why the landscape professions should collectively ensure that they do likewise.

3   The Landscape Institute should make specific provision for involving ecologists and managers in future Garden Festivals, since they represent major events for displaying the fruits of integration. The competitions at

Liverpool and Stoke-on-Trent disappointingly seem to have concentrated primarily upon design contributions.

4 When contemplating measures to bring about environmental improvement in the private sector, there are many grounds for recommending that the measures should be as broad-based as possible. We should avoid the single-minded, 'all-controls' or 'all-incentives' approaches and advance on as wide a front as possible, so that a mixture of measures can be selected to suit each individual case (Cobham *et al.* 1984).

In conclusion, we will make real progress in identifying the landscape problem areas and in solving them, only if we are resolved to make the triumvirate stronger than in the past and are thus fully committed to the integrated approach.

ACKNOWLEDGMENTS

The author gratefully acknowledges the assistance given by his colleagues Margery Slatter, Lois Bowser and particularly Suki Pryce in compiling Table 3.2. Special thanks are also due to Dr R. Bunce of ITE for the estimates of urban landscape areas.

REFERENCES

Beynon, R.B., Cobham, R.O. & Matthews, J.R. (1984). Kuwait effluent utilization project—concept and implementation. In *Proceedings of the 3rd Arab Water Technology Conference.* Dubai, United Arab Emirates, October 1984.

Chartered Institute of Public Finance and Accountancy (CIPFA) (1981). *Financial, General and Rating Statistics 1981–82.* CIPFA Statistical Information Services SIS Ref 41.82 CIPFA, London.

Cobham, R. O. (1983). The economics of vegetation management. In *Management of Vegetation* (ed. J.M. Way), pp. 35–66. British Crop Protection Council Monograph No 26. BCPC Publications, Croydon.

Cobham, R.O. (1984). Blenheim: the art and management of landscape restoration. *Landscape Research,* **9,** 4–14.

Cobham, R.O. & Hutton, P. (1983). Brown in memoriam: Blenheim park in perpetuity. *Landscape Design,* **146,** 11–12.

Cobham, R.O., Matthews, J.R., McNab, A., Stephenson, E. & Slatter, M.J.S. (1984). *Agricultural Landscapes Demonstration Farms.* CCP 170 Countryside Commission, Manchester.

Cobham Resource Consultants (1983). *Countryside Sports: Their Economic Significance.* Standing Conference on Countryside Sports, London.

International Labour Organisation (1968). *The International Standard Classification of Occupations.* International Labour Organisation, Geneva.

Landscape Institute (1983). *Register of Landscape Practices.* Landscape Institute, London.

Van der Post, L. (1983). *Yet Being Someone Other.* Hogarth Press, London.

Walker, S.E. & Duffield, B.S. (1983). Urban parks and open spaces—an overview. *Landscape Research,* **8** (2).

World Wildlife Fund *et al.* (1983). *The Conservation and Development Programme for the UK.* Kogan Page, London.

4. ECOLOGICAL ELEMENTS IN LANDSCAPE HISTORY

G. F. PETERKEN

Nature Conservancy Council, Northminster House, Peterborough, PE1 1UA

SUMMARY

The British landscape has developed continuously through millenia of fluctuations and developments in land use. In some periods the rate of change has been revolutionary, but mostly the landscape has been fairly stable, changing only in detail. Despite these changes some original natural features have survived. Elsewhere, despite the prevailing intensive use of land, natural vegetation has reappeared on unused patches, though the composition is unlikely to be identical to the original natural vegetation on these sites. The detailed distribution of plant and animal species is influenced by existing land use and the history of individual elements in the landscape. In any consideration of landscape management, ancient elements must be identified, retained and properly managed.

AN ENGLISH LANDSCAPE

My home is situated on a low spur of limestone and boulder clay between the river Nene and a small tributary, the Lyveden Brook. The main valley is intensively farmed, with carefully trimmed hawthorn hedges separating a mixture of arable and pasture fields. This is a predominantly nineteenth century landscape dating from inclosure in 1811, which still has a few hedgerow ash trees dating from the era of planned inclosure (Fig. 4.1). The gentle slopes beyond, rising onto the boulder clay plateau, were likewise enclosed from open fields—the ploughed-out ridge and furrow can still still be seen in the snow-melt patterns—but here the lacework of arable and hedges is relieved by small plantations sheltering Barnwell Castle. The horizon, however, is formed by a much older feature, Ashton Wold, a substantial wood which originated in or before the Middle Ages. Parts of the valley flood plain also appear to be well wooded, but this is an illusion created by a mixture of scrub willow fringing recent gravel workings, and regularly spaced tree willows planted along the various channels of the main river. To the north-

G. F. PETERKEN

FIG. 4.1. Panorama of Oundle and surrounding countryside from the south-west. Beyond the straight inclosure hedges and the contemporary ash tree (centre), Oundle occupies a slight elevation. The Nene flows from the right and bends right in front of the town. Trees along the valley bottom are mainly of recent origin around lakes left by gravel working, just visible to right of centre. Although Oundle contains a number of buildings originating before 1600, the oldest feature in the view is the woodland of Ashton Wold, on the skyline to the extreme right.

FIG. 4.2. The linear woodland along the Lyveden Brook, beyond which is the medieval Biggin Park Wood. The ground between was once covered by other medieval woods, but these were cleared sometime between 1565 and 1800, and the land is now divided by inclosure hedges.

east, on the next low ridge, lies Oundle, a small town built of limestone, whose density and variety of mature trees makes it one of the most arborescent features in the landscape. To the north, across the Lyveden Brook, stands the medieval Park Wood, situated within the perimeter of the fourteenth century deer park of Biggin (Fig. 4.2).

Although it is not obvious to a modern observer, this land has been intensively used for at least 2000 years. There is an important Roman settlement at Ashton and a high density of Roman and Romano-British occupation sites in and around the Nene valley. Fleeting crop marks on both the valley and the plateau indicate earlier inclosures and possibly cultivation (RCHM 1975). The valley of the Lyveden Brook was a notable centre for medieval pottery manufacture. All this is now invisible, so now the only features to pre-date the eighteenth century are the townscape of Oundle and the ancient woodlands of Ashton Wold and Biggin Park.

Millenia of cultivation have deprived this landscape of any direct contact with the original natural vegetation, and the combination of inclosure and modern agriculture has severed most links with the medieval landscape, but despite this, certain natural features survive. The Nene itself, still winding within natural meanders and divided channels, floods the view in most years.

The Lyveden Brook, though flowing in an artificially deepened bed, is lined with alder (*Alnus glutinosa*), willow (*Salix alba* and *S. viminalis*), sallow (*Salix cinerea*), elm (*Ulmus procera*), oak (*Quercus robur*), poplar (*Populus canescens*), blackthorn (*Prunus spinosa*) and other shrubs, all of which were probably constituents of the natural riparian woods (Fig. 4.2). The ancient woods were managed for centuries as coppice-with-standards and, though now unmanaged, still have this structure. Their mixture of pedunculate oak, ash, maple, hazel and hawthorn is typical of alkaline, poorly-drained clays in the East Midlands, and must be natural in some sense. The woods are well stocked with native woodland plants and animals. Ashton Wold is probably not primary woodland, but ancient secondary, originating on soils which were little altered by primitive cultivation in a landscape containing many patches of semi-natural vegetation.

This particular landscape is no more typical of Britain than any other, but it does illustrate some important general points.

1 Natural features have influenced the distribution of land use and the choice of trees for planting. For example, permanent pasture is confined to the flood plain. Woods are absent from the flood plain, where in the past land was more valuable as meadow. Willows have been planted on the river bank, but oak and ash have been planted on the clay plateau.

2 The landscape has evolved over a long period, but has experienced short periods of rapid change, most recently during early nineteenth century inclosure and the modern switch from pasture to arable.

3 Landowners have changed the landscape (or retained features) not only for utilitarian purposes, but also for aesthetic and recreational reasons.

4 Ancient elements survive, even in this perpetually changing landscape. The features we see date from a variety of historical periods.

5 Natural elements also survive, even in this intensively farmed landscape. Some, such as the composition of the ancient woods, have persisted from a remote period, but others, such as the young scrub round the gravel pits, are very recent in origin, and still rapidly developing.

6 Features which are both ancient and relatively natural contain both richer wildlife communities and rarer species. For example, Ashton Wold was until recently a haven for many butterfly species. Yellow archangel (*Galeobdolon luteum*) is just one example of a plant species which grows in the ancient woods, but not in those of more recent origin.

7 Natural communities can be created even after severe and prolonged disturbance, though they are usually stocked only with common and widespread species. This is true of the scrub, grassland and aquatic vegetation developing on the abandoned gravel pits, and would be true of any arable field reverting to pasture or woodland.

LANDSCAPE CHANGE

Certain features of landscape change can be illustrated with studies of woodland changes in three parts of the East Midlands (Peterken 1976). These areas entered the nineteenth century with very different densities of woodland (Table 4.1), but during that century their differences were reduced by substantial clearances in the well-wooded parts and patchy planting in the least wooded parts. This process of convergence was reversed in modern times, when many woods were converted from coppice to plantation—a process which favoured the larger and thus more economically managed woods. The small woods of sparsely wooded areas were vulnerable to clearance for cultivation. This broad pattern was complicated by open-caste ironstone mining in Rockingham Forest, which has accelerated the turnover of woodland sites but—due to land restoration with trees—not diminished the total woodland area. We see here a feature which I suspect is widespread; namely, that national economic, social and technical trends can have markedly different impacts on the local landscape, partly at least because the landscapes available for them to act upon differ substantially as a result of earlier historical phases. In this instance Rockingham Forest had remained well wooded because deforestation came as late at 1796, and parts of

TABLE 4.1. Changes in the area of woodland in Rockingham Forest, Central Lincolnshire and West Cambridgeshire since the early nineteenth century

	RF	CL	WC
Clearance rate (ha per year)			
1817–34 – 1885–87	−46·0	−19·1	−0·7
1885–87 – 1946	−6·3	−2·2	−0·7
1946 – 1972–73	−17·3	−4·3	−9·3
Afforestation rate (ha per year)			
1817–34 – 1885–87	+3·9	+5·4	+5·4
1885–87 – 1946	+16·5	+8·0	+5·3
1946 – 1972–73	+18·0	+13·3	+0·4
Net rate of change in woodland area (ha per year)			
1817–34 – 1885–87	−42·2	−13·7	+4·2
1885–87 – 1946	+10·0	+4·8	+4·6
1946 – 1972–73	−0·7	+9·0	−8·9
Extent of study area (ha)	41550	49950	42900
Woodland area in 1817–34 (ha)	7642	4247	981
Woodland area in 1972–73 (ha)	5525	3912	1234

Lincolnshire on marginal farmland had also remained well wooded. Indeed, the irregular pattern of woodland which influenced the changes in the nineteenth and twentieth centuries had been established by the eleventh century (e.g. Darby & Terrett 1954).

Even when landscapes appear to be unchanging in broad pattern, they are changing in important detail. For instance, many woods occupy the same land as they occupied in the Middle Ages, yet in this century their semi-natural mixture of native broadleaves treated as coppice-with-standards has often been replaced by evergreen conifers planted as even-aged high forest (Table 4.2): the impact on the landscape is considerable. More subtly, as Roberts (1959) showed in Snowdonia, changes in upland pastoral practices influenced the spread of bracken and the balance between various types of grassland. Even before the recent spate of hedgerow clearance, the number and form of hedgerow trees changed from time to time (Rackham 1977). The New Forest unenclosed woods, symbols of unchanging landscape, have in fact changed considerably in structure over the last 300 years (Peterken & Tubbs 1965).

The fine detail of landscape inevitably changes. As an illustration we can take part of the bounds of the manor of Maidensgrove in the Chilterns, as described in 1774 (Oxford Record Office VOR II/x/1/w):

'thence by the left hand hedge through Bishops Wood to the Freedom Wood gate;

thence by the left hand hedge to a cross (roads) on Bix Hill;

thence by a right hand hedge to a great yew tree;

thence forward to a cross (road) to the land;

TABLE 4.2. The fate of the 3726 ha of ancient woodland present in Central Lincolnshire in 1820, as determined in 1973

Actively coppiced	2
Neglected coppice	1018
Coppice promoted to high forest	82
Naturally regenerated broadleaved high forest	67
Replanted with broadleaved species	326
Replanted with coniferous species	824
Replanted with mixed conifers and broadleaves	216
Degenerated to scrub	1
Opened into parkland	7
Recently felled and not yet restocked	16
Cleared to agriculture	1168

Sources: fieldwork by P.T. Harding & G.F. Peterken.

then to the right up the land by the left hand hedge to a great beech in the lane;

then to the left over the hedge and across the field into Soundalls Wood to a ditch a little left of the gate;' etc.

Today, the woods and lanes are there, but the hedges have changed with the farming pattern. The great and no doubt prominent trees have gone. If they had survived, they would now be obscured by the change in the woods from coppice to high forest of beech.

ANCIENT ELEMENTS IN THE LANDSCAPE MOSAIC

The landscape we inherit has evolved over thousands of years into a complex patchwork of elements dating from different eras. This was demonstrated by Yates (1960) in his study of Harting and Rogate parishes in Sussex. His map (Fig. 4.3) indicates that early settlement and cultivation was concentrated at the foot of the Downs. Other areas were brought into cultivation at later dates. Some original vegetation survives in the form of woodland.

Hoskins (1955) and subsequent authors in the series on *The Making of the English Landscape* showed that numerous ancient features survive in the modern landscape, including village cores, old farmsteads, lanes and field walls, which are not the province of the ecologist. Indeed, for example, in the Penwith peninsula of Cornwall and the north-east of Norfolk, field patterns survive from Iron Age and Roman times, and many regions are dominated by an ancient inclosure pattern. Rackham (1976, 1980) has demonstrated the great antiquity of many of our woodlands. Even grass fields, which one might expect to be ephemeral habitats within the fluctuations of farming, include some examples which have existed continuously as grassland for centuries (Sheail & Wells 1969).

The units in the landscape may themselves be mosaics of different historical elements. In the somewhat exceptional circumstances of the Porton ranges, grasslands originating at different periods over the last 200 years (Fig. 4.4) are now clearly distinguishable on the ground by their soils and composition (Wells *et al.* 1976). Many existing woods are a mixture of ancient and more recent components (Fig. 4.5).

On a wider scale, some landscape features have expanded and contracted with fluctuating farming economics, but have never entirely disappeared. In recent studies of some national parks, Parry *et al.* (1982) have recognized and delimited moorland cores, where moorland has persisted throughout the last 100 years and probably for centuries before that. On the moorland fringes, however, land has slipped into and out of cultivation (Fig. 4.6), changing as

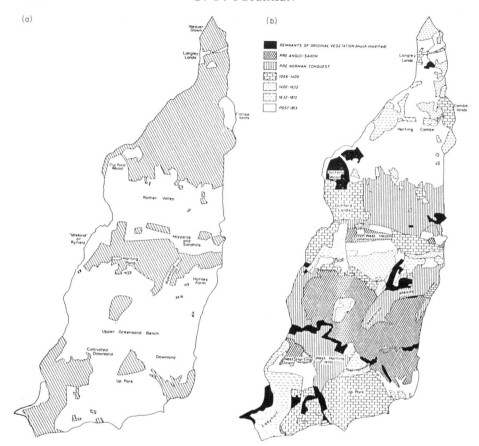

FIG. 4.3. Elements in the landscape of South Harting and Rogate parishes in West Sussex.
(a) Pre-1400 elements in the landscape (not shaded). (b) Land utilization: period of origin of
the existing enclosure pattern. Reproduced by permission of Dr E.M. Yates and the Royal
Geographical Society.

much as four times in 100 years. Much the same goes for many medieval forest
areas, which remain well wooded today.

LONG-TERM ECOLOGICAL RESPONSE TO
LANDSCAPE CHANGES

When a new habitat is created, e.g. when an arable field is planted with trees,
non-ecologists might tend to assume that the entire, original ecological
community of a natural woodland is restored. However, evidence from
woodlands suggests that the recovery is at best incomplete. Many woodland

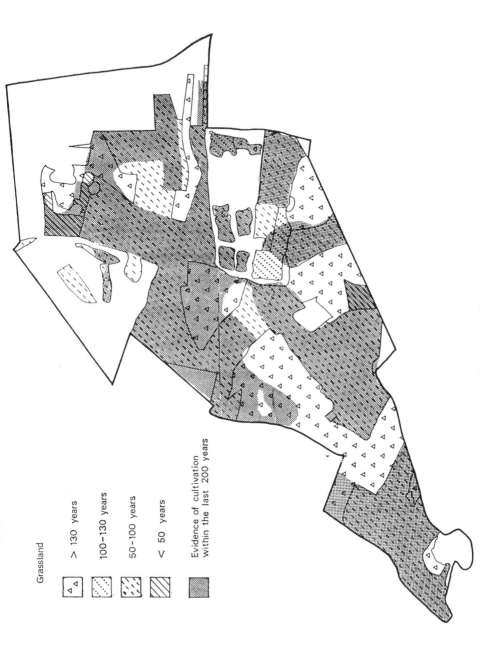

F𝙸𝙶. 4.4. Conjectured age of grasslands at Porton Ranges, Wiltshire, as indicated by documentary sources and evidence of former cultivation. Scale 1:50000. Reproduced by permission of Dr J. Sheail and Blackwell Scientific Publications.

Grassland

> 130 years

100–130 years

50–100 years

< 50 years

Evidence of cultivation within the last 200 years

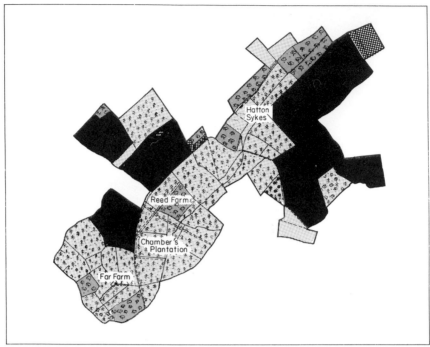

Chambers Farm Plantations, Lincolnshire
Periods of origin of existing woodland

◼ Ancient   ▨ 1920's

▨ Pre-19C secondary   ☐ 1943-1952

▨ 1820-1887

FIG. 4.5. Chambers Farm Plantations, central Lincolnshire, a patchwork of woodland originating at different periods. Most of the ancient woods are semi-natural, not plantations. Scale 1:50 000.

herbs are slow to colonize newly-available woodland in at least part of their range: these are the species found only or mainly in ancient woodlands. This pattern is well marked in eastern England (Rackham 1980; Peterken & Game 1984), where the woods have long been sparsely scattered and separated by cultivated land. The pattern is far less pronounced in historically well-wooded areas such as the Weald, and in uncultivated upland districts, but even here some specialized species are confined to ancient woods, e.g. some ferns and oceanic bryophytes (Ratcliffe 1968). Evidence that other groups include species which require not just the right habitat but also continuity of that habitat over several centuries at least, comes from the work of Boycott (1930)

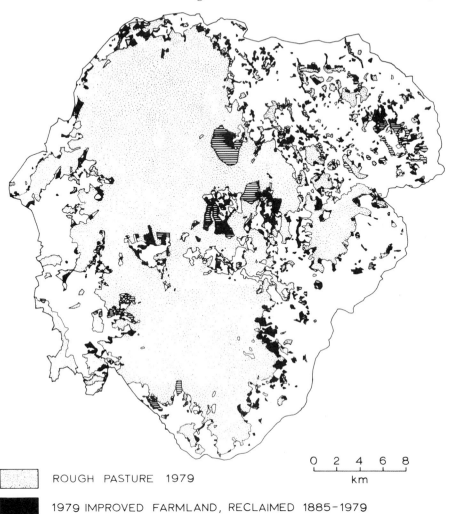

ROUGH PASTURE 1979

1979 IMPROVED FARMLAND, RECLAIMED 1885-1979

1979 WOODLAND, AFFORESTED 1885-1979

FIG. 4.6. Dartmoor National Park, Devon. Land use in 1979 showing the land which was converted to forestry or improved as farmland from rough pasture after 1885, and the remaining rough pasture. Reproduced by permission of Dr M.L. Parry and the Department of Geography, University of Birmingham.

on molluscs, Rose (1976) on epiphytic lichens, Hammond (1974) on beetles, and others.

Many factors seem to conspire to prevent some species recolonizing. Many woodland species have poor dispersal mechanisms, which are satisfactory in natural forest but unsuited to colonization across hostile cultivated land. Soils may become more fertile, more alkaline and better drained as a result of cultivation, so that when cultivation ceases and heathland, woodland or grassland return, the species present before cultivation may no longer be able to compete with the vigorous vegetation which has developed. Even if habitat and soil are suitable, the site may be too isolated from surviving populations of potential colonists.

Nevertheless, some rich and fairly natural habitats are associated with sites of early disturbance, and their very existence demonstrates a capacity for semi-natural communities to recover. Many chalk grasslands are associated with prehistoric fortifications and ancient field systems (Sheail & Wells 1969). Some herb-rich meadows overlie ridge-and-furrow. The Norfolk Broads, with their associated rich fen and aquatic communities, are long-abandoned peat diggings (Lambert *et al.* 1960). One might expect that such recovery will be equally possible in the future, but this seems unlikely given that the potential colonists are now so thinly scattered. In favoured areas such as the Sussex Weald, woodland herbs are rapidly colonizing the old commons as they revert to woodland. Here, the species are common in the vicinity in other woods and the dense network of hedges. But if land reverted to woodland in, say, Cambridgeshire, it is unlikely that many of the woodland herbs would even reach the new woodland from their remote refuges in the sparse scatter of ancient woods.

Ecologists are therefore concerned to identify surviving lammas meadows, long-established and unimproved pastures, medieval hedges, ancient woods and other habitats which have endured from much earlier times. These all tend to be rich in natural features, wildlife species and cultural associations, as has been fully demonstrated by Rackham's (1980) work on the ancient woodlands of East Anglia.

NATURAL DISTURBANCE AND LANDSCAPE DEVELOPMENT

Although ecologists stress the importance of long-enduring habitats, and particularly habitats which are direct survivals of primaeval vegetation (though in modified form), it is important to recognize that disturbance was part of the natural environment, and therefore that absolute stability of conditions may be an undesirable goal of management. In most circumstances, the effect of man has been to increase greatly the rate and variety of disturbance over the natural level, thereby favouring species characteristic of open conditions and early succession, notably many annual 'weeds'. Original natural broadleaf forests in lowland Britain were probably far less prone to natural catastrophes, such as fire and windthrow, than the original northern coniferous forests, yet even here 0.5–2.0% of the canopy trees probably died or toppled in an average year (Runcle 1982). Subsequent coppice management maintained a higher rate of canopy disturbance (8.3% per annum on a 12-year rotation). Neglect of coppicing in the present century has reduced disturbance rates to zero for many decades, and as a result many of the species of open habitats and early growth conditions have declined substantially.

LESSONS FOR LANDSCAPE DESIGN

When looking forward to future landscape designs, it is worth trying to learn some lessons from landscape history:

1 Much of our landscape was not intended to be beautiful, yet it is (or was). We should not presume to plan every landscape development.

2 Enduring and stable elements in the landscape should be retained. They give character and interest to an area, though from time to time they may be obscured by younger developments. These ancient elements support some of the richest and least replaceable plant and animal communities.

3 Absolute stability of the landscape cannot be achieved and would probably be undesirable if it could. Moderation of the pace of change, and change by piecemeal modification of existing features is the most desirable course from an ecological point of view. It is also in keeping with landscape development traditions.

4 Each area has its own distinctive features and responses to pressures for change. Landscape architects should not try to apply the same answers everywhere.

5 Species will not always respond to opportunities. We may provide the habitat, but many species will not colonize without assistance. This reinforces the value of stable and enduring elements in the landscape, but also indicates the need to consider whole communities when creating new habitats. The creation of new woodland is more than mere tree planting.

ACKNOWLEDGMENTS

I thank Dr John Sheail for comments on an early version of this paper.

REFERENCES

Boycott, A.E. (1930). The habitats of land Mollusca in Britain. *Journal of Ecology*, **22**, 1–38.

Darby, H.C. & Terrett, I.B. (1954). *The Domesday Geography of Midland England.* Cambridge University Press, Cambridge.

Hammond, P.M. (1974). Changes in the British Coleopterous fauna. *The Changing Flora and Fauna of Britain* (Ed. by D.L. Hawksworth), pp. 323–369, Academic Press, London.

Hoskins, W.G. (1955). *The Making of the English Landscape.* Hodder and Stoughton, London.

Lambert, J.M., Jennings, J.N., Smith, C.T., Green, C. & Hutchinson, J.N. (1960). *The making of the Broads. A reconsideration of their origin in the light of new evidence.* RGS Research Series, no. 3.

Parry, M.L., Mathieson, C.W., Bruce, A. & Harkness, C.E. (1982). *Changes in the extent of moorland and roughland on Dartmoor: a supplementary survey to 1979.* Department of Geography, University of Birmingham.

Peterken G.F. (1976). Long-term changes in the woodlands of Rockingham Forest and other areas. *Journal of Ecology,* **64,** 123–146.

Peterken, G.F. & Game, M. (1984). Historical factors affecting the number and distribution of vascular plant species in the woodlands of central Lincolnshire. *Journal of Ecology,* **72,** 155–182.

Peterken, G.F. & Tubbs, C.R. (1965). Woodland regeneration in the New Forest, Hampshire, since 1650. *Journal of Applied Ecology,* **2,** 159–170.

Rackham, O. (1976). *Trees and Woodlands in the British Landscape.* Dent, London.

Rackham, O. (1977). Hedgerow trees: their history, conservation and renewal. *Arboricultural Journal,* **3,** 169–177.

Rackham, O. (1980). *Ancient Woodland.* Arnold, London.

Ratcliffe, D.A. (1968). An ecological account of Atlantic bryophytes in the British Isles. *New Phytologist,* **67,** 365–439.

Roberts, R.A. (1959). Ecology of human occupation and land use in Snowdonia. *Journal of Ecology,* **47,** 317–323.

Rose, F. (1976). Lichenological indicators of age and environmental continuity in woodlands. *Lichenology: Progress and Problems* (Systematics Association Special, Vol. 8), (Ed. by D.H. Brown, D.L. Hawksworth & R.H. Bailey), pp. 279–307, Academic Press, London.

RCHM (1975). *An Inventory of the Historical Monuments in the County of Northampton.* I. Sites in North-east Northamptonshire. HMSO, London.

Runcle, J.R. (1982). Patterns of disturbance in some old-growth mesic forests of eastern North America. *Ecology,* **63,** 1533–1546.

Sheail, J. & Wells, T.C.E. (Eds) (1969). *Old grassland—its archaeological and ecological importance.* Monks Wood Experimental Station Symposium no. 5. Nature Conservancy Council.

Wells T.C.E., Sheail, J., Ball, D.F. & Ward, L.K. (1976). Ecological studies on the Porton Ranges: relationships between vegetation, soils and land-use history. *Journal of Ecology,* **64,** 589–626.

Yates, E.M. (1960). History in a map. *Geographical Journal,* **126,** 32–51.

II
ESTABLISHMENT

INTRODUCTION

The establishment phase is obviously critical to the success of any new landscape scheme. Failure, or only partial success at this point, can require costly extra work: if this has already been costed into the original estimates then these estimates will be unnecessarily high. But poor establishment is worse than this, because it puts the whole scheme into jeopardy from external factors, ranging from drought to vandalism. Plants that are growing well are not only more resilient and more able to cope with physical stresses, but also more quickly accepted by the public and less likely to be vandalized. What is surprising is that we take failure as a normal part of establishment, and so the general public accepts it as inevitable. If the job was done properly, failures would be minimal, just as they are in the products of modern manufacturing industries, but our expectations in establishment are wrong, and few people are concerned. Yet what would happen to a car manufacturer whose products had the same failure rate or poor early performance as newly planted trees?

Establishment is of course a very practical phase in which tradition and hard experience have ruled for centuries. This was possible when there was a large permanent labour force with continuity of expertise. At the present time, with a never-ending search for cuts in labour and overall costs, that source of expertise has been eroded and many operations are carried out blindly and sloppily, with resulting lack of success. Often we do not even know what went wrong.

This suggests that, to put establishment back on a firm footing, we must find out exactly what operations are necessary and what are crucial. At the same time we need to have a much clearer idea of potential sources of trouble, which involves going back to first principles. In doing this we will inevitably rediscover the importance of operations and ideas which were widely used in the past, as pointed out by Brian Clouston. Perhaps, because they were never properly recorded, we have been able to forget them.

The chase back to first principles is in progress, and its achievements are well indicated in this group of five papers. It is stimulating a great deal of new thinking and new principles, particularly from ecology. Some of these are exciting ideas which have still to be incorporated; others, such as the whole methodology of wild gardens, are fast becoming accepted garden systems. But some of this new work has direct relevance to mundane and practical operations such as tree planting. Errors in tree planting are now explicable in terms of basic principles; as a result there should be no excuse for failure in the future. Soils have equally received some radical thinking over the last few years. It is clear that we no longer need to specify topsoil blindly: we can achieve more, at lower cost, if we understand the soil-forming potential of materials and what limited operations are required to achieve this potential.

Ultimately, all this should shake us out of a conservative approach, requiring that we understand the basic principles involved, and have the courage to apply them. We will then see what can be achieved, as at Warrington New Town. Whether this is matched by landscape practitioners of the past is less important than the fact that we now have a coherent, scientifically-based approach to follow.

5. DESIGN CONSIDERATIONS AT ESTABLISHMENT

J. C. BAINES

Department of Planning and Landscape, City of Birmingham Polytechnic

SUMMARY

In the past 40 years we have lost a great deal of our rich natural heritage through agricultural improvement and urban expansion. The new recreational landscapes we have created as compensation have tended to be uninspiring by comparison. This paper suggests ways in which new landscapes can be manipulated, particularly at the capital-intensive establishment stage, to provide for greater opportunities for creative management.

INTRODUCTION

There seems to be increasing confusion, particularly within the landscape profession, about where design ends and management begins. The great joy of the living landscape is that it changes with time: to design with landscape is to harness that quality of change, to control and to direct it, the better to enjoy it; in other words, to celebrate those dynamic characteristics which make landscape architecture such a unique design discipline. Wherever land is being managed, the quality of that landscape is being continually influenced by the manager. Every decision will affect the pace and direction of change, and will consequently be a design decision (Baines & Smart 1984).

When a forest foreman selects a tall, straight stem in favour of a 'characterful' twisted one, this is a decision affecting the forest's design. More subtly, if woodland is cropped by felling in phased, small, sequential compartments, this will favour the conservation of woodland wildflowers and encourage the local migration of light-demanding species. Their increased presence will enhance the enjoyment of the woodland for visitors. If the manager chooses to leave occasional dead timber standing, or to stack some brashing within the wood, this will diversify the nesting habitats, and thus boost the amount of spring birdsong, once again contributing to the landscape's design.

In the more clinical world of the public open space, a park-keeper who raises the mower blades a centimetre or two will allow the creeping wildflowers in the green sward to flower and seed. Again, visitors will have their enjoyment enriched, this time with patches of speedwell and selfheal, and children will be able to make summer daisy chains.

Landscape design does not end with the signing of the drawings, or with the spending of the capital budget. It is true, however, that this initiation stage can have a disproportionate influence on the subsequent development of the dynamic landscape, and it is decision-making at this establishment stage which I will now address. Any managed ecosystem is a collection of responses to three inter-related groups of factors, as already argued by Bradshaw (Chapter 2): physical environment (geology and climate), management over time, and availability of species. Ironically, particularly in a country as rich and varied as Britain, the third factor is far less influential than the first two, and yet it is the factor with which landscape architects concern themselves most avidly. This preoccupation with plant selection and disregard for soils, climate and management regime inevitably results in landscapes which must be rigorously *maintained*. Such landscapes are impoverished by the continuous struggle to prevent the natural characteristics of the site from re-emerging and 'spoiling' the design.

ENVIRONMENTAL INFLUENCES

Few, if any, 'natural' landscapes have the blandness which we portray in our 1:5000 scale models. Even in landscapes such as saltmarsh or upland heath, there is punctuation: a creek, a ruined structure or a rocky tor. Significantly it is the 'wrinkles' in these landscapes which emphasize their character. In the enclosed landscape of woodland, the pleasure of the footpath is enriched by the wet, mossy bank, the small hollow filled with leafmould, or the patch of thinner soil which lightens the leaf canopy and lets in the sun. It is the wet bits in our otherwise well-drained meadows which harbour the taller summer wildflowers, survive the browsing of cattle (or mower) and provide food for the wintering snipe. The remarkable diversity of species to be found around our cities, almost entirely due to the great diversity of environments which has been left by past industrial activity, has been well illustrated recently (Teagle 1978).

For more interesting, long-term landscapes, then, we must look to the 'wrinkles'. Where they exist already, we must find ways of conserving them. We must welcome slopes steeper than one in three, for their instability will influence their vegetation; and they are fun to slide down. If there are wet patches in our school grounds, we should perhaps be draining the sports pitches into them and waiting for the orange-tip butterflies to lay their eggs on the cuckoo flowers. If there is rubble in our plantation, or the soil is too thin for trees, then this will be the open glade where speckled wood butterflies might feed on the bramble flowers.

If engineers must leave us with bland, smooth landscapes to 'tart-up', then we should have the guts to roughen these sites a little, to impede their immaculate drainage and create a modest wetland, to strip off some of their British Standard topsoil and expose the low fertility subsoil that diverse wild flower communities need. How much better, though, to work with the engineers, using their special skills to create a more imaginative structure for our landscapes from the beginning.

The question of soil fertility is worth particular attention. Millions of pounds are spent each year on importing topsoil on to amenity landscape sites. Not only is this practice outrageously expensive, it is almost always environmentally counter-productive. Where grassland is the prime community, topsoil will boost growth, favour coarse grasses, increase the frequency of mowing and mitigate against stress-tolerant colourful wild flowers. That may be fine for high-productivity agriculture, but in terms of landscape experience, richness is encouraged through poverty.

Where a woody canopy is intended, the topsoil will promote lush growth of herbaceous 'weeds' which will dominate the root zone, successfully outcompete the trees and shrubs for moisture, and generally lead to a high failure rate through drought stress in the first spring. The money saved by eliminating topsoil would be far better spent on skilful ground modelling to create some new 'wrinkles', and perhaps to pay for such apparent luxuries as pruning, mulching and post-planting irrigation to increase effective root establishment.

Where wetlands are concerned, physical site structure is especially influential. Aquatic plants are particularly sensitive to water depth. That hackneyed cross-section of the pond margin to which we all pay lip-service, actually does have serious implications for ground modelling on lake bottoms, river banks, etc. Emergents such as flag-iris and burr-reed will only tolerate a maximum of 150–200 mm of water, and *Phragmites* and reedmace will only stop spreading if the depth extends beyond a metre.

If you are planning a rich mosaic of varied emergent aquatics in the 'nature reserve' corner of your next water-ski arena, spend your money creating a criss-cross of ridge and furrow, and then tip in a lorry load of seed-laden pond-dredgings. The physical unevenness will quickly be reflected in the vegetation matrix, with each species occupying its appropriate niche.

More imaginative manipulation of physical site structure is relatively easy to achieve. It involves spending capital funds on orthodox items such as machinery and land drains, though there is a need for more emphasis on site supervision and perhaps rather less on drawings. Half an hour spent showing

Where shallow water margins are designed to provide habitats for species such as bog-bean, flag-iris, burr-reed and water mint, it is vital to avoid the early introduction of *Phragmites* or *Typha*. Both are invasive colonizers which will sweep through shallow open water. If the less aggressive plants are well established first, then when the windborne seed of reedmace and reed arrive and germinate, they are more likely to colonize without swamping the other plants.

With schemes of native vegetation, I believe we should allow for much more natural colonization. Instead of attempting to produce native woodland in one stage, I am convinced results would be far better if initially we established dense scrub, and so created a sheltered woodland environment at ground level. There is hardly a site in Britain that would not naturally acquire climax species, such as oak. Whether the jays deliver their acorns in year 5 or year 50 hardly matters if we are realistic about the time-scale of oak-woodland. If we must concertina the process, then a handful of acorns tossed into established scrub will succeed far better than the vulnerable nursery stock incorporated into most people's planting mixes.

If the practice of planting pioneer shrubbery to nurse longer-term species is developed a little further, then there is again scope for pre-determining future development at the establishment stage. If light-demanding pioneer shrubs such as gorse and broom are planted on the edge of plantations, they are likely to thrive and establish a viable colony even when the shady tree canopy has developed. If these same light-demanders are used within the plantation, then as the tree canopy grows up and closes over, the shade it casts will kill off the broom and gorse, making way for a shade-tolerant ground flora. Some shrub species, such as holly, hawthorn and dogrose, are tolerant of both open exposure and of shade. These can be employed as pioneer nurse species, but are likely to survive the closing tree canopy and remain as underscrub.

As landscape designers we tend to be simple-minded about communities. There is at least one senior practitioner who is happily establishing oakwoods by planting oak trees, ivy and bluebells, since discovering ecology. Among more experienced practioners, there must be an increasing awareness of the need to establish genuinely indigenous stock, and also to find ways of recolonizing sites with diverse invertebrate fauna. The litter techniques developed for healing pipeline scars on heather moorland are a start. Woodland litter is being introduced into plantations to help regenerate a more complex woodland community, and for many years conservationists have been transferring wetland communities from one pond to another in the form of sludge and waterweed.

Wherever possible, we must try to conserve pockets of established landscape to act as recolonizers after redevelopment. On landfill sites, for example, carefully designed phasing might allow for the retention of areas of silver birch. Since birch is regenerative within 5–6 years, and the seed are numerous and wind-dispersed, phase one of the landfill could be colonized by seed of indigenous birch before the parent colony is lost in the excavations of phase three. The same approach would work with willows. This practice would have the advantage of maintaining a population of truly indigenous genotypes within the site. In conserving relics of tree cover there would also be a greater chance of sustaining populations of a wider range of plant and animal species on site than it is possible to provide by new planting.

Again in the Blackbrook Valley, the Urban Wildlife Group has persuaded Coal Contractors to conserve one area of species-rich meadow whilst temporarily tipping on adjacent land. The meadow is rich in wildflowers such as yellow rattle, knapweed and centaury and the Group is collecting indigenous seed for revegetating the reclaimed tip site 2 years hence. By sustaining a viable population throughout the 2-year life of the tip, we hope that both animal and plant life will recolonize. There is a large colony of common blue butterflies, thriving on the birdsfoot trefoil of the low nutrient spoil heaps. Restoration of the stock-pile site will recreate conditions of low-fertility. Locally collected seed will be included in the sowing mixture, and in this case local schools will work with our ecology unit to monitor the rate of recolonization by invertebrates. It could be said that we are simulating a miniaturized version of the Pleistocene retreats of the Amazon basin; as building development on urban greenspace continues to be pursued so relentlessly, the landscape profession must vigorously champion the conservation of sanctuary communities, if its efforts in re-creation are to be more than naive cosmetics.

Viable, dynamic landscapes are never finished. On many sites the landscape architect will be spending the capital budget on changing the direction of landscape development, and on some occasions there may actually be a new landscape created. In either case, that establishment stage must be considered as part of a continuing process of creative development or management, but the pace and direction which the landscape is given at the start can have a very significant influence throughout the future life of the scheme.

REFERENCES

Baines, J.C. & Smart, J. (1984). *GLC Ecology Handbook*, No. 2. *A Guide to Habitat Creation*, Greater London Council Ecology Unit, London.
Baines, J.C. (1985). *How to Make a Wildlife Garden*. Elm Tree Books, London.
Baines, J.C. (1986). *The Wild Side of Town*. BBC Publications, Elm Tree Books, London.
Teagle, W.G. (1978). *The Endless Village*. Nature Conservancy Council, Shrewsbury.

6. THE ECOLOGY OF ESTABLISHMENT

P. J. GRUBB

Botany School, University of Cambridge, CB2 3EA

SUMMARY

Approaches to the study of plant establishment in the context of landscape design are briefly reviewed, and three issues are considered in more detail: the diversity of pioneers in relation to substratum type, the rarity of pioneer plants with nitrogen-fixing symbionts in relation to the increase in rooting density during succession, and the frequent importance of 'third parties' in controlling the balance between any two plants during establishment in gaps in an already complete vegetational cover. Examples of third parties are birch trees tipping the balance between two fir species during regeneration in a subalpine forest, voles making gaps in a heath layer in which birches may regenerate, and mycorrhizal fungi having favourable effects on some species of short-lived plants and deleterious effects on others.

INTRODUCTION

There are two kinds of working situation in which an ecologist may be confronted with landscape design. First, he may have to advise on the establishment of vegetation on totally bare sites. Secondly, he may have to draw up a scheme for the management of an area with a complete vegetational cover, where some are species are to be encouraged, others introduced and yet others ousted. A sound understanding of the ecology of establishment of different sorts of plants is essential for landscape design in both the first case (which we may call 'constructive') and in the second case (which we may call 'manipulative').

Among ecologists working with natural and semi-natural vegetation, much less attention has been paid to landscapes newly available to plants than to landscapes with a vegetation cover, albeit often subject to massive disturbance by fire, flood, windthrow or man's activities. Wherever ecologists have studied in detail the dynamics of vegetation, they have found that every plant species present has slightly different requirements for the best chance of producing a new adult individual. The differences between species concern the conditions that are most favourable for flower-formation, pollination, seed-set, dispersal, germination, establishment and onward growth; an extensive

83

review of this issue was provided not long ago (Grubb 1977). More recently, accounts emphasizing the different kinds of natural disturbance that favour particular species have been published for example, by White (1979), Pickett (1980) and various authors in the volume edited by West *et al.* (1981). A valuable alternative approach to the study of establishment has been the abstraction of major characteristics of whole groups of species with similar requirements. For example, Wells (1976) and Bazzaz (1979) have reviewed the differences in morphology and physiology between early successional and late successional species on forest clear-cuts and abandoned fields in the eastern USA, and Grime has produced his C-S-R model (see Chapter 11).

In the present paper I cannot attempt a comprehensive review, and instead I shall develop just three generalizing themes which have, in my opinion, received insufficient attention: (i) the diversity of pioneers that is to be understood in relation to the type of substratum colonized; (ii) the general failure of vascular plants with symbiotic nitrogen-fixing micro-organisms to be pioneers; and (iii) the frequent importance of 'third parties' (other plants, animals or micro-organisms) in determining the balance between any two plant species during establishment.

DIVERSITY OF PIONEERS IN PRIMARY SUCCESSION

A primary succession is said to occur where there is long-term directional change in the composition of the plant cover on a site initially devoid of vegetation and without a developed soil. A 'pioneer,' in the strict sense, is one of the first plants to become established at a particular site; to be effective it must also be able to continue development to the stage of reproductive adult. 'Establishment' is taken to mean the attainment of a recognizable juvenile vegetative stage; it is not possible to be more precise without referring to the type of plant concerned (lichen, moss, tree, herb, etc.). Pioneers may be transitory or persistent; their essential feature is their ability to become established before other plants.

I shall confine myself to plants which are pioneers on relatively stable terrestrial sites; the special problems of pioneers on screes and sand dunes, salt marshes and lake edges are reviewed by Ellenberg (1982). The kind of landscape which has been open to pioneer plants on much the largest scale in the last million years is that which has been newly freed from glacial cover, and it is convenient to start by considering the pioneers found these days where a long glacial tongue has retreated from well within the range of present-day forest. Examples of such landscapes have been described for western North America (Cooper 1916, 1923), Scandinavia (Faegri 1933), the European Alps (Richard 1975) and New Zealand (Wardle 1980). In parallel with the range of topographic site-types we find a remarkable range of pioneers (Fig. 6.1) with distinctive distributions (Fig. 6.2). Little precise information is available on their physiological characteristics, but a tentative analysis is included in Fig. 6.2. The growth rate is consistently inversely proportional to the length of life, but the relationship between final size and length of life is more complicated as (in general) the smallest live longest, the tallest next longest and the plants of intermediate size live the shortest time.

Among the pioneers, the growth form and growth rate match the capacity of the substratum to supply water and mineral nutrients. I suggest that the differentiation of plant types may have more to do with nutrient supply than water supply. Of particular interest is the differentiation between trees and herbs. There is increasing evidence that in forested country in more mature landscapes an abundance of herbs in the field layer is characteristic of relatively nutrient-rich sites, and an abundance of shrubs and young trees is characteristic of relatively infertile soils, and this is true of both dry areas (where the herbs are often grasses) and moist areas (where they are usually broad-leaved dicotyledons). References are given by Grubb (1984).

The relationship between the type of pioneer and the type of substratum shown in Fig. 6.2 applies not only to deglaciated landscapes but also to landslip scars, large outcropping rock-formations, and volcanic debris. The importance of a comparatively poor nutrient supply in favouring trees as pioneers is clearly brought out by considering the plants of landslip scars in the wetter regions of the world. It is often seen that trees are pioneers on the rockiest parts of the scars, where the nutrient supply is poorest, while faster-growing shrubs and herbs are found on the mixed debris at the lower ends of the landslips. Wardle (1980) has described how, in the Westland of New Zealand where the annual rainfall is up to 5000 mm and there is no dry season, the trees *Metrosideros umbellata* and *Weinmannia racemosa* are the pioneers on upper, rocky parts of landslips, whiie shrubs such as *Coprosma rugosa* and *Hebe salicifolia* are the pioneers on the lower parts composed of rock debris including a certain amount of fine earth. Trees can also be the pioneers on bare rock in seasonally dry climates, like *Oedomatopus ovata* on the convex surfaces of granite 'Inselbergen' in the Orinoco basin, where various drought-tolerant herbs (incuding the strange *Vellozia tubiflora*) are the pioneers on ledges where weathered rock accumulates, and where grasses dominate on the deepest soil-accumulations (P.J. Grubb, pers. obs.).

In Japan, lavaflows and 'mudflows' (often including aggregations of huge rocks) are widespread and very varied in texture. Yoshioka (1974) has emphasized that trees and shrubs are the usual pioneers at rocky sites, and has illustrated the contrast with fine-textured mudflows on which grasses form a closed cover that impedes establishment of woody plants.

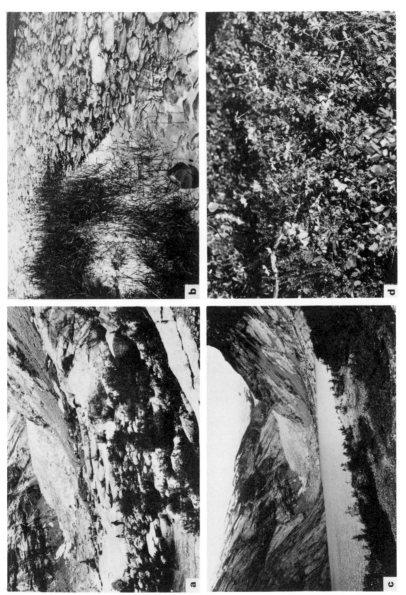

FIG. 6.1. (a) Area beneath snout of the Nilgarbreen Glacier in Norway in August 1983, with a little morainic material and much bare rock. *Salix* species (with mosses) were the major pioneers in rock crevices, while grasses (with mosses) were the major pioneers on gravel and silt. (b) Grasses as the major pioneers on fine-grained material at the margin of a small lake (with floating ice in August), a little further from the glacier than the foreground of (a). (c) Moraine further from the glacier than the foreground of (a), showing mixture of woody and herbaceous plants in developing birch-wood. Note the dark crowns of the nitrogen-fixing *Alnus incana* to the right, characteristically larger (faster-growing) than *Betula* or *Salix* but not a pioneer. (d) The major herb with nitrogen-fixing symbionts, *Lotus corniculatus*, in the succession at Nilgarbreen; characteristically found from mid- to late-successional stages, and not as a pioneer.

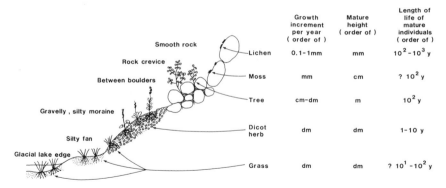

FIG. 6.2. The types of pioneer plant characteristic of different types of substratum in a recently deglaciated area within the range of present-day forest, and selected properties of the plants.

In fact, many different kinds of trees have been described as pioneers of rocky landslip sites in temperate and tropical regions, e.g. in Chile *Nothfagus* spp. (Veblen & Ashton 1978), in the Himalayas *Pinus* spp. (Stainton 1972), in Malaya *Macaranga* spp. (Whitmore 1975), and in Central and much of northern South America *Cecropia* spp. (Cuatrecasas 1958). In North America, trees such as *Betula papyrifera* have been described as pioneers on both large-block talus slopes (Cooper 1916) and on the rockiest parts of landslip-scars (Flaccus 1959), and in Norway *Betula* spp. can be seen behaving similarly on talus. Because talus slopes and landslips are rare in Britain it is in abandoned quarries that we most often see trees as pioneers—usually species of *Salix*, *Betula* or *Quercus*.

I do not mean to suggest that all woody pioneers can tolerate nutrient-poor sites. In fact, many are characeristic of sites with an abundance of fine earth. Such trees and shrubs may be found all the way from the lowlands of the tropics, e.g. *Eucalyptus deglupta* and *Octomeles sumatrana* in Papua New Guinea (Whitmore 1975), to the mountains of the temperate zone, e.g. *Myricaria germanica*, *Hippophae rhamnoides* and *Alnus incana* in the European Alps (Ellenberg 1982).

Contrasting strongly with pioneer trees of nutrient-poor sites are the fast-growing herbaceous pioneers of drift lines along the sea shore and along major rivers. They are peculiar 'pioneers' in that they are never followed by a second wave of invaders because each year the old drift is washed away and a new lot of drift is deposited; their communities are thus 'Dauer-Pioneer-Gesellschaften' (Tüxen 1975). Examples for the Rhine and its tributaries are given by Lohmeyer (1971) and summarized by Ellenberg (1982). They include many of the herbs found these days in fertilized fields, on roadsides and in other nutrient-rich habitats, e.g. species of *Chenopodium, Matricaria, Papaver* and *Polygonum*. In contrast, the tree species of rocky landslip scars spread characteristically to road cuttings and fields that have been abandoned because of their infertility (cf. Whitmore 1975).

I suggest that the correct world perspective is to see trees as generally the pioneers of sites with more than enough mineral nutrients for lichens and bryophytes but insufficient for herbs. Where the growing season is too short and/or the temperatures are too low to support woody plants, then the pioneers at the corresponding sites are probably mostly bryophytes.

The physiological basis for the advantage of woody plants over herbs on nutrient-poor sites is only partially understood. It has long been known that woody plants tend to have lower maximum relative growth rates than herbs (Coombe 1960), and that this means that woody plants are prone to be eliminated by interference from herbs on more fertile sites. The lower relative growth rates result from lower maximum photosynthetic rates in the leaves (Jarvis & Jarvis 1964), and these may be linked to greater mean life-lengths for leaves on woody plants (cf. Orians & Solbrig 1977). I have suggested elsewhere (Grubb 1984) that the longer leaf-life may involve the use of a greater proportion of the leaf's nitrogen in 'defence' so that less is available for production of photosynthetic machinery.

The maximum development of dicotyledonous herbs on sites supplying somewhat less nutrients than those occupied by grasses, shown in Fig. 6.2, is also probably an observation that can be widely generalized. It is a fact that the nitrogen-starved 'Xerobrometum' communities of Central Europe often lack grass (*Carex humilis* taking its place; Ellenberg 1982), and the failure of grasses (as opposed to various shrubs, dicotyledonous herbs, orchids and sedges) to become established on extremely chalk-rich ex-arable in southern England as a result of nitrogen-shortage was elegantly demonstrated by Lloyd & Pigott (1967). Of course there are bound to be exceptions; *Sesleria albicans* is the most convincing case of a pioneer grass able to withstand severe nitrogen shortage in northern Europe (Dixon 1982). The physiological basis for the advantage of dicotyledonous herbs (and certain grass-like plants such as *Carex* spp.) over grasses on nitrogen-poor sites is for the moment unknown.

RARITY AMONG PIONEERS OF PLANTS WITH NITROGEN-FIXING SYMBIONTS

It is generally accepted that even the most favourable of the microhabitats shown in Fig. 6.2 is short of nitrogen (Lawrence 1958), and yet it is characteristic of deglaciated landscapes that the vascular plants with nitrogen-fixing symbionts are not among the pioneers, but among the second or third wave of invaders, e.g. the legumes *Astragalus* and *Hedysarum* in North America (Viereck 1966) and *Lotus* and *Trifolium* in Europe (Ellenberg 1982), species of *Alnus* and *Dryas* in both continents (Cooper 1923; Faegri 1933; Viereck 1966; Ellenberg 1982), and *Carmichaelia* and *Coriaria* in New Zealand (Wardle 1980). On recent moraines on tropical mountains there are commonly no vascular plants known to have nitrogen-fixing symbionts; this is apparent from the accounts of Hedberg (1968) for East Africa and Hope (1976) for West Irian, and is also true of the Andes (P.J. Grubb, pers. obs.). However, it is possible that many pioneer vascular plants have effective parasymbiotic nitrogen-fixers in the rhizosphere, as is believed by some to be the case for marram grass (*Ammophila arenaria*) on dunes (Huiskes 1979). Shortage of nitrogen seems to be particularly severe for a group of pioneers not so far considered—epiphytes (Grubb, Flint & Gregory 1969; Grubb & Edwards 1982). In their case there is evidence that the shortage of nitrogen is slightly relieved through significant nitrogen-fixation by free-living micro-organisms in tree crowns (Jones *et al.* 1974), and by related organisms associated with epiphytic bryophytes (Edmisten 1970).

I suggest that the most probable explanation for the common failure of plants with nitrogen-fixing symbionts to be pioneers is that it is impossible to produce an exceedingly light and outstandingly dispersible seed which also contains sufficient 'starter nitrogen.' It is well established that the nitrogen capital in the relatively large, nitrogen-rich seeds of legumes is needed for investment in nodule-formation before any net return of nitrogen accrues to the angiosperm partner, and that the establishment of leguminous plants is often nitrogen-limited (Sprent & Thomas 1984). Other possible explanations could be a need for an increased phosphate-supplying power to develop in the soil by weathering (many legumes are strongly promoted by addition of phosphate), or a low probability of invasion together by vascular plant and symbiont. The two latter explanations seem less probable as there are a few cases where legumes are found as pioneers, e.g. *Astragalus alpinus* on base-rich till eroded by river action in northern Sweden, and *Lupinus densiflorus* var. *aureus* on steep roadside banks in the Coastal Range of California (P.J. Grubb, pers. obs.). In such cases the pioneers invade over distances of only a few metres, and the sites invaded are often below the source of seed.

A quite different approach to the problem is to suggest that a plant with nitrogen-fixing symbionts comes to have a strong competitive advantage only when the rooting density has reached a rather high value, and the effective shortage of nitrogen has increased relative to other limiting factors. Nitrate is highly mobile in the soil (Nye & Tinker 1977) and it can be taken up by any given root from a large volume of soil around it. In contrast phosphate, also in short supply in most raw soils, is immobile and can be taken up from only a narrow sleeve of soil around each root. A simple way of showing the increasing importance of nitrate-limitation relative to phosphate-limitation as rooting density increases is to grow plants at the rate of one per large pot of soil, and to use soil which is (in general terms) low in both nutrients; additions of the nutrients are then made to separate plants. At least with some species it is found that at first only phosphate-addition will increase growth, and later only nitrate-addition (Fig. 6.3). It can easily be shown that this result is

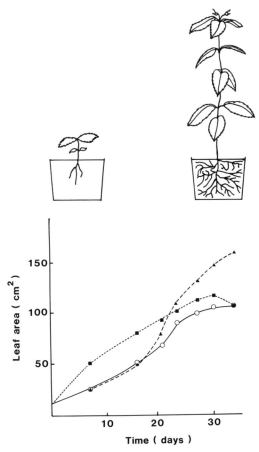

FIG. 6.3. Two stages in the development of a plant in a pot of soil, and the change-over with time from phosphate-limitation to nitrate-limitation; based on results of Peace & Grubb (1982) for *Impatiens parviflora* grown on a woodland soil derived from calcareous boulder clay. O, no addition; ■, phosphate added; ▲, nitrate added.

dependent on rooting density and not on some other time-dependent factor by growing plants at a high rooting density from the beginning, e.g. ten in a small pot; then nitrate limits growth from the start (Peace & Grubb 1982).

The time is ripe for a critical investigation of the nitrogen relations of natural pioneers, and particularly the importance of nitrogen-shortage relative to other limiting factors at the pioneer stage and in the later phases of succession.

IMPORTANCE OF THIRD PARTIES IN DETERMINING THE BALANCE BETWEEN COMPETITORS DURING ESTABLISHMENT

Other plants as third parties

Weatherell (1957) was one of the first to illustrate the importance of the third party during establishment, when he showed that *Calluna vulgaris* was able to suppress severely establishment of *Picea sitchensis* but not of that of *Larix leptolepis* in the Scottish uplands. An even more arresting example in a wholly natural system has been found by Kohyama (1984) in an area of Japanese subalpine forest, a part of which is subject to wave-patterns of regeneration driven by wind (Fig. 6.4). The two most abundant trees are species of *Abies*, both relatively shade-tolerant, but one (*A. mariesii*) more shade-tolerant and slower-growing than the other (*A. veitchii*). Where a windthrow area is created, leaving advanced growth of the two *Abies* spp. exposed, and there is no invasion by *Betula*, the less shade-tolerant *Abies* outgrows the other and forms a majority of the trees in the forest regenerated on the site. But if *Betula* invades, it grows faster than either of the *Abies* spp. and forms a mixed canopy with the *Abies* species. This mixed canopy tends to break up in an irregular fashion. Under these conditions the more shade-tolerant *Abies* species is favoured; it builds up a 'sapling bank' as opposed to a 'seedling bank', and individuals eventually grow up through the broken canopy. The frequency of invasion of gaps by *Betula* is such that the mean proportion of the more shade-tolerant *Abies mariesii* in the canopy over a wide area is about 0·6; without the intervention of the *Betula* it would certainly be rarer, and might be eliminated.

Animals as third parties

Remmert (1984) has emphasized the role of animals in controlling the balance between plant species in many kinds of landscape. A striking case that I recently studied was in northern Sweden near Abisko (68°N). The subalpine forest there has one principal tree species, *Betula pubescens* ssp. *tortuosa*

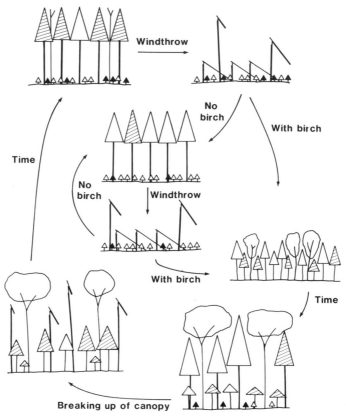

FIG. 6.4. A diagrammatic representation of the two major routes of regeneration in a subalpine forest in Japan studied by Kohyama (1984). The conifers with filled crowns are *Abies mariesii* (more shade-tolerant), those with open crowns are *A. veitchii*, and the broad-leaved trees are *Betula ermanii*.

(Rune 1965), and this forms multi-stemmed individuals that may be very long-lived. On acidic soils the field layer is composed of a dense 10-cm tall growth of *Empetrum hermaphroditum* and *Vaccinium* spp., in which seedlings of *Betula* cannot become established. However, every few years there are eruptions in the local populations of the vole *Clethrionomys ruficanus*, and areas of 0·5–4 m² in the heathy field layer are destroyed (Callaghan & Emanuelsson 1985). It seems that only in these gaps can the birch regenerate (Fig. 6.5).

Micro-organisms as third parties

It is a striking fact that very few species of short-lived plants can maintain populations in continuous short turf on chalk-derived soils in northern

(a)

(b)

(c)

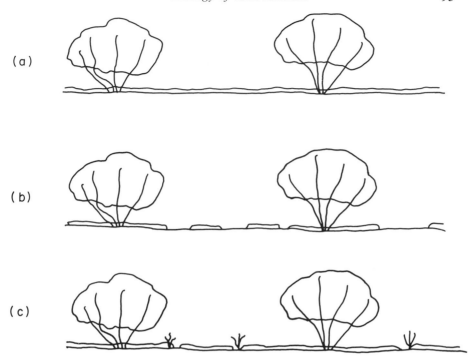

FIG. 6.5. Profiles of a relatively open, ridge-top facies of heath-type subalpine birch forest in northern Sweden, showing the creation of gaps by voles, and the apparent dependence of birch on these gaps for regeneration.

Europe, even when the turf is grazed to a height of only 2–10 cm. Common annual and biennial 'weeds', if sown into such turf, germinate but fail to become established (Gay *et al.* 1982). Fenner (1978) showed experimentally that common colonists of bare ground are highly susceptible to below-ground interference. He made gaps in an artificial sward of *Festuca rubra* kept to a height of 1 cm by regular cutting, and sowed various species; the colonists failed to grow where turf-inhabiting perennials succeeded in becoming established (Fig. 6.6). At least for *Arenaria serpyllifolia* addition of phosphate (but not nitrate) will relieve seedlings of the inhibition by a dense grass turf (P.J. Grubb, unpubl. data). It is therefore of particular interest that the most widespread biennials of chalk turf (*Linum catharticum* and various Gentianaceae) have peculiar mycorrhiza (Gay *et al.* 1982), almost certainly involved in phosphate uptake (Harley & Smith 1983). Taylor (1982) has shown experimentally that one of the Gentianaceae concerned (*Centaurium erythraea*) is very strongly dependent on mycorrhizal infection for development. Even more remarkable, she found in the same experiment that infection of nearby

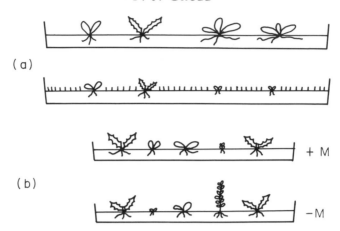

(a)

(b)

+ M

−M

FIG. 6.6. Diagrams of (a) the experiment of Fenner (1978) showing differential suppression of 'turf-incompatible' plants, and (b) the experiment of Taylor (1982) showing differential effects of mycorrhizally infected inoculum plants on seedlings of various species grown in sterilized calcareous sandy soil.

'inoculum' plants of *Centaurium erythraea* and *Leontodon hispidus* with mycorrhiza caused a dramatic reduction in the growth of seedlings of the 'turf-incompatible' *Arenaria serpyllifolia* and *Verbascum nigrum* (Fig 6.6). It has yet to be proven that mycorrhiza can tip the balance in the field, and not merely under laboratory conditions, but the results are highly suggestive.

The issue of 'turf-compatibility' concerns not only short-lived herbs but also shrubs and trees. For example, in northern Europe and in the eastern United States species of *Crataegus* are especially effective in invading derelict grassland, whereas species of *Cornus* and *Viburnum* are very effective in invading bare soil left after arable cultivation, or where rabbits have had their burrows. The late P.S. Lloyd suggested that differences in root morphology were of key importance in this context (Ellenberg 1982) but I suspect that the nutrient relations of the whole plant and the mycorrhizal type, may well prove to be of equal importance.

CONCLUSIONS FOR LANDSCAPE DESIGN

Natural pioneers are enormously varied, and much of this variation can be related to the extent of the nutrient supply. Of particular interest, not least in the applied context, is the balance between trees and herbs. If, for example, it is desired to cover motorway cuttings with trees, it might be better not to cover the slopes with fertile topsoil which will encourage the growth of grasses that can suppress any trees that are planted, but to plant trees (or scatter seeds) on the infertile subsoil and so save a good deal of money.

Legumes and other plants with nitrogen-fixing symbionts are undoubtedly important in encouraging growth of plants in young systems on infertile soils, e.g. conifer plantations on old sand dunes, but the nitrogen relations of the very earliest stages in primary succession, as in old chalk-pits and other quarries, need to be re-examined critically.

Species of later stages in succession differ among themselves, not only in their requirements regarding physical aspects of the environment which are relatively easily measured, e.g temperature and moisture supply, but also in the degree to which their juveniles are dependent on third parties for successful establishment. This element of dependence is important naturally in determining which species can grow in a community and their relative abundance. Therefore, if it is desired to manipulate a landscape, and change the balance of species, the inter-relations between the juveniles of those species and various third parties in the ecosystem should be considered most critically.

ACKNOWLEDGMENTS

I am particularly indebted to Drs T.V. Callaghan, U. Emanuelsson and T. Kohyama for showing me their papers before publication, and to Dr Pamela Taylor (née Gay) for permission to quote her unpublished results; her research was supported by the Natural Environment Research Council. I thank Professor P.W. Price (Northern Arizona University) for encouraging my interest in the importance of 'third parties'.

REFERENCES

Bazzaz, F.A. (1979). The physiological ecology of plant succession. *Annual Reviews of Ecology and Systematics*, **10**, 351–371.

Callaghan, T.V. & Emanuelsson, U. (1985). Population structure and processes of tundra plants and vegetation. *The Population Structure of Vegetation* (Ed. by J. White), pp. 399–439. Junk, Dordrecht.

Coombe, D.E. (1960). An analysis of the growth of *Trema guineensis*. *Journal of Ecology*, **48**, 219–232.

Cooper, W.S. (1916). Plant successions in the Mount Robson region, British Colombia. *Plant World*, **19**, 211–238.

Cooper, W.S. (1923). The recent ecological history of Glacier Bay, Alaska. *Ecology*, **4**, 93–128, 223–245, 355–365.

Cuatrecasas, J. (1958). Aspectos de la vegetacion natural de Colombia. *Revista de la Academia de Ciencias Naturales de Colombia*, **10**, 221–262.

Dixon, J.M. (1982). Biological Flora of the British Isles: *Sesleria albicans* Kit. ex Schultes. *Journal of Ecology*, **70**, 667–684.

Edmisten, J.A. (1970). Preliminary studies on the nitrogen budget of a tropical rain forest. *A Tropical Rain Forest* (Ed. by H.T. Odum & R.F. Pigeon), pp. H-211–H-215. US Atomic Energy Commission, Oak Ridge, Tennessee.

Ellenberg, H. (1982). *Vegetation Mitteleuropas mit den Alpen in ökologischer Sicht*, 3rd edn. Ulmer, Stuttgart.

Faegri, K. (1933). Über die Langenvariationen einiger Gletscher des Jostedalsbre und die dadurch bedingten Pflanzensukzessionen. *Bergens Museums Årbok*, **1933** (7), 1–255.

Fenner, M. (1978). A comparison of the abilities of colonizers and closed-turf species to establish from seed in artificial swards. *Journal of Ecology*, **66**, 953–963.

Flaccus, E. (1959). Revegetation of landslides in the White Mountains of New Hampshire. *Ecology*, **40**, 692–703.

Gay, P.E., Grubb, P.J. & Hudson, J.H. (1982). Seasonal changes in the concentrations of nitrogen, phosphorus and potassium, and in the density of mycorrhiza, in biennial and matrix-forming perennial species of closed chalkland turf. *Journal of Ecology*, **70**, 571–593.

Grubb, P.J. (1977). The maintenance of species-richness in plant communities: the importance of the regeneration niche. *Biological Reviews*, **52**, 107–145.

Grubb, P.J. (1984). Some growth points in investigative plant ecology. *Trends in Ecological Research for the 1980s* (Ed. by J.H. Cooley & F.B. Golley), pp. 51–74. Plenum Press, New York.

Grubb, P.J. & Edwards, P.J. (1982). Studies of mineral cycling in a montane rain forest in New Guinea. III. The distribution of mineral elements in the above-ground material. *Journal of Ecology*, **70**, 623–648.

Grubb, P.J., Flint, O.P. & Gregory, S.C. (1969). Preliminary observations on the mineral nutrition of epiphytic mosses. *Transactions of the British Bryological Society*, **5**, 802–817.

Harley, J.L. & Smith, S.E. (1983). *Mycorrhizal Symbiosis*. Academic Press, London.

Hedberg, O. (1968). Taxonomic and ecological studies on the Afroalpine flora of Mt. Kenya. *Hochgebirgsforschung*, **1**, 171–194.

Hope, G.S. (1976). Vegetation. *The Equatorial Glaciers of New Guinea* (Ed. by G.S. Hope, J.A. Peterson, I. Allison & U. Radok), pp. 113–172. Balkema, Rotterdam.

Huiskes, A.H.L. (1979). Biological Flora of the British Isles: *Ammophila arenaria* (L.) Link. *Journal of Ecology*, 363–382.

Jarvis, P.G. & Jarvis, M.S. (1964). Growth rates of woody plants. *Physiologia Plantarum*, **17**, 654–666.

Jones, K., King, E. & Eastlick, M. (1974). Nitrogen fixation by free-living bacteria in the soil and in the canopy of Douglas Fir. *Annals of Botany*, New Series, **38**, 765–772.

Kohyama, T. (1984). Regeneration and coexistence of two *Abies* species dominating subalpine forests in central Japan. *Oecologia*, **62**, 156–161.

Lawrence, D.B. (1958). Glaciers and vegetation in southeastern Alaska. *American Scientist*, **46**, 89–122.

Lloyd, P.S. & Pigott, C.D. (1967). The influence of soil conditions on the course of succession on the chalk of southern England. *Journal of Ecology*, **55**, 137–146.

Lohmeyer, W. (1971). Über das Polygono-Chenopodietum in Westdeutschland unter besonderer Berücksichtigung seiner Vorkommen am Rhein und in Mündungsgebiet der Ahr. *Schriftenreihe Vegetationskunde, Bonn*, **5**, 7–28.

Nye, P.H. & Tinker, P.B. (1977). *Solute Movement in the Soil–Root System*. Blackwell Scientific Publications, Oxford.

Orians, G.H. & Solbrig, O.T. (1977). A cost-income model of leaves and roots with special reference to arid and semi-arid areas. *American Naturalist*, **111**, 677–690.

Peace, W.J.H. & Grubb, P.J. (1982). Interaction of light and nutrient supply in the growth of *Impatiens parviflora*. *New Phytologist*, **90**, 127–150.

Pickett, S.T.A. (1980). Non-equilibrium coexistence of plants. *Bulletin of the Torrey Botanical Club*, **107**, 238–248.

Remmert, H. (1984). And now? Ecosystem research! *Trends in Ecological Research for the 1980s* (Ed. by J.H. Cooley & F.B. Golley), pp. 171–191. Plenum Press, New York.

Richard, J.L. (1975). Dynamique de la végétation au bord du grand glacier d'Aletsch (Alpes suisses). *Sukzessionsforschung* (Ed. by W. Schmidt), pp. 189–206. Cramer, Vaduz.

Rune, O. (1965). The mountain regions of Lappland. *Acta phytogeographica suecica*, **50,** 64–77.

Sprent, J.I. & Thomas, R.J. (1984). Nitrogen nutrition of seedling grain legumes: some taxonomic, morphological and physiological constraints: opinion. *Plant, Cell and Environment*, **7,** 637–645.

Stainton, J.D.A. (1972). *Forests of Nepal*. Murray, London.

Taylor, P.E. (1982). *The development and ecological significance of mycorrhiza in contrasted groups of chalkland plants*. Ph.D. dissertation, University of Cambridge.

Tüxen, R. (1975). Dauer-Pioniergesellschaften als Grenzfall der Initialgesellschaften. *Sukzessionsforschung* (Ed. by W. Schmidt), pp. 13–19. Cramer, Vaduz.

Veblen, T.T. & Ashton, D.H. (1978). Catastrophic influences on the vegetation of the Valdivian Andes, Chile. *Vegetatio*, **36,** 149–167.

Viereck, L.A. (1966). Plant succession and soil development on gravel outwash of the Muldrow Glacier, Alaska. *Ecological Monographs*, **36,** 181–199.

Wardle, P. (1980). Primary succession in Westland National Park and its vicinity, New Zealand. *New Zealand Journal of Botany*, **18,** 221–232.

Weatherell, J. (1957). The use of nurse species in the afforestation of upland heaths. *Quarterly Journal of Forestry*, **51,** 298–304.

Wells, P.V. (1976). A climax index for broadleaf forest: an *n*-dimensional, ecomorphological model of succession. *Central Hardwood Forest Conference* (Ed. by J.S. Fralish, G.J. Weaver & R.C. Schlesinger), pp. 131–176. Department of Forestry, Southern Illinois University, Carbondale, Illinois.

West, D.C., Shugart, H.H. & Botkin, D.B. (Eds) (1981). *Forest Succession: Concepts and Application*. Springer, New York.

White, P.S. (1979). Pattern, process, and natural disturbance in vegetation. *Botanical Reviews*, **45,** 229–299.

Whitmore, T.C. (1975). *Tropical Rain Forests of the Far East*. Clarendon, Oxford.

7. THE SELECTION AND MANAGEMENT OF SOILS IN LANDSCAPE SCHEMES

R. D. ROBERTS AND J. M. ROBERTS

Department of Biology, University of Essex, Colchester, Essex

SUMMARY

Landscaping is carried out for a variety of reasons and in a variety of ways. Landscape architects are particularly adept and proficient at selecting the most appropriate plant materials according to the needs of individual schemes. In contrast, flexibility is only rarely exercised when selecting and specifying soil materials. The usual approach is to specify topsoil as defined by BS 3882:1965—the original layer of grassland or of cultivated land. This approach may be appropriate for some schemes but, more frequently, it is inadequate and produces disappointing results, which can be expensive to rectify.

The belief, inherent in BS 3882 and most variants developed from it, that surface layers from grassland or cultivated land are appropriate for all plant materials and all landscape schemes, is fallacious. For this and other reasons, it is argued, by reference to specific examples, that the present approach is inadequate and must be replaced if the landscape profession is to maintain customer satisfaction at reasonable costs. A more flexible approach, based on selection and management of a much broader range of soil and soil-like materials, is suggested. Appropriate soil assessment and management methods are reviewed and their application to landscape schemes assessed. It is concluded that, given appropriate treatment, there are many materials other than topsoil that can be used in landscape schemes with greater success and at lower cost.

INTRODUCTION

Soil conditions are critical in almost all landscape schemes, since they determine which species of plants are able to survive and to what extent they grow. However, although there are some notable exceptions, little attention is paid to soil conditions in most landscape schemes. The usual approach is to rely on the use of topsoil as defined by BS 3882:1965, combined occasionally with applications of fertilizers.

The concept of topsoil as embodied in BS 3882, has some advantages and applications. It may be useful in agricultural or garden situations where fertility and plant growth is maintained by continuous cultivation and fertilization. Indeed, the standard was initially conceived at a time when our understanding of soils was derived principally from agriculture and when the landscape industry was concerned primarily with agricultural or garden situations.

More recently, ecological approaches to the selection and management of soils have begun to be introduced into landscape schemes, partly due to a growing realization that the traditional concept of soils, based on BS 3882, is unable to meet many requirements of the present-day landscape profession.

This paper attempts to explain the basis of this view and to demonstrate how the problem can be overcome. A change in approach is important if the landscape industry is to meet, in its usual imaginative and creative way, the challenges inherent in current land-use patterns and the aesthetic, recreational and environmental values of present-day society.

THE PRESENT POSITION

BS 3882:1965 is not a standard but rather contains recommendations for the classification and use of topsoil. It defines topsoil as being the original surface layer of grassland or cultivated land (specifically excluding natural or semi-natural areas of vegetation) and describes it, in general terms, in relation to colour, organic matter content, structure, porosity and weed content. For certain specialized (but unspecified) purposes or for more intense cultivation, it suggests that a more precise specification of texture, acidity/alkalinity and stone content may be desirable and provides details of soil-testing methods and classification systems for these properties. The content of BS 3882 is non-specific; most soil properties are covered in general terms only. For example, 'Topsoil tends to be friable and shows some degree of porosity' is the only reference to structure and drainage characteristics. The only firm, quantitative recommendation is that, normally, stones should be less than 50 mm in any dimension and should not exceed 20% of the soil weight.

Clearly, BS 3882 is imprecise, incomplete and refers only to cultivated soils. It is not a standard. Nevertheless, because of the wording of the document, it can be interpreted as a standard and is often used in this way. This is unfortunate. BS 3882 and variants currently used by the British landscape profession have advantages and applications but they do not meet the requirements and challenges of today's landscapes.

The following examples show the limitations of the present approach. In some cases the type of soil which is used is inappropriate. In other cases, the right type of soil is used, but because the wrong soil properties are assessed, very poor survival and growth of plants are achieved. In other cases, even when good growth is obtained, topsoil can be much more expensive than alternative materials.

Wrong soil types

Landscape schemes involve a variety of after-uses. These can be grouped into the following five broad categories:
    (a) arable agriculture and horticulture;
    (b) grazing of improved pastures;
    (c) formal recreation involving sports facilities;
    (d) informal recreation and amenity;
    (e) landscape improvement and wildlife conservation.

Arable agriculture requires soil conditions which promote maximum yields of a large variety of crop species and are amenable to mechanical cultivation. Grazing also requires high yields but of a smaller range of plant species—predominantly grasses—and less intense cultivation. For formal recreation, the major requirements are for soils that are able to withstand trampling and to support the growth of a relatively small number of plant varieties bred for use as sports turf. The remaining after-uses typically involve tree, shrub or herbaceous species, all of which thrive naturally in much poorer soil conditions than would be acceptable for arable, grazing or formal recreation uses.

Selection of appropriate soils is critically important. A soil which may be suitable for informal recreation may be too infertile to promote maximum growth of agricultural grasses and crops and, consequently, may be unsuitable for arable or grazing uses. Conversely, a soil suitable for grazing may be too productive for informal recreation. The fertile conditions will promote high growth rates and tall herbaceous vegetation which can be unattractive and a fire hazard unless regularly mown. Agricultural grasses and weed species are often best adapted to these conditions. Consequently, they may invade the sward and because they tend to be aggressive they may ultimately dominate the sward at the expense of the species originally sown or planted.

This point is illustrated by the results of field experiments at three landfill (refuse disposal) sites in Essex. At each site, eight different soil or soil-forming materials were placed at varying thicknesses, onto the surface of the refuse in $2 \times 1 \cdot 5$ m plots. Each material was then seeded with four grass species and the biomass (above-ground yield), amount of ground cover and species composition recorded after 9 months.

Conventionally, all three sites would be restored by importation of topsoil.

However, higher yields (i.e. agricultural potential) were obtained using river dredgings and pulverized fuel ash (PFA) mixed with sewage sludge than using topsoil (Fig. 7.1). The topsoil plots were also unsightly and ecologically uninteresting: they consisted mainly of agricultural weed species whose propagules were present in the original material. In contrast, all other materials contained few indigenous propagules and the sward consisted of the sown species together with, in some cases, a number of self-sown wildflower species (Fig. 7.2).

Prior to harvesting, the plots were inspected by a group of professional landscape architects, each of whom allocated two marks to each material

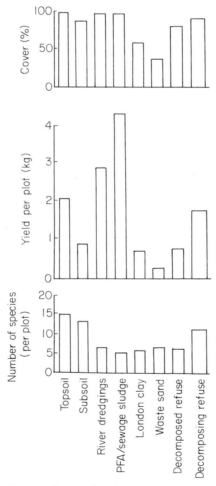

FIG. 7.1. Vegetation performance in topsoil and seven non-topsoil materials used to cover refuse in a landfill site. (PFA = pulverized fuel ash).

FIG. 7.2. Landfill (refuse disposal) sites are traditionally restored by burying the refuse with topsoil. (a) But invariably this produces a very poor, monotonous sward dominated by a few species of grasses. Many other materials could be used. (b) In these trial plots 8 different soils or soil-like materials are being used to cover the refuse: all of them support some growth without being given any management. (c) Given appropriate management some, such as waste sand, will produce extremely attractive swards which will require little, if any, long-term maintenance.

according to the acceptability of the sward for (i) a picnic area and (ii) livestock grazing. The individual scores were then collated and the materials arranged into ranking tables in order of acceptability (Table 7.1). Decomposed and decomposing refuse scored the highest marks for the picnic afteruse, followed by subsoil. All other materials scored notably lower marks for a variety of reasons including too much growth (dredging, PFA/sewage sludge), unsightliness (topsoil), too little growth (sand) or too muddy (clay). In contrast, the scores for grazing ranked PFA/sewage sludge and dredgings first, followed by decomposing refuse and top soil, with other materials scoring low, primarily due to poor yield.

Subsequently, the plots were split into two, one half fertilized and the other left unfertilized. Highest yields after 3 years of growth were obtained from river dredgings. Topsoil, subsoil and decomposed refuse all responded to fertilizer addition and could be used for grazing. Unfertilized topsoil remained reasonably productive and was still dominated by agricultural grasses and weeds. In contrast, fertilized sand and clay and unfertilized dredgings, subsoil and decomposed refuse all supported low-growing, visually-attractive swards and were much better suited than the topsoil for informal recreation, amenity and landscape areas.

Clearly topsoil, if it is obtained from grassland or cultivated land, must be the right type of soil for many agriculturally-orientated landscape schemes. However, because of its relative fertility and weed content, it is fundamentally wrong for many other schemes.

The problem lies with the belief that there is some easily recognizable standard type of soil that can be used in all landscape schemes. There is,

TABLE 7.1. Comparison of visual acceptability of topsoil and non-topsoil materials for productive (pasture) and low-maintenance (picnic area) uses when used as final covering of three landfill sites in Essex

		Picnic	Pasture
Acceptability score	0%	Waste sand Topsoil London clay PFA/sewage	Waste sand Decomposed refuse Subsoil, London clay Decomposing refuse Topsoil
	100%	Subsoil Decomposing refuse Decomposed refuse	Dredgings PFA/sewage

however, no standard landscape scheme, no standard design which simply needs arranging to fit the space available, and there is a rich variety of plant material provided by nature and selective breeding. In response to all this, landscape architects have become extremely adept and proficient at selecting the most appropriate designs and materials according to the needs of individual schemes.

Equally, however, there is a rich variety of soils and soil-forming materials available. Cultivated topsoil provides only a limited range of the materials available and is the optimum growing medium for relatively few plant species. We must select soil materials according to the particular requirements of each scheme, not on the basis of what is easy to specify.

Poor soil quality

Soils ideally suited to a particular scheme are rarely available to the landscape architect. More typically, the soils that are available will require some maintenance or treatment at the time of establishment and/or during a period of after-care. Unfortunately, experience suggests that this is often overlooked or that attention is paid to the wrong soil properties.

In order to assess the quality of soils used in restoration schemes, samples of topsoil were collected from twenty-four restored landfill sites and their fertility compared to that of the refuse buried beneath the topsoil. As restoration was initially to grassland, topsoil (i.e. original layers from grassland or cultivated land) should provide an appropriate type of soil. However, in all cases, the quality was either poor or atrocious. Of the topsoils sampled, 70% were able to support agricultural grass species but the amount of growth was no better, and often less than that obtained from the refuse underlying the topsoil (Fig. 7.3). The remaining 30% were so badly contaminated as to be completely unsuitable for landscaping schemes. Chemical analysis suggested that poor growth could be attributed to nutrient deficiency (Table 7.2). In subsequent pot experiments, notable increases in growth were achieved by addition of nitrogen or nitrogen and phosphorus.

Similarly depressing results have been reported by Bloomfield *et al.* (1981) for topsoils used in various parts of Merseyside, Greater Manchester and Cheshire. Of forty-four topsoil samples, none approached the quality of good garden soil acting as a control. Half of them were no better than brick waste from an urban clearance site, one-fifth were no better than raw colliery waste and one prevented growth altogether. Again, analysis suggested that this poor performance could be attributed mainly to nutrient deficiencies, particularly lack of nitrogen.

BS 3882 concentrates on the physical properties of soils at the expense of

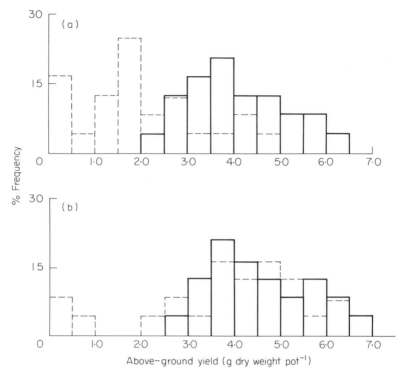

FIG. 7.3. Growth of ryegrass (*Lolium perenne*) in topsoils (---) and refuse (—) from twenty-four and landfall sites. (a) Without fertilizer addition, (b), plus N, P, K (100, 50, 50 kg ha^{-1}) and lime (1000 kg ha^{-1}).

chemical and biological properties. Nutrients are ignored completely. Physical properties, such as texture, structure and porosity, are very important. They determine how easily a soil can be cultivated, what its drainage properties will be; whether it is liable to erosion, waterlogging, poor aeration and nutrient losses through leaching. They provide the framework or skeleton of a soil and determine how much response will be obtained from addition of fertilizer. But the data in Table 7.2, together with that reported by Bloomfield *et al.* (1981), suggest that the amount of growth achieved in practice is not determined by physical properties but rather by the chemical characteristics, in particular nutrient supply.

There are three main causes of this problem: (i) cultivated soils do not necessarily contain adequate reserves of nutrients to support maximum agricultural yields in the absence of fertilizers or manures; (ii) lifting, storage and spreading of soils causes several changes, many of which lead to reduced

TABLE 7.2. Chemical analysis of topsoils and surface refuse materials from twenty-one different restored landfill sites in SE England. (μg g^{-1} except pH; mean $\pm$ standard error; minimum–maximum values in parentheses)

	Topsoils	Surface refuse
pH	7·38 ± 0·05	7·21 ± 0·06
	(7·2–7·5)	(6·9–7·5)
Total N	1879 ± 662	5197 ± 869
	(546–4745)	(2133–9735)
Extractable N	57·6 ± 12·6	66·3 ± 8·42
	(24·8–102)	(23·8–114)
Mineralizable N	20·9 ± 4·02	31·6 ± 4·19
	(11·7–34·6)	(11·5–58·8)
Total P	469 ± 46·3	1558 ± 241
	(286–647)	(370–2765)
Extractable P	9·46 ± 1·43	49·7 ± 7·01
	(3·42–14·5)	(22·0–88·4)
Total K	3693 ± 582	1801 ± 174
	(1984–6033)	(699–2701)
Extractable K	109 ± 23·3	263 ± 30·7
	(51·7–277)	(129–486)

nutrient availability; (iii) materials sold as topsoils sometimes include a large proportion of subsoil. The topsoils used in restoration, consequently, rarely contain an adequate supply of nutrients. If topsoil is being used in the expectation of high productivity, then it clearly cannot be relied upon, and the deficiencies must be repaired if adequate growth is to be achieved. Soil selection and management must take into account a much broader range of soil properties than those currently included in BS 3882.

The cost of topsoil

Table 7.3 lists estimated costs for a landfill restoration scheme based on local quotations. Two comparative costings are provided, one based on topsoil delivered to the site, the other on waste sand obtained from a local borrow pit. The waste sand incurs costs for lifting as well as carting and spreading, together with higher costs and lower returns during the after-care period. Nevertheless, restoration costs using waste sand are approximately half those for topsoil and afford savings of approximately £6000 ha^{-1}. In this particular instance, purchase costs for topsoil were just below £2 m^{-3}. In some other parts of the UK purchase costs exceed £5 m^{-3}. This would increase restoration costs to around £25 000 ha^{-1} compared to £6200 using non-topsoil materials.

TABLE 7.3. Comparative costs for restoration of a landfill site to pasture using topsoil and locally-available waste sand

		Topsoil	Waste sand
Costs (£ ha^{-1})			
Provision and spreading of soil		12 434	5486
Cultivation and establishment		234	234
After-care	Year 1	424	424
	Year 2	99	99
	Year 3	424	424
	Year 4	33	99
	Year 5	33	33
Fencing		254	254
Total		£13 935	£7053
Income (£ ha^{-1})			
After-care	Year 1	365	320
	Year 2	75	62
	Year 3	365	330
	Year 4	75	70
	Year 5	75	75
Total		955	857
Balance		£12 980	£6196

Similar large savings have been indicated for establishment of grass in urban areas (Dutton & Bradshaw 1982).

Traditional topsoil (i.e. the original layer of grassland and cultivated land) is a limited (and declining) resource. Waste materials from the mineral extraction operations are plentiful (and accruing). In a recent survey of sand and gravel pits in East Anglia, 36 out of 57 were found to contain unwanted material which, given appropriate after-care, was potentially usable as pasture soil; 45 out of 57 had materials potentially suitable for amenity, informal recreation and wildlife conservation after-uses (S. Gregson & R.D. Roberts, unpubl. data).

AN ALTERNATIVE APPROACH

Clearly the present system is unsatisfactory. It must be replaced by one which is more precise, based on soil properties additional to texture, reaction and stone content, and which encompasses a larger range of soils and soil-forming materials than 'the original surface of grassland or cultivated land'. Such a system is possible. It is based on the belief, in contrast to BS 3882, that virtually no soil material can be used in a landscape scheme without receiving some treatment but that, given appropriate treatment, many materials could be suitable.

Of the materials used in the landfill site experiments of Fig. 7.1 (for example), river dredgings, topsoil, subsoil and decomposed refuse could all be developed into grazing soils, given appropriate treatments. Subsoil, waste sand, dredgings and decomposed refuse could also provide conditions suitable for informal recreation, amenity, landscape improvement and, in some cases, wildlife conservation. All these materials, other than the topsoil, were either available free of charge in the locality of the landfill sites or were delivered to the sites as wastes for disposal. All required some treatment, including the topsoil. The challenge then is to understand what treatment is required and how much this may cost.

The selection of soils in landscape schemes should commence with an analysis of the soil conditions required for a particular scheme. Several or all of the soil-forming materials available within an area should then be tested and the necessary treatments identified. These can be costed and the materials compared on a cost/benefit basis in relation to the requirements of the landscape scheme. Where suitable materials are not available, or cannot be manufactured at reasonable cost, the detail of the landscape scheme must be changed. This may only require exchanging the desired plant material for other species better suited to the soils that are available. In severe cases it may, however, be necessary to radically rethink the scheme. It is important, therefore, that consideration is given to soils at an early stage in the decision process and certainly not as an afterthought. Ideally, the sequence summarized in Fig. 7.4 should be followed. This approach does, however, require a thorough understanding of the biology of soils and the ecology of plant materials. In particular, a knowledge of soil assessment methods and soil improvement techniques is essential.

SOIL CHARACTERISTICS

Properties to be assessed

Table 7.4 lists the major soil properties which may need to be tested as a prerequisite for the selection of soil materials or soil-forming materials in landscape schemes. Details of analytical methods and fuller discussions of each property are available in a number of publications (including Black 1968; MAFF 1973; Allen *et al.* 1974; Fitzpatrick 1974; Russell 1974; Courtney & Trudgill 1976; Foth 1978).

Soil properties other than those listed in Table 7.4 may also need to be

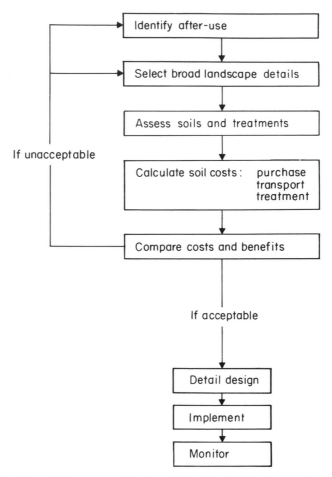

FIG. 7.4. Decision-making sequence for the selection of soils in landscape schemes.

tested in special circumstances, but these should be rare. Not all the tests will
be needed in every situation. In general, texture, structure, pH and nutrients
are the most useful: many other properties can be inferred from these by an
experienced soil biologist. Nevertheless, the tests need to be chosen according
to the nature of the soil materials under investigation and the objectives and
content of each landscape scheme.

     The agricultural grade of the land and the quality of existing crops should
be important indicators. The species composition and quality of vegetation in
non-agricultural sites can be good indicators if examined by an experienced
ecologist. A large number of plant species thrive only in very specific soil

TABLE 7.4. Major soil properties which may need to be considered in the selection of soils

Physical properties	texture
	structure
Chemical properties	pH
	lime requirement
	nutrient capital
	extractable nutrients
	toxins
Biological properties	organic matter content
	nitrogen mineralization
	plant propagules: weeds
	beneficial species

conditions (Bradshaw & Chadwick 1980). Their presence, or absence, in particular sites can be very useful indicators of soil conditions and, consequently, of the tests that may be required. Visual inspection of the vegetation should also reveal any obvious signs of variability. Where variability occurs over relatively large areas, each area must be sampled and tested, and the soils regarded as separate materials at least until the cause of the variability has been identified.

Physical analysis

The two most important physical properties of soils are texture and structure. The mineral particles in a soil can be divided into several site categories: sand (0·06–2·0 mm in any dimension), silt (0·002–0·06 mm) and clay (< 0·002 mm). Texture refers to the relative proportions of each particle size and is determined either by subjective inspection or sieving combined with hydrometry. Soil structure refers to the aggregation of individual soil particles into larger compound units and can be assessed quickly by subjective inspection or more precisely by density measurements.

The importance of physical properties has already been emphasized in this paper. They influence a number of soil conditions which can affect the ease with which particular species can be grown. They particularly determine the mechanical strength of a soil and how easily it can be cultivated. Consequently, they are particularly important in agricultural, garden and intensive recreational situations.

For other after-uses, the physical properties may be less critical. A broader range of textures may be suitable for public open space or amenity than for intense agriculture or recreation (Fig. 7.5). An even wider range of textures

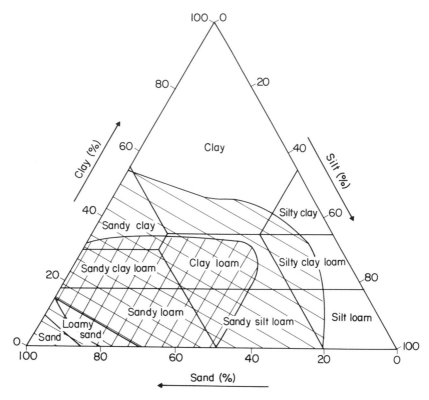

Fig. 7.5. Soil texture triangle demonstrating the textural classes that may be suitable for arable agriculture (▨) and amenity (⟍).

could be used for wildlife conservation, provided that appropriate species of plants are selected.

Chemical analysis

Past experiences of landscape schemes suggest that, however, for a broad range of soils, physical properties rarely limit the species composition or quality of vegetation. In general, the mineral constituents provide the skeleton of a soil, but it is the chemical and biological processes which determine actual plant growth.

Soil reaction (pH) and nutrient supply are the most important chemical properties. Toxic materials may also need to be considered, but these should rarely be a problem. Although in agricultural soils, nutrients are now usually added as fertilizer in amounts proportional to the quantities required by crops, for many other after-uses, the nutrients can only be obtained from the capital within the soil. Consequently, analysis for nitrogen, phosphorus, potassium is often essential, together with calcium, magnesium and trace elements where appropriate.

The total amount of each nutrient, together with an indication of its availability for plant uptake, can be obtained from many organizations together with recommendations as to whether the values obtained are satisfactory or how they should be augmented. ADAS has a coding system relating analytical data to fertilizer requirements for specific crops. For tree-planting schemes, values at the lower end of the ADAS range may be acceptable provided there is sufficient capital to maintain the extractable values for a sufficient period of time. For wildlife conservation and some amenity purposes, values of half to one-third of those for trees could be appropriate depending upon other soil conditions and the particular species to be used.

Nitrogen may require special attention. It is critically important for plant growth but is deficient in most soils. Most nutrients are derived from the minerals present in the parent rock material. Soil minerals and solid rock contain almost no nitrogen (Wlotzka 1978). The capital contained within topsoil has been accrued over many years from inputs in rain and dust and by the activity of a small number of nitrogen-fixing organisms which are able to absorb and use nitrogen from the atmosphere (Black 1968; Marrs *et al.* 1983). This nitrogen eventually accumulates in the soil as organic matter. The organic matter is subsequently broken down by soil animals and micro-organisms and the nitrogen liberated in forms that can be taken up by plants. The total amount of nitrogen in a soil is a guide to what is ultimately available, but does not indicate the extent to which organic matter decomposition and nutrient release is occurring. This can be estimated by a nitrogen mineralization test in which soils are incubated at $25°C$ for 2 weeks (Keeney & Bremner 1967). The increase in extractable nitrogen over this period provides an indication of the ability of the soil to provide nitrogen. Depending upon the after-use, values of $30–100$ μg N g^{-1} soil may be acceptable.

Biological analysis

In the final analysis, whether a soil is satisfactory or not is indicated by its ability to grow plants, and its ability to grow the right plants. This may be best indicated by a direct test using an appropriate plant as a bioassay. Typically a standard plant, such as ryegrass, radish or tobacco, is grown in a greenhouse on replicated samples of the possible materials, together with replicated samples of a good quality soil to act as a control (Bloomfield *et al.* 1981). If deficiencies are expected, various amendments, such as additions of organic matter and/or nutrients, can be included. If these have effects, they will indicate not only the nature of the deficiency, but also the amount of amendment necessary to overcome it (Roberts *et al.* 1983).

Topsoils, in particular, often have problems because of the seed or fragments of obnoxious and troublesome weeds that they contain, such as couch grass (*Agropyron repens*) and docks (*Rumex obusifolius* and *R. crispus*). While it may be possible to recognize vegetative fragments without a test, it is rarely possible to assess the presence of dormant seed directly. If information on potential weed problems is wanted, a simple and effective test is a bioassay of the soil material by keeping it in shallow trays in a greenhouse, carefully watered, for about a month and observing what germinates. One very great advantage of subsoils is, by contrast, their almost complete absence of weeds.

Costs

The costs of soil testing will vary according to the particular tests and the organization carrying them out. Invariably, costs per sample also decrease as the number of samples and tests increases. At current rates costs are typically in the order of £10–50 per sample. But it must be remembered that soil analysis can easily save its costs several times over, and also improve the end product. Tree planting, for example, in Essex typically costs around £4500 ha^{-1}. A 10% improvement in survival could be sufficient to offset the cost of analysis.

SOIL IMPROVEMENT TECHNIQUES

Treatments

The first stage in the selection of soils in landscape schemes should be to assess the materials already existing at the site. Although the top soil may have been lost or damaged, many indigenous materials can be used. They may have one or more limitations but frequently these can be overcome. Table 7.5 lists some of the potential problems for a range of soil and soil-like materials. Whether they require treatment, and the degree of treatment, will depend upon the content of each landscape scheme. In general, arable agriculture and sports facilities require more demanding conditions than other after-uses. Consequently, soils to be used for these purposes may suffer greater limitations and require more intense treatment.

Potential remedial treatments are summarized in Table 7.6 and are discussed more fully in Bradshaw & Chadwick (1980) and Roberts *et al.* (1983).

TABLE 7.5. Potential major limitations to plant growth in soil and soil-like materials

	Texture	Structure	High temperature	Instability	pH	Nutrients	Toxicity	Micro-organisms & soil fauna
Clay subsoil	*	*	−	*	−	*	−	*
Sandy subsoil	+	*	−	*	+	*	−	*
Urban clearance site material	−	*	−	−	−	*	−	*
Sand and gravel extraction wastes	+	*	−	+	−	*	−	*
China-clay wastes	*	*	−	*	*	*	−	*
Ironstone spoil	*	*	−	+	−	*	−	*
Slate wastes	*	*	*	*	−	*	−	*
Coal spoil	*	*	*	+	+	*	−	*
Pulverized fuel ash	+	*	+	*	−	*	+	*
Decomposed domestic refuse	+	*	−	−	−	+	+	−
River dredgings	−	*	−	−	−	+	+	*

− Rarely a serious problem.
+ Can be a problem with some materials.
* May be a serious problem.

TABLE 7.6. Potential soil treatments

Problem	Solution
Poor texture	None
Poor structure	Cultivate; organic matter; vegetation
High surface temperature	Vegetation; mulch
Instability	Vegetation; mulch; stabilizer; modify slope
pH	Lime
Nutrients	Fertilizers; manures; legumes
Toxicity	Tolerant plants; bury
Micro-organisms	Vegetation
Soil fauna	Inoculate

Physical improvement

It is almost impossible to alter the texture of a soil. But where materials of contrasting texture are available, it may be possible to achieve some mixing of them if appropriate plant is available on site, or at least produce a layered soil. Some of the effects of poor texture can be ameliorated by the accumulation of organic matter, but this is rarely a complete solution. Hence, it is critically important that material of the correct texture is selected in the first place. If soils of the correct texture cannot be obtained at an acceptable price, the original design must be modified in line with the soils that are available.

Poor structure should be prevented before it occurs, by proper handling (McRae 1983). However, it can be improved by a number of methods including cultivation, addition of organic matter or growth of plants, especially legumes. The technology of cultivation systems is now highly developed (Binns 1983) and can be used to reduce compaction, improve aeration and prevent waterlogging. Poor texture and structure sometimes combine with exposed position on steep slopes, resulting in soil loss by rain-drop impact or wind erosion. This can be overcome in many cases by establishment of deep-rooted vegetation combined, where necessary, with mulching until the vegetation becomes established.

Chemical improvement

Unfavourable pH values can be improved by simple addition of lime. The amount of lime required is determined not just by the pH of the soil but also by its buffering capacity. The latter can vary considerably between soils. Typically, it is high in clay-rich or organic soils and low in sandy soils. However, the amount that is required can be estimated relatively easily in the laboratory from the results of a lime requirement test. Many soils have a natural tendency to become acidic; single applications of lime may not be sufficient to overcome the problem.

Nutrient deficiency can be overcome by the application of fertilizers or manures but, for many landscape schemes, nutrients should be added not simply according to the immediate requirements of the vegetation, otherwise it will be necessary to return annually. The challenge must be to replenish the nutrient capital so that ultimately the soil is able to provide available nutrients at levels sufficient for plant uptake without the need for fertilizers.

For nitrogen, a soil capital of over 1000 kg N ha^{-1} has been shown to be required. Since mineral wastes and subsoils typically contain almost no nitrogen, they start with a nitrogen deficiency of at least 500 kg N ha^{-1} (Marrs et al. 1983). At normal fertilizer application rates of 50–100 kg N ha^{-1} y^{-1} the application must be repeated for at least 10 years to replenish the capital, even assuming that all the fertilizer nitrogen is retained within the soil–plant system. Although topsoils should start with an adequate capital, in practice this is not necessarily the case and therefore they may require repeated fertilizer applications over several years.

Two other solutions may be possible: (i) a heavy treatment of nitrogen-rich organic material such as farmyard manure or mushroom compost can add as much as 500 kg N ha^{-1} in a single application; (ii) nitrogen-fixing legumes such as white clover (*Trifolium repens*) or tree lupins (*Lupinus arboreus*) can add 100–200 kg N ha^{-1} y^{-1} (Dancer *et al.* 1977a, b; Skeffington & Bradshaw 1980). Many types of nitrogen-fixing plants are available and can be used to suit different purposes (Jefferies 1981).

In contrast to nitrogen, we know considerably less about the replenishment of soil capital for other nutrients. They are derived from the mineral constituents of soils and, therefore, tend to be less deficient than nitrogen. The most common strategy appears to be to add fertilizers to the soil at the time of establishment and subsequently at intervals dictated by conditions for up to 5 years. This approach appears to work, although we do not as yet understand how effective it is in restoring the soil capital and maintaining extractable levels in the long term. The main reason for this is that many other nutrients occur in soils in several different chemical and physical forms, each of which can release or bind nutrients depending upon conditions in the soil. The matter is discussed in the various books on soil properties already mentioned. But we need to know more about phosphorus and potassium nutrition in landscape soils, and about the formation of the various nutrient forms during natural soil development and as a result of specific management treatments.

Biological improvement

Unwanted weed species already present in the soil material can be controlled by mowing, herbicides or cultivation. There is now a wide range of very effective herbicides. However, the weeds may reinvade the site, particularly if soil conditions are very fertile. Aggressive species can often become a problem partly because they may be unsightly but, more importantly, because they may interfere with the desired species. Their control through the use of selective herbicides and growth retardants is possible, but may be expensive and have a temporary effect only. Continuous cropping and burning to remove nutrients (See Green, Chapter 12) deserve further investigation as possible long-term methods of reducing fertility and, therefore, controlling the aggressiveness of perennial weeds. Another approach may be to add chemicals such as iron or aluminium compounds or lime to soils to reduce the availability of phosphorus. Nevertheless, the simplest way to control aggressive weeds is to select soils which do not possess them and have the appropriate fertility in the first place. In these two respects non-topsoil materials are ideal.

The converse of the weed problem—the absence of propagules of appropriate species—is much easier to resolve since seeds, cuttings or young plants are available commercially for many species of plants. Similarly, once a vegetation cover has become established, other organisms such as fungi, bacteria, invertebrates and small mammals tend to colonize the site relatively easily (Table 7.7) (Neumann 1973; Lanning & Williams 1979; Roberts & Mellanby 1983; S. Gregson & R.D. Roberts, unpubl. data). Some invertebrates and small mammals, however, can take many years to develop a stable population. Woodlice and earthworms are often slow to colonize areas, and yet both are extremely important to organic matter decomposition—and, hence, soil structure and nitrogen supply—in normal soil–plant systems. In extreme cases, these and other species may have to be introduced. Some startling work has been carried out with earthworms on newly reclaimed polder soils in Holland (Hoogerkamp *et al.* 1983). The role of invertebrates in landscape schemes and techniques for their cultivation require urgent study.

THE NEED FOR MONITORING AND AFTER-CARE

The selection and management of soils in landscape schemes is a developing and, at the moment, incomplete science. Soils and organisms are complex and can behave in unexpected ways. Consequently, consideration of soils should not end with the establishment of the vegetation. Indeed, in many ways this is only the beginning. The real challenge is to develop the conditions which are required for long-term trouble-free maintenance of a self-sustaining ecosystem, as argued by Bradshaw (Chapter 2). This can rarely be achieved instantly, and may require several years of inspection, testing and retreatment. Plant

TABLE 7.7. Development of invertebrate populations at Pitsea landfill site (Roberts & Mellanby 1983)

Selected taxa	Common description	Age since tipping (years)					Control grassland
		1	3	4	15	25	
Annelida	Segmented worms	+	+	+	+	*	+
Mollusca	Snails, etc.	−	+	+	+	+	+
Collembola	Spring tails	+	+	+	*	+	*
Orthoptera	Grasshoppers, etc.	+	+	*	+	+	+
Hymenoptera	Bees/wasps/ants	+	+	*	*	*	*
Coleoptera	Beetles	*	*	*	*	*	*
Acari	Mites/ticks	−	−	+	+	*	+
Lepidoptera	Butterflies/moths	+	+	*	*	+	+
Myriapoda	Millipedes/centipedes	−	+	+	*	+	*

− Absent.
+ Less than 10 individuals trapped.
* More than 10 individuals trapped.

material in landscape schemes is frequently cared for over a number of years; soils need similar care and attention.

A very precise after-care system was developed by MAFF a long time ago in relation to the reclamation of ironstone working. Several mineral operators and MAFF are now developing after-care systems for present mineral extraction restoration schemes. In these, each site is inspected at least on an annual basis for up to 5 years: soil samples are collected and analysed, and treatments for the following year developed on the basis of the test results. Such a system could usefully be applied to many landscape schemes.

APPLICATION OF A NEW APPROACH

During recent years, there has been a marked increase in the use of more imaginative, ecologically-based approaches than BS 3882 to the selection and management of soils in landscape schemes. There are numerous schemes illustrating the success that can be achieved at costs comparable to, or less than, traditional approaches. The following examples have been selected to illustrate what can, and has been, achieved.

The first example refers to the use of topsoil. This may seem rather strange in a paper which has been arguing against topsoil. Nevertheless, there are occasions where the original surface soil is undoubtedly the most appropriate material. One such instance is where an area contains ecologically-interesting natural or semi-natural vegetation, and the aims of the landscape scheme are to recreate that vegetation following a temporary disturbance. Here, the original soil will be extremely useful and should be preserved at all cost, since it will contain useful plant propagules and more or less correct conditions for growth of those species. It is, however, rarely sufficient to strip, stockpile and respread the soil. Invariably, some propagules and nutrients are lost and these must be replaced if the scheme is to be successful.

One of the best examples of this approach can be found amongst the sand dunes on the east coast of Australia; these are covered with an attractive heathland vegetation which must be preserved, but the dunes contain valuable reserves of zircon, rutile and ilmenite. These are currently being mined by passing the complete sand dune through a processing plant. Immediately before an area is mined, the topsoil is removed very carefully and then respread after mining (Brooks 1976). A seed mix containing a nurse crop, a perennial grass and some native species is then sown, the area covered with brush to reduce erosion and fertilized. After periods of 18 and 48 months, the species composition is assessed, and supplementary nursery-raised seedlings added if required. Within 5–8 years the original species composition and balance is re-established, including the dominant shrub and tree species. Similarly impressive schemes have been carried out in this country, particularly in relation to re-establishment of semi-natural vegetation following pipeline burial (Gilham & Putwain 1977).

The second example refers to the opposite extreme: a soil material which has virtually none of the properties of topsoil. Mining of china-clay waste in Cornwall produces large amounts of waste materials which are deposited as steep-sided tips. The waste consists almost entirely of sand-sized quartz particles with very few silt or clay particles. It would be extremely difficult and very expensive to cover these heaps with imported soil. As a result, the waste itself is being treated and vegetation established directly upon it. The waste is slightly acid, very deficient in most nutrients and has poor structure and moisture-retaining properties (Table 7.8). The major limitation to growth, however, is nitrogen deficiency. Provided the wastes are sown with a legume and/or given nitrogen fertilizer, excellent growth can be achieved (Fig. 7.6). Many legumes are very sensitive to phosphorus supply and soil reaction. In the case of china-clay wastes, both lime and phosphorus must be added if many herbaceous or woody legume species are to survive (Fig. 7.7). Single applications are rarely sufficient, since calcium ions are rapidly leached, and much of the phosphorus added as fertilizer becomes bound in the soil in unavailable forms. Both lime and phosphorus must be added every 2–3 years for as long as the legume component is required (Roberts et $al.$ 1983).

Many china-clay tips have now been restored using this approach. The nitrogen contents of a number of sites of different ages have been assessed by analysis of surface soils (0–30 cm), vegetation and litter. Mean net accumulation rates were calculated as 108 kg N ha^{-1} y^{-1} (Marrs et $al.$ 1983). This compares with estimated inputs of nitrogen from fertilizer and rainfall of 81 kg N ha^{-1} y^{-1}. The remaining 27 kg N ha^{-1} y^{-1} must be derived from legume fixation and provides a minimum estimate of the effects of legumes. When various losses from the system are taken into account, the fixation rates may be as high as 63 kg N ha^{-1} y^{-1}. This range of values (27–63 kg N ha^{-1} y^{-1})

TABLE 7.8. Typical nutrient analyses of china-clay sand wastes and adjacent agricultural pasture soils

	pH	CEC (mequ. 100 g^{-1})	Loss-on-ignition (%)	Total nutrient (μg g^{-1})					
				N	P	K	Ca	Mg	Zn
Sand wastes	4·1	1·4	<1	29	20	1100	40	50	5
Pasture soils	5·8	12	18	2400	980	3410	9120	4110	130

FIG. 7.6. China-clay waste has virtually none of the properties associated with topsoil. Nevertheless, provided that legumes are sown and nitrogen, phosphorus and lime added, a very productive sward can be obtained. This site is now grazed by sheep from the surrounding open moorland and supports a much greater density of animals than adjacent land.

compares well with estimates of fixation (49 kg N ha^{-1} y^{-1}) by white clover (*Trifolium repens*) under field conditions using acetylene reduction techniques. The necessary nitrogen capital should, therefore, be achieved within 10 years if the nitrogen fertilization is continued and the necessary occasional lime and phosphorus treatments provided to maintain the legume.

There are many examples of innovative landscape schemes which fall between the two extremes of the Australian sand dunes and china-clay wastes. At the Mucking landfill site in Essex, one area has been restored to cattle grazing by burying the refuse beneath 50 cm of subsoil. The subsoil was ploughed, disced, seeded with a grass/clover mix, fertilized and harrowed. Nitrogen was reapplied at the end of the first year and both nitrogen and phosphorus during the second year. The site now supports an excellent sward, and nutrient capitals and soil structure have improved dramatically. The sward is actively grazed by cattle and currently supports a greater density of animals than adjacent undisturbed fields. A landfill site in mid-Wales has similarly been restored using sand-rich overburden, and a site in the Midlands using clay-rich subsoil. They were given appropriate after-care and subsequently leased to local tenants with specific after-care clauses. Both are now being used very successfully for grazing and hay cropping.

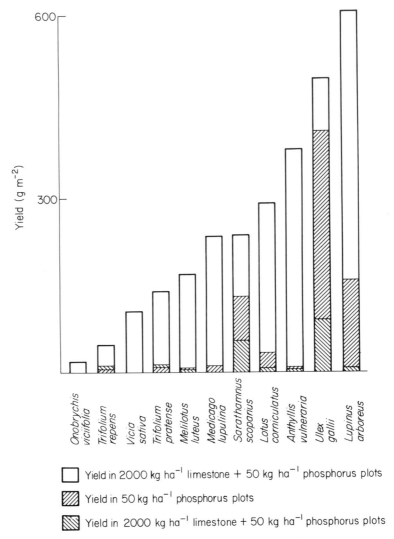

FIG. 7.7. Effects of lime and phosphorus addition on the growth of legumes in china-clay wastes.

A number of roadside verges have been landscaped without the use of topsoil. One of the most notable examples occurs on the Ellesmere Port motorway, where one part of the verge was seeded with grass and clover directly on to subsoil. The subsoil material had a sandy clay loam texture, and poor nitrogen and relatively low phosphorus levels. The site was given unusually high nitrogen and phosphorus applications in the first growing season but relatively little after-care. The fertilizer rates were designed to encourage rapid establishment and initial growth of both the clover and the grass in order to stabilize the surface, but not to promote vigorous growth in the long term. As a result, the sward is now relatively short and contains small open patches which have allowed invasion of a number of slow-growing wildflower species. In contrast, control plots where topsoil was included have encouraged vigorous growth of grasses and agricultural weeds. This has produced a very monotonous sward, dominated by red fescue, whose only relieving features are a number of tall-growing noxious agricultural weed species.

An attractive sward, similar to that obtained on the subsoil plots, has also been encouraged on a restored slate tip in North Wales. The slate tip was reshaped by scraper and bulldozer and the surface layers crushed by a vibrating roller. The area was then seeded with a grass/clover mixture and given fertilizer additions at establishment and subsequently. Again, nutrient additions were designed to encourage rapid initial growth but subsequent slow growth and an open sward. The site now contains an extremely attractive sward and supports in excess of fifty plant species, many of which are important food plants for a number of butterfly and bird species.

A further development of the concepts behind the roadside verges and slate schemes might be to attempt to create specific wildlife habitats, such as heaths or reed beds, in areas which currently support agricultural or amenity uses. The resulting habitats would have a general ecological value by protecting genetic resources and providing sites for public enjoyment in order to reduce pressures on other areas. They may even develop important nature conservation value of their own, as has happened in areas of derelict land (Bradshaw & Chadwick 1980).

CONCLUSIONS

BS 3882:1965 has many advantages and applications, but it also has failings since it is inappropriate for many landscape schemes. A development of BS 3882 seems timely, as has been suggested by Bradshaw (1983) and Holden (1983). These authors recognize the importance of nutrients and the need for detailed assessment. The three-tier system proposed by Bradshaw (1983) has the additional advantage that it accepts all soils are likely to be poor in quality and require treatment. It also recognizes that there are other potentially-useful materials, in addition to topsoil from grassland or cultivated land.

Nevertheless, such a system will still have limitations. On the one hand, any grouping of soils or soil-like materials into distinct categories of quality assumes that discrete soils exist: this is rarely the case; there is continuous variation amongst soils. On the other hand, a grouping alone will not deal adequately with the requirements of particular landscape schemes and individual sites.

What seems to be needed is a more flexible approach, based upon assessment of soil and soil-like materials available within the area of the scheme, and their evaluation in relation to the landscape scheme and the costs and practicalities of soil management techniques. When carried out properly, this approach can produce very successful results, alleviating some of the problems experienced with more traditional approaches. It can also lead to major cost reductions (Fig. 7.8).

However, a flexible approach requires a thorough understanding of soils, plants and animals and their inter-relationships, i.e. a knowledge of biology, and particularly ecology, is required. The re-creation of habitats and communities in landscape schemes offers a considerable, but exciting, challenge to the ecologist and landscape architect working in close collaboration. We now know a great deal about the biology of many systems and we can design exciting schemes. But we need to explore in much more detail how to bring those schemes about more effectively, and more economically, by the management and manipulation of starting materials. We do not yet know all the answers, but the achievements so far suggest that major developments are possible.

FIG. 7.8. Considerable costs are incurred in the purchase and spreading of topsoil. The use of other materials combined with restoration of soil nitrogen supply is often overlooked, even though such treatment would cost very little by comparison with the cost of topsoil.

REFERENCES

Allen, S.E., Grimshaw, H.M., Parkinson, J.A. & Quarmby, C. (1974). *Chemical Analysis of Ecological Materials.* Blackwell Scientific Publications, Oxford.

Binns, W.O. (Ed.) (1983). *Reclamation of Mineral Workings to Forestry.* Forestry Commission, Edinburgh.

Black, C.A. (1968). *Soil–Plant Relationships.* John Wiley, New York.

Bloomfield, H.E., Handley, J.F. & Bradshaw, A.D. (1981). Topsoil quality. *Landscape Design,* **135,** 32–34.

Bradshaw, A.D. (1983). Topsoil quality—proposals for a new system. *Landscape Design,* **141,** 32–34.

Bradshaw, A.D. & Chadwick, M.J. (1980) *The Restoration of Land.* Blackwell Scientific Publications, Oxford.

Brooks, D.R. (1976). Rehabilitation following mineral sand mining on North Stradbrook Island, Queensland. In *Landscaping and Land Use Planning as Related to Mining Operations.* Australian Institute Mining and Metallurgy, pp. 93–104. Aust. Inst. Min. Metall. Adelaide.

Courtney, F.M. & Trudgill, S.T. (1976). *The Soil: An Introduction to Soil Study in Britain.* Edward Arnold, London.

Dancer, W.S., Handley, J.F. & Bradshaw, A.D. (1977a). Nitrogen accumulation in kaolin mining wastes in Cornwall. I. Natural communities. *Plant and Soil,* **48,** 153–167.

Dancer, W.S., Handley, J.F. & Bradshaw, A.D. (1977b). Nitrogen accumulation in kaolin mining wastes in Cornwall. II. Forage legumes. *Plant and Soil,* **48,** 303–314.

Dutton, R.A. & Bradshaw, A.D. (1982). *Land Reclamation in Cities.* HMSO, London.

Fitzpatrick, E.A. (1974). *An Introduction to Soil Science.* Oliver and Boyd, Edinburgh.

Foth, H.D. (1978). *Fundamentals of Soil Science* (6th edn). John Wiley, New York.

Gilham, D.A. & Putwain, P.D. (1977). Restoring moorland disturbed by pipeline installation. *Landscape Design,* **119,** 34–36.

Holden, R. (1983). Topsoil specifications, or beware of BS 3822. *Landscape Design,* **141,** 34–35.

Hoogerkamp, M., Rogaar, H. & Eijsackers, H.J.P. (1983). The effect of earthworms (Lumbricidae) on grassland on recently reclaimed polder soils in the Netherlands. In *Earthworm Ecology* (Ed. by J.E. Satchell), pp. 85–104. Chapman and Hall, London.

Jefferies, R.A. (1981). Legumes for the reclamation of derelict and disturbed land. *Landscape Design,* **134,** 39–41.

Keeney, D.R. & Bremner, J.M. (1967). Determination and isotope ratio analysis of different forms of nitrogen in soils. 6 Mineralizable nitrogen. *Proceedings of the Soil Science Society of America,* **31,** 34–39.

Lanning, S. & Williams, S.T. (1979) Nitrogen in revegetated china clay sand waste. I. Decomposition of plant material. *Environmental Pollution,* **20,** 147–159.

Marrs, R.H., Roberts, R.D., Skeffington, R.A. & Bradshaw, A.D. (1983). Nitrogen and the development of ecosystems. In *Nitrogen as an Ecological Factor* (Ed. by J. A. Lee, S. McNeill & I.H. Rorison), pp. 113–136. Blackwell Scientific Publications, Oxford.

Ministry of Agriculture, Fisheries & Food (1973). *Analysis of Agricultural Materials.* Technical Bulletin 27. MAFF, London.

McRae, S.G. (1983). Good practice in the agricultural restoration of sand and gravel workings. *Landscape Design,* **141,** 29–32.

Neumann, U. (1973). In *Ecology and Reclamation of Devastated Land* (Ed. by R.J. Hutnik & G. Davis), Vol. 1, pp. 335–348. Gordon and Breach, New York.

Roberts, R.D., Marrs, R.H., Skeffington, R.A. and Bradshaw, A.D. & Owen, L.D.C.O. (1983). The importance of plant nutrients in the restoration of china clay and other mine wastes. *Transactions of the Institute of Mining and Metallurgy,* **A91,** 42–49.

Roberts, T.E. & Mellanby, K. (1983). Ecological monitoring of landfill sites. In *Reclamation 83,* pp. 492–502. Industrial Seminars Ltd., Kent.

Russell, E.W. (1974). *Soil Conditions and Plant Growth* (9th edn). Longman, London.
Skeffington, R.A. & Bradshaw, A.D. (1980). Nitrogen fixation by plants grown on reclaimed china clay waste. *Journal of Applied Ecology*, **17**, 469–477.
Wlotzka, F. (1978). Nitrogen. In *Handbook of Geochemistry* (Ed. by K.H. Wedepohl), Vol. II, 7B M. Springer-Verlag, Berlin.

8. CAUSES AND PREVENTION OF ESTABLISHMENT FAILURE IN AMENITY TREES

H. INSLEY[1] AND G. P. BUCKLEY[2]

[1]*Forest Research Station, Alice Holt Lodge, Wrecclesham, Farnham, Surrey and* [2]*Department of Horticulture, Wye College (University of London), Nr Ashford, Kent*

SUMMARY

The survival and growth of young trees are affected by (i) their physiological condition at the time of planting and (ii) the ecological situation, determined by the plant communities and site conditions, into which they are introduced. These influences were investigated using examples of young broadleaved tree-planting on motorway verge sites in England and Wales.

Exposure of bare-rooted young trees during handling and transplanting reduced their survival and subsequent growth. Tolerance to exposure varied with species and was related to the interaction of root morphology, water loss and reductions in carbohydrate, although root cell structure and root:shoot ratios were probably also implicated. Sampling of transplants delivered from the nursery to the motorway in 1978 and 1979 showed that up to 10% of the trees had already lost more than 50% of their contained water and were unlikely to survive.

The road verges were inherently fertile, although soils were thin and nitrate and phosphate levels were universally low. The swards, already established before tree-planting in the majority of cases, were derived from a standard seed mixture dominated by *Lolium perenne*. Control of these swards using herbicides demonstrated the strong competition exerted by grass cover, the response of tree growth being proportional to the area of sward removed. Glasshouse experiments with *Betula* and *Lolium* seedlings suggested that this competition took the form of more efficient nutrient capture by grass when water was not limiting. However, in the field the capacity of a grass sward *per se* to reduce soil moisture to wilting point in the upper part of the profile would have impaired the ability of trees to take up nutrients and obtain water.

Large improvements in the success of tree-planting schemes can thus be achieved by the simple expedients of transporting bare-root stock inside closed polythene bags to reduce water loss, and by maintaining sward-free zones around the transplants for 2–3 years after planting. These precautions will often outweigh the disadvantages of poor site conditions and allow successful tree and shrub establishment.

INTRODUCTION

In the late 1970s an estimated £54m per annum was spent on amenity tree and shrub stock, a figure which would have been easily doubled by the further costs of establishment (Insley 1982a). Simple observation suggests that much of this expenditure was redundant because poor handling, bad planting and inadequate maintenance caused numerous losses. Brief examination of the number of planting replacements required for motorway and trunk road schemes initiated by the Department of Transport shows that considerable opportunities for improvement and cost-saving exist (Table 8.1).

Attempts have been made to overcome or to avoid the problems of transplanting small tree stock caused by root exposure. Direct seeding onto sites is such an avoidance mechanism which has recently been advocated in Britain (Luke *et al.* 1982; La Dell 1983). While this technique undoubtedly overcomes the problems induced by plant handling, others are introduced, such as achieving germination in hostile conditions, preventing competition by 'nurse' or indigenous species and, if successful, re-spacing and protecting crowded saplings. Moreover, direct seeding into already existing grass swards sown for the purposes of stabilization, the context considered here, is unlikely to be successful. Elsewhere, its suitability depends on subtle understanding of a number of environmental and ecological factors which may be difficult to transmit to landscape contractors.

At the other extreme, many landscape architects are attracted by the immediate effect of introducing standard or heavy standard trees (British Standards Institution, 1966). Unsightly dead or dying trees, still neatly staked and tied, are almost a trademark of urban and roadside planting schemes and

TABLE 8.1. Total number of plants and the proportion of beat-ups (replacements) used for motorway planting during the period 1976–81 (Insley, 1982a)

Planting season	Total plants used	Replacements (%)
1976–77	2651 775	37·6
1977–78	2069 705	26·4
1978–79	1942 029	36·1
1979–80	1377 927	39·6
1980–81	1267 477	33·7

are a direct result of nursery production techniques which encourage low root:shoot ratios. Subsequent transplanting shock (which is very much greater in larger than smaller stock) is exacerbated by planting into vigorous young grass swards, but the weeding of large trees is frequently overlooked.

Another much-favoured solution is to change from bare-rooted to container-grown or containerized stock. In some circumstances containers do enable the planting season to be extended, but the transporting and handling costs are usually much greater and there may be problems of root deformation and future instability (Insley 1982b). In addition, relatively small differences in survival and growth of bare-root and container-grown stock have been reported on motorways (Insley 1981).

These findings have, therefore, led to the present investigation into the precise effects of plant handling and of sward control techniques convention-ally used on roadsides in Britain. From these it was hoped that the benefits and cost-effectiveness of improved techniques could be evaluated.

EFFECTS OF PLANT HANDLING

Smith (1977) reported drastic effects in the survival of oak seedlings after reductions in fresh weight of 15–20%. In an experiment under prevailing nursery conditions, we exposed dormant broadleaved seedlings of eight different species* for periods of up to 5 days before replanting. These desiccation treatments produced a wide range of species responses with some, such as *Tilia*, *Aesculus* and *Robinia*, showing only minor reductions in subsequent survival after the maximum exposure period, compared with massive losses in *Betula* and *Nothofagus*.

Further experiments under controlled-environment conditions were used to examine the main location of water loss in the root and to determine relative loss rates within and between species. The two species chosen for these experiments were selected for their contrasting root structures: *Fraxinus*, in which seedlings are characterized by a dominant tap-root and comparatively coarse minor rootlets; and *Betula*, the seedlings of which have strongly-branched, fine root systems with no well defined tap-root. For example, the fine lateral roots of samples of *Betula* seedlings from a local nursery were all less than 0·1 mm in diameter, whereas those of *Fraxinus* had an average diameter of 0·24 mm.

A comparison of the root size distribution of the two species (Fig. 8.1)

---

* The species used in the different plant handling experiments described in this paper were: *Quercus petraea, Nothofagus obliqua, Acer platanoides, Aesculus hippocastanum, Betula pendula, B. pubescens, Crataegus monogyna, Robinia pseudoacacia, Tilia cordata, Fraxinus excelsior* and *F. angustifolia*.

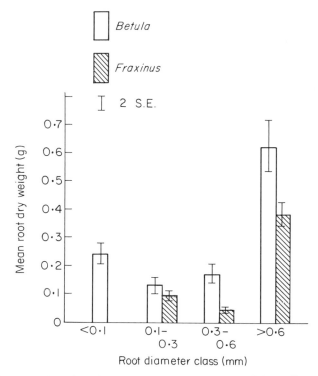

FIG. 8.1. The proportion of *Betula pubescens* and *Fraxinus angustifolia* seedling root systems in designated diameter classes.

shows that although *Fraxinus* seedlings are smaller than *Betula*, a relatively much greater proportion of the roots are in larger diameter classes. The more finely structured root system of the *Betula* seedlings might be expected to increase their susceptibility to drying out during handling. To test this, the drying gradients of samples of seedlings were examined by exposing them under laboratory conditions of approximately constant temperature and humidity for periods of up to 72 hours. After exposure the plant roots were divided according to their natural branching habit into laterals and tap-root (Fig. 8.2) and oven-dried in order to determine the moisture content (as percentage of dry wt). In general, moisture losses with increasing exposure took an exponential form, the drying rates being rapid at first but declining rapidly, such that

$$\% \text{ moisture} = AB^t$$

where A = initial moisture content of the seedlings, B = rate of change of moisture content and t = time. The fitted curves (Fig. 8.3) showed that the

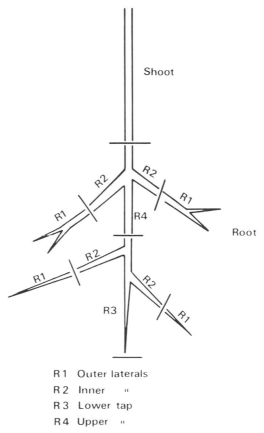

R 1 Outer laterals
R 2 Inner   "
R 3 Lower tap
R 4 Upper   "

FIG. 8.2. Diagram to show the plant subdivisions used to measure and compare drying gradients.

initial moisture content of the shoots of both species was much lower than in the roots, but the latter lost water more quickly. Within the roots, the outer laterals of smallest diameter had the highest moisture contents and dried out quickly in comparison with the root axis. Between-species differences were less obvious, but statistical comparisons suggested that for whole-root systems the rate of moisture loss was similar in each species although they had differing initial moisture contents. However, the large roots and shoots of *Fraxinus* took longer to dry out than those of *Betula*.

These results confirm that small roots are the most susceptible to desiccation and that short periods of intense drying are sufficient to reduce their moisture content to fatal levels. Several authorities have, in addition, shown that poor survival and slow growth of nursery stock follows prolonged soil moisture content under an area sown with a standard motorway seed mixture (Department of Transport 1976) at Alice Holt Research Station. The soil was a silty loam with flints, overlying Gault Clay and had good moisture retention. Field capacity was 40% (w/w), with a permanent wilt point determined as 13% (Briggs & Schantz 1912). After establishment of the sward, replicated 4 × 4 m plots were tested as follows:

(a) left uncut (control);
(b) mown regularly to 30 mm;
(c) killed with herbicide and thereafter maintained bare;
(d) covered with bituminized felt as a sheet mulch;
(e) covered with pine bark chippings to 150 mm depth.

Five gypsum soil moisture probes were inserted below each plot to depths of 50, 100, 200, 300 and 400 mm and continuous recordings of soil moisture were taken for two growing seasons. In general, in the experimental area soil water only fell below field capacity during the summer after periods of low rainfall (Fig. 8.5). The least drying out occurred under the two mulch treatments and was most severe under the uncut sward, which dried almost to permanent wilting point to 200 mm depth after a period of low rainfall in June 1980. Drying was slightly less severe under the mown sward at first, but this increased compared with the uncut sward as a mat of dead vegetation began to develop in the latter. Herbicide-treated plots in contrast dried only super-ficially, extending to depths of 200 mm only after prolonged dry spells during spring 1980 and summer 1981, while the soil moisture content below remained at field capacity.

These results allow some perspective on the use of water by grass swards and confirm that, in comparison with either mown or neglected swards, herbicides and mulches are important means of establishing newly-planted trees. However, the form of sward control imposes its own characteristics, and different surface treatments might be expected to influence other soil factors such as temperature, aeration and mineralization rates as well as moisture content (Buckley et al. 1981).

Even in the absence of moisture stress, nutrient capture by the sward species appears to be more efficient than that of young trees. A pot experiment, in which Lolium perenne and Betula pubescens seedlings were grown together in sand culture under glasshouse conditions, demonstrated that the domination of the nutritional state of the trees by the sward could only be moderated slightly by increased inputs of water and nitrogen. The total nutrient application during the course of the experiment was equivalent to 27·5 and 55 kg ha^{-1} of N at the two levels used, including 5·6 kg ha^{-1} of P and approximately 27·3 kg ha^{-1} of K. Nutrient capture by Betula seedlings

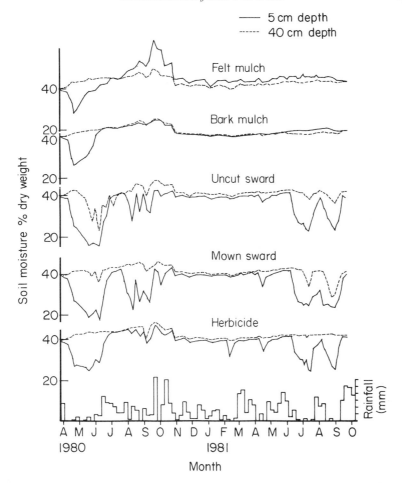

FIG. 8.5. Soil moisture contents at 5 and 40 cm under five surface treatments and weekly rainfall during the period April 1980–September 1981.

grown with and without *Lolium* and by the *Lolium* seedlings themselves is illustrated in Table 8.4.

The actual effect of the *Lolium* sward was to reduce the nitrogen capital of *Betula* seedlings by about 75% and that of P and K by approximately 70% in each case. This success of *Lolium* in capturing nutrients can be gauged by observing the resource partitioning of available nutrients between the two species. At the lower level of nitrogen, *Lolium* had accumulated 6·9, 5·2, and 26·9 times the amount of nitrogen, phosphorus and potassium respectively in *Betula* seedlings at the end of the experiment.

TABLE 8.4. Mean plot nutrient content of *Betula* and *Lolium* (elemental wt in g × 100) grown
for 99 days with two levels of nitrogen in substrate

| | Betula | | | | Lolium | |
| | Without Lolium | | With Lolium | | (With Betula) | |
Nutrient/treatment	N_1	N_2	N_1	N_2	N_1	N_2
N	5·21	8·30	1·33	2·11	9·19	12·20
P	0·84	1·12	0·25	0·38	1·29	1·61
K	2·28	3·48	0·67	0·93	18·04	22·62

In order to determine the integrated effects of sward competition on tree
growth in the field, an experiment was carried out on the verges of the A610 at
Ripley, Derbyshire, in which successively larger circular areas of sward were
removed from around newly-planted transplants of *Alnus cordata*, *Crataegus
monogyna* and *Acer pseudoplatanus*. The four levels of weed control removed
respectively 0, 12·5, 25 and 50% of the vegetation present on the plots and
these weeding treatments were combined with or without a dressing of urea
giving 77 kg N ha^{-1}.

Analysis of height and stem diameter growth after three growing seasons
showed that growth responses improved with increasing weed control. Height
increments of trees given the maximum herbicide treatment were 50% greater
than in control plots, while basal area increments doubled in *Crataegus* and

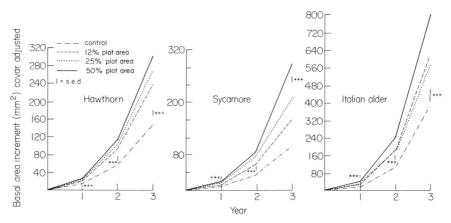

FIG. 8.6. Change in mean basal area increment with four different levels of sward control
over a 3-year period at Ripley, Derbyshire. * Significance of difference between treatments.

Alnus and increased fourfold in *Acer* (Fig. 8.6). Growth was mostly exponential or linear for each separate treatment over 3 years and the trees in weedier treatments showed no sign of gaining on those enjoying better sward control. Previous workers have shown similarly clear-cut tree responses to weed control. Moreover, Buckley *et al.* (1981) have demonstrated that responses by *Acer platanoides* transplants and whips are proportional to both the area and diameter of the weed-free zone provided around the trees, as was also the case in the Ripley experiment, implying that optimal growth responses may be achieved at well over 50% of sward control at these spacings.

At Ripley, there was a marked difference in response between *Crataegus* and *Acer* and the nitrogen-fixing *Alnus*. The former both responded positively to additions of nitrogen, but when these were discontinued after the second year both species showed some fall-off in height though not in basal area increments. In contrast, the height and basal area increments of *Alnus* were not improved by fertilizer, suggesting that nitrogen fixation was satisfactory in this species. Interactions between sward removal and nitrogen additions occurred only in *Acer*, where fertilized trees gave relatively greater basal area responses with increase in the herbicide-treated area in the final year.

DISCUSSION

These simple investigations raise fundamental questions about the current specifications of many tree- and shrub-planting schemes and their practical implementation. The current work done on plant handling suggests some of the reasons for planting failures as well as means of preventing them. Relatively short periods of exposure clearly have the capacity to destroy the fine lateral roots of transplants, impair survival and reduce subsequent growth rates. This is verified by other recent work (Dutton & Bradshaw 1982). Conversely, simple techniques such as sealing plants in polythene after lifting, soaking root systems prior to planting and ensuring good soil:root contact in the planting pit are likely to be effective unless the desiccation process has proceeded too far. Further work on the desiccation tolerance of other common broadleaved species and their root size distribution in relation to current nursery practices would be of considerable benefit.

The Code of Practice prepared by the Joint Liaison Committee for Plant Supplies (JLCPS 1980), of which the Landscape Institute is a member, was much modified from its original draft and now suits the nursery industry at the expense of the product. The size and type of plant stock frequently used for landscape schemes could often be modified to reduce costs and often to speed up results. Redundant overspecification is nowhere more apparent than in the use of standard trees for immediate visual effect, although these are frequently outgrown within a few years by transplants or whips. Similarly, the whole concept of importing topsoil for planting sites may sometimes be questioned. In some cases, the same result could undoubtedly be achieved by planting into existing media modified by the addition of nutrients at lower cost, as is discussed by Roberts (Chapter 7). Moreover, topsoil spread during civil engineering projects then requires stabilization by fast-growing grass swards which are detrimental to tree growth and establishment. Simple attention to post-planting maintenance would achieve greatly enhanced growth rates where the use of topsoil and swards cannot be avoided. However, all too frequently swards are kept mown for appearance everywhere except within the areas planted with trees and shrubs, precisely where sward removal is most desirable.

ACKNOWLEDGMENTS

We thank the following members of our research organizations who contributed to the work described: Roger Boswell, Terry Davis, Roger Warn, Peter Howard, Joe Gardiner, Sonia Wright, Ziad Al Barazi and John Williams. Much of the research described was carried out under Department of the Environment contract No. 483/7 and the assistance of Anthony Dunball and his staff at the Department of Transport in providing sites and facilities is gratefully acknowledged.

REFERENCES

Aldhous, J.R. (1972). Nursery practice. *Forestry Commission Bulletin, 43,* London.
Bloomfield, H.E., Handley, J.F. & Bradshaw, A.D. (1981). Topsoil quality. *Landscape Design,* **135,** 32–34.
Briggs, L.J. & Schantz, H.L. (1912). The wilting coefficient for different plants and its indirect determination. *USDA Bureau of Plant Industry Bull.,* **230,** 1–77.
British Standards Institution (1966). *Recommendations for transplanting semi-mature trees. BS 4043:1966* British Standards Institution, London.
Buckley, G.P., Chilton, K.G. & Devonald, V.G. (1981). The influence of sward control on the establishment and early growth of ash (*Fraxinus excelsior* L.) and Norway maple (*Acer platanoides* L.). *Journal of Environmental Management,* **13,** 223–240.
Coutts, M.P. (1981). Effects of root or shoot exposure before planting on the water relations, growth and survival of Sitka spruce. *Canadian Journal of Forest Research,* **11,** 703–709.
Department of Transport (1976). *Specifications for road and bridge works.* HMSO London cl. 611, 32–33 & cl. 2616, 150.
Dunball, A.P. (1978). Establishing trees on difficult sites—motorway planting. *Arboricultural Journal,* **3,** 273–280.
Dutton, R.A. & Bradshaw, A.D. (1982). *Land Reclamation in Cities.* HMSO, London.
Insley, H. (1980). Wasting trees—the effects of handling and post-planting maintenance on the survival and growth of amenity trees. *Arboricultural Journal,* **4,** 65–73.

Insley, H. (1981). Roadside and open space trees. *Research for Practical Arboriculture*, pp. 84–92. Forestry Commission, Farnham.

Insley, H. (1982a). *The effects of stock type, handling and sward control on amenity tree establishment*. Ph.D. thesis, Wye College, University of London.

Insley, H. (1982b). The use of container-grown broadleaved trees. *Landscape Design*, 137, 38–40.

JLCPS (1980). *Code of practice for plant handling*. Joint Liaison Committee on Plant Supplies. HTA Reading.

La Dell, T. (1983). An introduction to tree and shrub seeding. *Landscape Design*, 144, 27–31.

Luke, A.G.R., Harvey, H.J. & Humphries, R.N. (1982). The creation of woody landscapes on roadsides by seeding—a comparison of past approaches in West Germany and the United Kingdom. *Reclamation & Revegetation Research*, 1, 243–253.

Marrs, R.H., Roberts, R.D., Skeffington, R.A. & Bradshaw, A.D. (1981). Ecosystem development on naturally colonized china clay wastes. II. Nutrient compartmentation. *Journal of Ecology*, 69, 163–169.

O'Flanagan, N.C., Walker, G.J., Waller, W.M. & Murdoch, G. (1963). Changes taking place in topsoil stored in heaps on opencast sites. *NAAS Quarterly Review*, 62, 85–92.

Smith, D.J. (1977). *Root:shoot relations in establishing transplanted trees*. Ph.D. thesis, Wye College, University of London.

Widholm, J.M. (1972). The use of fluorescein diacetate and phenosafranine for determining viability of cultured plant cells. *Stain Technology*, 47, 189–194.

FIG. 9.1. The derelict Royal Ordnance Factory. (Photograph: John Mills Photography Limited.)

OAKWOOD LANDSCAPE STRUCTURE

The creation of a bold new landscape structure based on extensive woodland planting, including a variety of parks and the careful management of existing areas of ecological interest, was seen as an essential element in the redevelopment of Oakwood housing area (Fig. 9.2). Four main stages are involved in the creation and care of the new woodland structure.

Planning the woodland structure

Prior to redevelopment, Oakwood was an open space within a wooded enclosure. This is reflected, on a smaller scale, in the physical planning of the new landscape structure. A web of woodland belts, linked to the existing woods on the margins of the area, is arranged to enclose 'cells' for residential developments, schools, playing fields and other land uses. The belts vary from 10 to 40 m in width and from their edges, 'fingers' extend to enclose small groups of houses, car parking areas and play areas. Continuity and linkage of the woodland belts defines the extent, individuality and 'territory' of residential areas, and reduces the perceived scale of development. The new woodland structure will also create shelter from wind and help filter gaseous

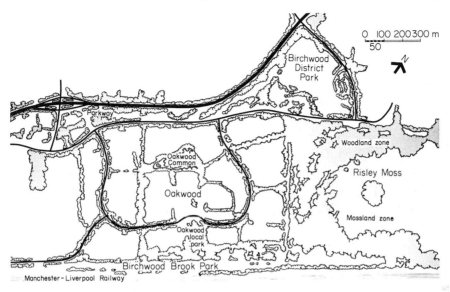

FIG. 9.2. The structure of linked woodland belts in Oakwood and the surrounding parks and open spaces which comprise Birchwood Forest Park.

and particulate air pollution. Perhaps most importantly, the belts with their links to the surrounding countryside can be seen by people as a continuous thread of nature through their residential environment (Fig. 9.3).

Advance planting and establishment of the woodland structure

The earliest woodland belts were established 3 years before the first housing completions and played a major role in creating a green environment for the first residents. The benefits of advance planting are financial as well as social. Low cost whips have, with good ground preparation and maintenance, caught up with standard trees within 3 years, and they show much healthier growth and more natural form. A variety of techniques have been used in the establishment of the plantations. Observations indicate that on compacted, disturbed subsoils the most important operations to ensure good growth are deep ripping, adequate drainage and mulching. The incorporation of locally available peat into the subsoil, which improves organic matter content and soil structural deficiencies, and the use of a high phosphate/slow release nitrogen fertilizer in initial years such as Enmag, have played a major part in establishment and growth rates. Experience has clearly shown that the costly importation of topsoil is not a prerequisite for good tree growth, and may even

FIG. 9.3. A linear 'greenway' connecting residential areas to the more natural landscapes of Birchwood Brook Park. (Photograph: Robert Tregay.)

be a hindrance due to the higher maintenance to deal with excessive weed growth.

Woodland planting design, maintenance and management

The new woodland structure has been established using indigenous species typical of the region. The technique involves planting all the woodland species, excluding herbaceous plants, at the outset; principal (canopy) trees, understorey species, light-demanding species, and edge shrubs. The structure and species composition of the woodland belts varies with location, width, function and the interaction of environmental factors. Narrow belts, for instance, consist essentially of two woodland edges backing onto one another. This reflects the emphasis placed on robust, diverse and visually interesting transition zones between communities, i.e. ecotones. Larger areas have more complex mosaics of vegetation associations and structures. The main aim is the creation of woodlands that will be dynamic. Therefore, a well planned and carefully implemented management programme is essential to their development.

Early plantations made only a simple distinction between 'woodland' and 'edge' mixes. Furthermore, all species were planted randomly at 1-m spacings and the nurse crop occupied 50–65% of the woodland mix. While initial visual impact was impressive, these plantations needed major thinning after only 3 years, and slow species such as *Quercus robur* were quickly suppressed by vigorous species such as *Alnus glutinosa*, *Betula pendula* and *Rosa canina*. It was also found that the critical edge zones were slow to establish and could not provide early protection to the woodland mix behind. In the light of this, five basic mixes are now employed (Table 9.1). To avoid the problems of suppression and the need for critically timed thinning operations, the nurse crop, in the true sense, is usually omitted in more recent designs, and most species are planted in individual groups of 5–100 plants. Edge shrub species are planted at 500-mm centres to produce an impenetrable thicket within 2–3 years, limiting access points to the plantations, channelling public pressure and creating quiet areas for wildlife.

Short-term maintenance generally involves keeping the ground weed-free for the first 3 years in order to reduce competition with the new planting and, ultimately, costs. The need for rapid establishment is considered to be the most critical factor, and the natural condition of young trees and shrubs growing up through a grass and herb sward has only been used on a few early experimental areas. After this initial maintenance phase, the herb layer is allowed to develop freely, but selective herbicides are used often for a further 2 years to combat noxious weeds. Within about 3 years, the canopy closes and the management phase begins. Management involves the traditional woodland techniques of thinning and coppicing to avoid leggy growth and the suppression of slower growing species, and to create multistemmed bushy plants. Coppicing is thus used both to remove unwanted growth and also to establish structural diversity and robust plants at an early stage.

The objective of management in the medium term is to create a diverse, uneven-aged and dynamic type of structure which differs from the traditional vision of a rural woodland in that the emphasis placed on a predominantly open canopy. This allows sufficient light into lower levels, to permit the development of a dense, coppiced shrub layer and a diverse field layer. Occupying larger breaks in the canopy are light-demanding groups, regenerating trees, pockets of scrub and herb-rich swards. The long-term management will continue the practice of coppicing, and some trees may be felled before maturing to create light pockets and suitable conditions for the continuation of woodland succession. It is intended that rather than developing into mature shady woods with a continuous high canopy, the plantations will retain much of the open and dynamic character of early successional phases. It is believed that, by achieving this woodland structure,

TABLE 9.1. Warrington structure planting species mixes

Main features of mix	Percentage composition	Species in mix
Woodland		
1 m centres; group planted with trees at	25	*Quercus robur*
10–50 per group and shrubs at 5–10 per	12·5	*Fraxinus excelsior*
group; *Alnus* and *Corylus* planted randomly	5	*Prunus avium*
with occasional group.	2·5	*Pinus sylvestris*
	2·5	*Ulmus glabra*
	25	*Corylus avellana*
	5	*Ilex aquifolium*
	2·5	*Sambucus nigra*
	20	*Alnus glutinosa*
Tall edge/Hedgerow		
1 m or 0·75 m centres; group-planted at	42·5	*Crataegus monogyna*
5–50 per group; *Lonicera* planted randomly;	17·5	*Corylus avellana*
frequently forms edge to planting; percentage	15	*Prunus spinosa*
of *Crataegus* increased when used as hedge.	10	*Alnus glutinosa*
	5	*Acer campestre*
	5	*Sambucus nigra*
	2·5	*Lonicera periclymenum*
	2·5	*Salix caprea*
Scrub		
0·75 m or 0·5 m centres; main species		⌈ *Crataegus monogyna*
group-planted at 5–30 per group; additional	75–100	⎰ *Prunus spinosa*
shrubs and trees randomly planted when		⎱ *Rosa canina*
included; forms blocks detached from other		⌊ *Ulex europaeus*
planting mixes.		⌈ *Corylus avellana*
		⎰ *Ilex aquifolium*
	0–15	⎱ *Sambucus nigra*
		⌊ *Viburnum opulus*
	0–10	Tree species
Light-demanding		
1 m centres; all species group-planted at 5–100	17·5	*Alnus glutinosa*
per group; may occasionally form edge to	17·5	*Betula pendula*
plantation; mostly group-coppiced on	12·5	*Sorbus aucuparia*
rotation.	5	*Populus tremula*
	2·5	*Pinus sylvestris*
	22·5	*Corylus avellana*
	10	*Acer campestre*
	7·5	*Sambucus nigra*
	10	*Acer campestre*
	7·5	*Sambucus nigra*
	5	*Ilex aquifolium*

Low edge
0·75 m centres; all species group-planted at
5–30 per group; percentages and combinations
may vary widely depending upon required;
small-scale variation important.

50	{ *Rosa canina* *Rosa arvensis*
30	{ *Crataegus monogyna* *Corylus avellana* *Prunus spinosa*
20	{ *Cornus sanguinea* *Ilex aquifolium* *Rosa pimpinellifolia* *Ulex europaeus* *Viburnum opulus*

General notes
1. Mixes are for guidance only but adaptations should reflect local conditions and function.
2. Single-species mixes may be included in any location.
3. Planting centres are dependent upon time available for establishment and importance of barrier function for edge mixes.
4. Size of planting groups will vary to increase variety and in relation to percentage of species included in mix.
5. *Rubus fruticosus* not planted as it will rapidly colonize suitable areas.
6. Planting percentages do not represent ultimate composition of mixes.
7. Above mixes cannot show full complexity of 'natural' planting but do illustrate the basic plant associations which provide potential for successful development.

it is possible to combine the needs of people for exciting, usable green space with assured cover and quiet areas for wildlife.

While management costs are relatively low, careful timing and supervision of the operations, based on detailed ecological knowledge, are clearly required. This is accomplished at present by the use of a contractor working with the landscape manager/ecologist or landscape architect. But the real need is for a specialist team of trained operatives who can work to a general brief, and who have not only some ecological expertise but also an eye for landscape.

Environmental education and involvement with residents

A good deal of contact with new residents and local schools is made through an environmental education team, park rangers, news letters and personal contact by landscape staff. This 'after sales service' is considered essential to protect the investment, by gaining the respect of the users. Further, it helps the community to use and enjoy their parks to the full and ensures that the potential for education is fully realized.

BIRCHWOOD BROOK LINEAR PARK

This 10-ha park occupies the former site of the Royal Ordnance Factory railway sidings, which were constructed of cinders up to 2 m deep, and also incorporated 5 ha of hummocky peat terrain, various tips of red shale and other debris, as well as a small area of clay subsoil. This diversity of difficult soils clearly posed some problems in the establishment of trees and shrubs. But it was considered that the site should be exploited rather than uniformly blanketed with an imported soil, as so often occurs in reclamation work. The basic design philosophy was, therefore, to retain the existing landscape character typified by *Betula* woods on peat, *Salix* scrub on mixed fill materials and areas of young regenerating *Betula* on deep beds of cinder. In addition, it was planned to exploit the variety of acidic, low-nutrient soils in order to create a landscape best described as heathland. The preliminary task was to clear the dense stands of fire-prone *Pteridium aquilinum* (bracken) which dominated the peat area. This was accomplished by two grading operations and an Azulox spray, the opportunity being taken to create a more pleasing landform and bury various derelict concrete structures. The main design principles can be summarized as follows:

(a) Whip and shrub planting employed a limited number of species, in particular *Betula pendula* and *Betula pubescens*, *Sorbus aucuparia*, *Salix caprea*, *Crataegus monogyna*, *Sambucus nigra*, *Prunus spinosa* and *Ulex europaeus*, with some *Pinus sylvestris*. All plants were pit-planted, mulched and fertilized in the area of the old railway sidings.

(b) Various grassland mixes, each linked to a particular soil and maintenance regime, were employed. These were carefully designed to respond to variations in landform, disposition of existing woodlands and clearings, and also to expected patterns of recreational use.

(c) On small grassed areas, expected to take heavier use, topsoil was used and the appropriate hard-wearing seed mix was sown.

(d) So that meadow areas on the soft peat would be capable of taking some wear and tear whilst retaining a heath-type flora, a low-nutrient sandy subsoil (produced as excavated waste from factory foundations nearby) was spread 100 mm deep. This was then seeded with a *Festuca–Agrostis* mix.

(e) Some areas of peat, expected to receive only moderate wear, were seeded without the use of imported soil, although ground magnesium limestone, differentially-graded phosphate and a pre-seeding fertilizer were applied in advance to assist initial establishment.

(f) Another *Festuca–Agrostis* mix was developed for the creation of diverse meadows on clay subsoil.

(g) Existing woodland areas were thinned and coppiced to increase light at ground level and encourage natural regeneration of trees and herb species. This was supplemented by underplanting of *Quercus robur*, *Corylus avellana*, *Sorbus aucuparia* and *Acer campestre* in clearings on the heavier soils.

(h) Several open peat areas, cleared of *Pteridium aquilinum*, were left untreated to allow natural regeneration to occur.

The park has now been established for 6 years. The meadow areas, after several cuts in the first year to assist grass establishment and reduce ruderal species, have developed well. A diverse flora of 70 species including, e.g., *Vicia cracca* (tufted vetch), *Vicia sepium* (bush vetch) and *Lotus corniculatus*, (birdsfoot trefoil), have now colonized on sandy subsoil areas; greater diversity is apparent on the clay subsoil which is richer in nutrients but low in nitrogen. Various maintenance regimes, involving two, three, six and twelve cuts per year are employed, and these have created a diverse pattern of grassland types (Fig. 9.4). The new woodland plantings show good initial growth, with up to 700 mm annual increments being recorded for *Betula pendula* and *Sorbus aucuparia* whips planted into the old railway sidings. Some natural regeneration is apparent in both woodland and open areas, the species

FIG. 9.4. Birchwood Brook Park: a colourful meadow achieved through natural diversification in only 4 years. (Photograph: Robert Tregay.)

including *Digitalis purpurea* (foxglove), *Silene dioica* (red campion), *Betula* spp., *Salix* spp. and *Calluna vulgaris* (ling). The development of *Betula–Salix* scrub is curtailed in some areas in order to assist the spread of *Erica* and *Calluna* spp. Certain herb species such as *Dactylorhiza fuchsii* (common spotted orchid found on clay soils only) were transplanted from areas nearby which were to be affected by housing development.

The cost of managing the landscape in Birchwood Brook Park (£6000 per annum) has remained constant, even taking into account inflation, demonstrating the fact that the ecological management of such a site can be done more economically than conventional horticultural management. The cost of managing Birchwood Brook is approximately 1/6 of the cost per hectare of maintaining other conventional parks in the New Town.

GORSE COVERT LANDSCAPE STRUCTURE

A system of landscape belts within this housing area, devised at the planning stage, have been used to connect the housing developments with the larger scale landscape on the periphery of the area to link with Birchwood's landscape structure. These belts have more recently been adapted and expanded at the east-end of Gorse Covert to include areas of tipping and demolition rubble, which proved uneconomic to remove or to build over (Fig. 9.5).

As Gorse Covert was already well served with a range of relatively controlled open spaces (e.g. District Park, Risley Moss) the opportunity was taken to develop its structural landscape as a naturalistic framework available for a wider range of uses.

Diversity

Diversity is one of the key elements in the creation of a more natural landscape and is achieved at three distinct stages of the design:
  (a) retaining and manipulating landform, soil and water;
  (b) planting and seeding;
  (c) management.
Small-scale variations in landform are typical of derelict land and are one of its main attractions for children's play. Combined with variations in soil type and an impervious subsoil, such as the clay in Gorse Covert, an uneven landform created potential for landscape diversity. Attempts to produce diversity without stage (a) have been much less successful and are more dependent upon management. If the time-scale was of no consequence, diversity could have been achieved through stages (a) and (c) alone.

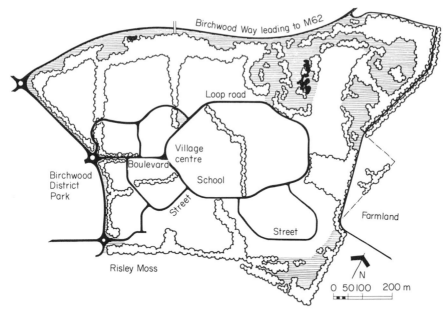

FIG. 9.5. The landscape structure of Gorse Covert, covering 35 ha and linking the area with the open space beyond.

Landform and vegetation structure were designed to relate to the new function of the site with frequent and complex transition zones used whenever possible. Coppice and scrub are frequently used as the basic structure, with trees introduced for particular emphasis. Variations in plant groupings and spacings also offer greater opportunities for creative work at the management stage. Management of the dynamic vegetation structure is seen as a continuation of the design process.

Respecting the site

The design began with an attempt to read the visual, scientific and historical factors which in combination, create the site's uniqueness. Only if these factors are fully understood can the site be respected. It is this information which provided the clues for appropriate design solutions and suggested any adaptations which were necessary for the site to fulfil its new function.

Establishing landscape

Gorse Covert's landscape structure is being created on both the clay of the site and on mounds of demolition material up to 16 m high which have been covered by a minimum of 1 m of clay. Most of the methods which are used to establish the vegetation are not new, being based on well known, though often ignored, techniques. However, the basic techniques and some of the most important adaptations are worth outlining.

Ground-shaping

The existing landscape has indicated the diversity which can result from the smallest changes in level. Even where there is less than 1 m to work with, it has been possible to introduce variety through small-scale landform. Cut-and-fill operations are therefore used to exaggerate, rather than to simplify, the topography.

Plantations

Due to its success throughout Warrington, the strategy of planting some 3 years in advance of building works was also accepted in Gorse Covert. This has provided the time-scale necessary to establish plantations from 600–900 mm transplants and 450–600 mm shrubs.

Ameliorants

The clay in Gorse Covert is typical of boulder clays in many parts of Britain, being difficult to work when wet and cracking badly when dry. It has a pH of around 6·5, rising to 7·5 where mortar is present; it contains good levels of almost every plant nutrient except nitrogen and totally lacks organic matter and structure. Severe compaction has resulted from the demolition process and subsequent engineering works.

The compaction of the clay subsoil has been relieved by ripping, and the soil structure improved by a bulky organic ameliorant. The local source of raw peat has now been exhausted but even imported peat has proved to be cheaper and more successful than topsoil. On areas where topsoil was available this has been incorporated instead of the peat but, because of its finer texture and lack of bulk, it has not been so successful in improving aeration, drainage and structure.

Plant mixes

The plant mixes developed during the earlier work in the New Town have been expanded to reflect a wider range of vegetation structures. *Alnus glutinosa* and

Betula pendula are now usually included in mixes at less than 20%, whilst other fast-growing species such as *Sambucus nigra*, *Populus tremula* and *Salix* spp. are only included in very small numbers. The plants are now planted in single-species groups which vary in size from 5 to 100 plants, according to the size of the planting area and the number of the species in the mix. This grouping allows greater freedom in the timing and approach to coppicing operations.

Maintenance

A clean ground maintenance policy is followed throughout the establishment period, which may include up to four contact sprays combined with a residual herbicide annually and selective herbicides if necessary. Herbaceous competition for moisture and nutrients is removed and fertilizer can be applied to the base of individual plants each spring.

These maintenance programmes have played an essential role in obtaining growth rates of up to 1·5 m per season in *Alnus*, *Betula*, *Prunus*, *Sorbus* and in reducing plant losses to below 5% (Fig. 9.6). By shortening the establishment period the plants are brought quickly through their most vulnerable stage. This creates earlier impact and allows a rapid transition to the management phase when a more natural character can begin to be developed.

Management

The early management has been aimed at further diversifying the range of forms established at planting. With fast-growing species such as *Alnus glutinosa* and *Betula pendula* it has been possible to commence the coppicing at the end of the second growing season. Delayed coppicing results in a plantation with a high canopy, straight trunks, the loss of lower branches and little vegetation at ground level.

Thinning operations are relatively cheap if carried out at an early stage when a hand-held rotary-cutter can be used. The choice of material to be coppiced responds to the design, the use of the site and the plant growth. The coppicing is usually carried out in blocks of different sizes relating to the planting groups and at varying densities. Where heavy coppicing has been undertaken, total weed control has been continued until the end of the fourth season. In the following season the selective treatment of only unwanted species has commenced and the desired herb layer, together with colonizing plants such as *Rubus fruticosus*, encouraged to develop. On some areas this has been assisted by the introduction of bulbs and seeds.

Fig. 9.6. Three-year-old plantation established from transplants on adapted clay and maintained with a 'clean ground' policy—markings on pole at 0·5 m intervals. (Photograph: Roger Greenwood.)

Species-rich grassland

Within a decade of the demolition of the Royal Ordnance Factory a range of attractive plants have colonized the exposed clay, including *Centaurium erythraea* (common centaury), *Lathyrus pratensis* (meadow vetchling), *Lotus corniculatus* (birdsfoot trefoil), *Prunella vulgaris* (self-heal), *Vicia cracca* (tufted vetch) and *Viola tricolor* (wild pansy). A smaller number of less common species are also present, e.g. *Blackstonia perfoliata* (yellow-wort), *Dactylorhiza fuchsii* (common spotted orchid), *Dactylorhiza purpurella* (northern marsh orchid) and *Lathyrus nissolia* (grass vetchling). It has occasionally been possible to include one of these species-rich areas into the new landscape structure, but they have been of greater value in providing a local seed source more likely to be genetically suited to conditions in Gorse Cover. With such a plentiful supply of seed in the area, the approach has been to create conditions suited to the plants and to allow them to colonize naturally.

Preparation

The basic requirement for the creation of low-growing, species-rich grassland is a low nitrogen level to reduce competition from taller growing plants. The retention of a compacted substrate also assists by preventing deep root penetration and therefore vigorous growth. An acceptable seed bed can be produced from the clay subsoil using standard cultivation techniques when the clay is neither too wet nor too dry; deep cultivation is avoided. The cost of the cultivation may be marginally higher than normal owing to the strict programming necessary and the psychological impact on a contractor of having to work on rubbish. It is not unusual to find contractors offering to import topsoil!

Seeding

The purpose behind the use of sown grasses is to form an open sward which wild flowers will colonize. The grasses give some protection to the surface from wear, provide cover during the colonization period and act as a background to the eventual community. The grasses present in the existing communities (e.g. *Anthoxanthum odoratum*) are not available commercially in large quantities and so modern turf grass varieties are used in their place. Clovers are excluded as increased nitrogen levels are not required and the commercially available cultivars quickly dominate the sward.

Germination and establishment with fine grasses on subsoil is often slow, therefore early spring and late autumn are avoided for sowing. A pre-seeding fertilizer helps to give the grasses an initial boost.

Maintenance

Newly-sown areas are maintained as a short sward for a minimum of two seasons to encourage the establishment of the sown grasses and to cut off any tall, unwanted species. This may involve up to six cuts during the season.

Management

The management of grassland sown directly onto the clay subsoil has consisted of two and sometimes three cuts per year. Most of the attractive low-growing herbs are summer flowering and the main cut is therefore timed for late summer to allow them to flower and seed. A cut is also carried out early in the season to prevent any coarse grasses present from seeding. If the growing season has been a long one, a third cut is sometimes carried out in late autumn. These cuts are carried out at around 100 mm high and the cuttings are spread evenly on site. It would be most desirable to remove the cuttings but this will not be practical on a large scale until a forage harvester, or similar machine, is developed, suitable for use on this type of area. Unwanted species are removed using a selective herbicide. Colonization has been assisted on some areas by sowing locally collected seed and implanting turves.

More frequent cutting can be used to make simple tracks through the grassland and narrow zones along the edge of paths and roads, creating visual contrast and pattern and indicating that the longer grass is there through design. In general, the approach to grassland management should be flexible enough to respond to vegetation developments.

Water

Areas with a high water-table provide exciting potential for the inexpensive creation of waterbodies. These features do not have to be large: ponds, ditches, shallow scrapes and even damp hollows can all add richness to the landscape. They also form one of the most valuable natural education resources.

Shaping

Ditches have been used in place of land drains. In addition to being cheaper and more attractive, these systems offer the potential for exploitation by a range of flora and fauna which would otherwise be very limited. Small ponds have sometimes been excavated at the end of ditches which help to keep the water level topped up. Where open water is required, a minimum depth of 1·5 m is created to prevent total colonization by *Typha latifolia* and *Phragmites australis*. The maximum water level is easily fixed by the installation of a land drain fitted with some form of debris guard. The shape of the ponds has been made as complex as possible to increase the length of the important edge zone. To increase safety and develop a wide belt of marginal vegetation the pond banks are gently shelved. Where access to open water is required a simple timber decking has been installed.

Maintenance

Initial maintenance usually consists of monitoring water levels and assisting vegetation to establish by transferring marginal and submerged plants (later followed by invertebrates) from established ponds. Such ecological systems take some time to stabilize and are sometimes hindered by fertilizer and herbicide run-off from adjacent planting areas.

Management

If the minimum depths have been achieved and the water inflows are constant, clean and relatively free of silt, then no major problems should arise. If too much planting has been carried out around the pool, some will have to be removed to avoid excessive shading and leaf decay in the water.

The biggest problems on small ponds usually arise through heavy use, particularly from fishing. An annually renewable lease is currently being drawn up with the local angling club, for the use of a pond in Gorse Covert, as an experiment in managing use and self-supervision. Ditches and other shallow areas will have to be cleaned periodically and colonization allowed to recommence. It is desirable to leave small sections as a source for the colonizing plants and to retain some cover.

CONCLUSION

Current maintenance costs in Oakwood and Gorse Covert indicate that the cost of managing the new 'ecological' landscapes can be dramatically reduced in comparison to conventional horticultural landscapes. If, however, established vegetation structures are to be developed to the full, they cannot be considered to be entirely self-sustaining. Operations are less frequent and intensive but are more dependent upon skilled and timely management decisions. The Development Corporation will only be responsible for the management of the landscape during its formative years, after which it will be handed over to a local authority. It is, therefore, essential that full advantage is taken of this initial period to help guide the long-term development of the landscape.

With the increasing and challenging workload generated by new town landscapes, local authorities will have to employ a level of expertise which does not currently exist, e.g. expertise in the management of natural habitats. One of the biggest problems local authorities will have to face will be training staff in different methods and convincing them that some landscape management should have an ecological rather than an horticultural basis.

There is an obvious need for new landscape management courses and it is gratifying to learn that universities are considering this. Horticultural colleges need to follow suit, as several of their courses are hopelessly out of date, emphasizing theory rather than practice.

FURTHER READING

Greenwood, R.D. (1983). Gorse Covert, Warrington: creating a more natural landscape. *Landscape Design*, **136**, 35–38.

Greenwood, R.D. & Moffat, J.D. (1982). In *An Ecological Approach to Urban Landscape Design* (Ed. by A. Ruff & R.J. Tregay), pp. 40–59. Department of Town and Country Planning, University of Manchester.

Tregay, R.J. & Gustavsson, R. (1982). *Oakwood's New Landscape: Designing for Nature in the Residential Environment.* Sveriges Lantbruksuniversitet and Warrington & Runcorn Development Corporation, Warrington.

Tregay, R.J. & Moffatt, J.D. (1980). An ecological approach to landscape design and management in Oakwood, Warrington. *Landscape Design*, **134**, 33–36.

Scott, D. (1983). *Handover of open space to local authorities.* Paper given to JCLI Conference, Liverpool, July 1983.

III
MANAGEMENT PRINCIPLES

INTRODUCTION

The success of any landscaping scheme will depend on whether clear objectives are identified at the outset. Equally, it will depend on adequate thought being given to the kind of management necessary to fulfil these objectives. In the long term, success may well depend as much on the effectiveness of management as on the quality of design: the designer must, therefore, be aware of different management options and their suitability. But here there is a real problem. We are dealing with a multi-disciplinary subject in which valid contributions can be made from a whole range of different specializations. Inevitably, what happens in practice is that many of the possible options are never considered because they lie outside the boundaries of a given designer's experience. So there is a perception gap. What may be obvious to an ecologist, familiar with the nutrient balance of a chalk grassland, may be totally unperceived by a designer looking at the same scene from a fundamentally different standpoint. These is no doubt that ecological understanding can offer a range of management options, often at a subtle level, which may not at first be appreciated.

But the ecological dimension is only one aspect of management. There is a whole spectrum of possible options and constraints, especially the level of resources available for management. When using unconventional methods it may even be questionable whose responsibility it is to do the work. The following six papers explore some of these aspects of management. Some deal with the application of ecological knowledge to specific problems, whilst others consider how alternative approaches might be developed to accommodate more varied methods of management. In all of this there is a reliance on management principles as a basis for achieving overall objectives.

So far as ecological understanding is concerned, there is a considerable body of existing knowledge. In many cases this has been developed in response to the needs of land managers, as in the case of rehabilitation studies, and much is of direct relevance to landscape design. But the information is not always readily available and there is a clear need for more synthesis of the available information by ecologists. This applies particularly to fundamental principles which will otherwise remain the province of the ecologist. Grime illustrates the point in demonstrating that a broad understanding of processes, such as the ecological strategies of different plants, can be applied to the management of vegetation. But it requires synthesis and interpretation before it can be applied effectively.

Ecological principles can also provide a basis for developing alternative methods of managing landscapes which, for socio-economic reasons, are no longer managed in traditional ways. Or, conversely, the traditional forms of husbandry may be applied in novel circumstances to meet a changing need, as in the case of introducing hay-meadow management to urban parks. But it is in these circumstances that the problem may arise of who should do the work. Again it seems that a continuing input of ecological knowledge is necessary for such alternatives to be successful.

The detailed examples give an insight into the range of opportunities which could be offered by a greater integration between ecology and landscape design. But they also show that implementation will not be without its attendant difficulties. It requires greater willingness on the part of ecologists to make relevant information more readily available, and adjustment on the part of designers to accommodate these novel approaches. Success will also depend on whether the person responsible for day-to-day maintenance understands the subtleties of ecological management sufficiently well to make it work.

10. MANAGEMENT OBJECTIVES AND CONSTRAINTS

V. G. KIRBY

Yorkshire Dales National Park Authority, Yorebridge House, Bainbridge, N. Yorkshire

SUMMARY

The need for land management has arisen through conflicts of interest in man's intervention in the land. Approaches to landscape management have traditionally been simple or sectoral, but now tend to be guided by a more complex, integrated or holistic approach.

The sectoral approach was adequate until technological advance accelerated the deterioration of landscapes. Post-war legislation excluded agriculture and forestry from planning control. In the last 20 years a systematic approach to landscape planning and management has been developed. In a time of limited capital resources this has enabled environmental improvement to continue at low cost.

The definition of management aims must be amplified by a list of objectives drawn up in accordance with certain principles, but reviewed regularly. It is essential that the objectives are those of the project, not of the professionals involved.

INTRODUCTION

Land must be managed in the best interests of society as a whole, but deciding what those best interests are is a complex task. Landscape architects and ecologists are frequently called upon to advise on the management of land and are asked to set out objectives and consider constraints without necessarily understanding what these terms mean. The purpose of this paper is to define the role that objectives and constraints play in landscape management, to trace the evolution of approaches to management, to explain the use of objectives as part of the management process and lastly to comment on some professional issues.

THE NEED FOR MANAGEMENT

Man has been managing land either directly or indirectly for thousands of years. Peterken (1983) refers to the early exploitation of Britain's forests at the end of the Atlantic period some 3000 years ago (see Chapter 4). Since then the scale and rate of intervention has increased so that now nowhere in the British Isles can be called truly wilderness. The demands on land are so intense that conscious efforts have to be made to define priorities for the use of every hectare. Yet there remain at national level intense conflicts between the aims of different agencies with responsibility over land. These have repercussions at all levels, right down to the daily decisions that professional landscape architects and ecologists have to take in relation to individual projects. Land is a scarce resource: it has to be looked after carefully. Using it insensitively and selfishly can damage it irretrievably, and taking actions without understanding their implications is equally undesirable.

Landscape management is the process whereby the intentions of those who control land and the activities thereon are put into practice. It is a continuous activity, following on in some cases from the implementation of a design or plan. But more often the introduction of management can be traced back to no fixed point in time, reflecting tradition or habit sometimes going back for hundreds of years. Given the many pressures which can operate on any one piece of land today, there are strong arguments for saying that we can no longer afford to have land management policies and practices whose aims are not explicitly stated and justified. To be fully effective the approach to landscape management needs to be explicit, systematic and organized. The use of aims and objectives, setting out the overall purposes of management and the detailed intentions of a management programme, involve the identification of constraints which will operate upon the system or site in question. No comprehensive approach to management which has any hope of continuing relevance over a period of time, will omit a consideration of the constraints or problems which may prevent the successful achievement of the stated purpose.

APPROACHES TO LANDSCAPE MANAGEMENT

There are two basic and contrasting ways in which landscape management can be approached. The first is the more traditional and can be labelled the simple or sectoral approach. This attitude to land-use assumes that there is one principal use for any one area, and that any other function must be distinctly subsidiary. An example is the assumption that the countryside is primarily for the production of food and that any other potential use must defer to the needs of agriculture. This view is challenged by those who support the alternative approach, which can be summarized as integrated and holistic. This is a complex concept, which implies that not only can each piece of land perform different functions, but that the essential character of a landscape is more than simply the sum of its parts. A field may be a part of an agricultural enterprise, a wildlife habitat, an archaeological site, a recreational resource, a potential source of mineral wealth, or it may contain a water supply or buildings of architectural interest. A full understanding of all its attributes is necessary before deciding how best to manage it so that it performs the desired functions whilst retaining its *genius loci* or essential unique character.

Supporting the integrated approach to landscape management does not deny the existence in many cases of one primary purpose behind management policies. It does imply, however, that decisions affecting the management of that land will not be taken without first examining all the attributes of and pressures on the land in question and assessing their importance. Once this is done, objectives can be confirmed which reflect in broad outline the programme of management. They should be ranked in order of priority, and can then form the basis of management policies and detailed prescriptions for action.

LANDSCAPE MANAGEMENT IN BRITAIN: A REVIEW

Implications of the sectoral approach

For thousands of years the basic aim of survival, combined later with economic gain, underpinned landscape management practices everywhere. Whilst the population remained small and technology was reliant on man and horse power, the fact that decisions were made in an *ad hoc* way and that plans were rarely written down, did not really matter. As the rate of change was slow, wildlife could adapt to changing circumstances, and traces of previous regimes were frequently relatively undisturbed. Certainly the slow demise in our primary woodland cover was recognized hundreds of years ago (Evelyn 1664), but (i) without adequate understanding of the means to rectify the loss, and (ii) without any agency to direct a large-scale programme of works, little could be achieved to remedy the situation. The first comprehensive management plans were produced by foresters in Europe in the nineteenth century (Simpson 1900). Although still principally sectoral, in that timber production was the overriding aim, these early forest plans are a landmark in the development of landscape management as a subject in its own right. Some English and Scottish estates followed the European example, and the Forestry Commission encouraged the practice during the second and third decades of this century (Forestry Commission 1933), but the practice was not extended to other land uses. Instead, attention was focussed on the effects of over 100 years of industrialization, the growth of cities, and environmental pollution. Over the next decade demands grew for control over *ad hoc* development and change, particularly within built-up areas. This culminated in the introduction of a mandatory planning system which was a major feature of the post-World War II boom in interventionist legislation.

Effects of post-war legislation

The legacy bequeathed by the legislators of this era is of immense value since, without the structures and principles established by the 1947 Town and Country Planning Act and the 1949 National Parks and Access to the Countryside Act, the prospects for a logical, socially responsible approach to landscape management would have been poor indeed. However, these Acts now seem in some respects excessively naive. For example, National Parks were to be set up for the purposes of providing opportunities for outdoor recreation and for the enhancement of natural beauty, and it was not thought necessary to define which purpose should take precedence. Similarly, the classic approach to town planning was that advocated in 1915 by Geddes (1968). His three basic steps of survey, analysis, plan, implied that once a master plan had been prepared, implementation along the prescribed lines would follow with relative ease. Although the early development plans were to be reviewed every 5 years, the emphasis was very much on the original master plan. The need to think about the day-by-day management of urban land was not appreciated. In the countryside as a whole there was an even greater lack of attention to management. Most rural land was designated as 'White Land' where no significant changes were proposed and where, as the production of food and timber was an overriding national priority which was not seen as causing environmental conflict, agriculture and forestry could operate largely without reference to planning law.

Conflicts in the landscape

By the 1960s the inadequacies of the sectoral approach and the problems of inadequate control over change in the countryside were becoming evident. This change of approach resulted from a realization that towns did not necessarily develop along the lines laid down in a two-dimensional plan. In addition, environmental problems were arising which had previously not been perceived or had not hitherto existed. Various threads from diverse sources began to be woven together: they resulted eventually in an appreciation of the need to produce plans which encompassed both urban and rural land, although for some years the need to concentrate on the long-term processes of land management, as opposed to the short-term processes of intensive development, was not fully appreciated. Town planners who had previously concentrated on the processes involved in producing plans, began to develop a methodology which took much more seriously the need to look at the complex inter-relationships of the elements in towns and cities, and looked critically at the implementation of plans over time. They learnt from the cybernetic approach pioneered by Wiener, who stressed that many phenomena—whether they are social, biological or physical in character—can usefully be viewed as complex interacting systems (Hall 1975). A systems view of planning was developed (McLaughlin 1969) which demanded at an early stage the definition of overall aims and objectives. Although there are many different models of the planning process, they have in common the principle that the object of study is changing continuously; therefore, throughout the process of plan formulation and implementation, there should be continuous feedback and analysis of the state of the system, so that changes can be made to the plan where necessary.

There were also changes in attitudes towards landscape and environmental planning. Dower (1965) appreciated the impact which the growth in leisure time would have on the countryside. He expressed this as the fourth wave, following on from industrialization, the spread of the railways and the growth of car-based suburbs. The changing pressures on the countryside were analysed by many writers (Fairbrother 1970; Patmore 1970) and the metamorphosis of the National Parks Commission into the Countryside Commission as a result of the 1968 Countryside Act was yet another expression of changing priorities. Ecologists were also being awakened to the growing complexity of environmental problems. Warren & Goldsmith (1974), writing on the development of the conservation movement, commented on the use of a unifying concept such as systems analysis or systems management in the solving of such problems.

Despite the introduction of the new system of structure plans under the 1967 Planning Act, which were to encompass the countryside as well as towns, the practice of landscape and countryside planning changed slowly. For some years there continued to be a reliance on site-specific solutions to problems, as witnessed by the emphasis on Country Parks and Picnic Sites in the 1968 Countryside Act.

Emergence of management planning

It is perhaps not surprising that few serious attempts were made to apply the systematic approach to landscape planning and management until the world recession of the 1970s reduced the funds available for new capital-intensive, site-specific projects. The Countryside Commission (1973) justified the need for the introduction of conscious landscape management by referring to 'the dispute between a more intensive agricultural industry and a public demanding increased access to and conservation of the countryside'.

Since that date, the work sponsored by the Countryside Commission has espoused landscape management on an ever-increasing scale. Relatively early projects such as the *New Agricultural Landscapes* report (Westmacott & Worthington 1974) and the *Lake District Upland Management Experiment* (Countryside Commission 1976) were based on this new, integrative approach, focussing attention on far wider tracts of countryside than earlier site-specific projects had done.

New Agricultural Landscapes

This study is a classic in its own right: it marked the first major departure of the Commission into a study of landscape conservation without a direct link with recreation. The brief was to discover 'how agricultural improvement can be carried out efficiently but in such a way as to create new landscapes no less interesting than those destroyed in the process' (Westmacott & Worthington 1974). The study did not define its own objectives, but contained a detailed survey of the changes in the landscape in seven lowland counties. The care with which the survey was undertaken and the understanding that is evident in the analysis have served as standards for those attempting subsequent landscape conservation projects. Of particular value are the comments on the interests and biases of the farmers whose land was surveyed, and the conclusions which stress the need to find the common ground between farmers, visitors and conservationists.

The Lake District Upland Management Experiment

This experiment was introduced because it had become 'increasingly difficult for farmers to maintain in good repair many features of the landscape which contribute so much to the scene enjoyed by visitors' (Countryside Commission 1976). The Commission wished to test the hypothesis that 'if small amounts of public money could be made available it might be feasible to encourage skilled persons to carry out small works which would produce a high quality landscape and afford better recreational access'. The formal objectives set for the two stages of the project were:

'(i) to test a method of reconciling the interests of farmers and visitors in the uplands by offering financial encouragement to farmers to carry out small schemes which improve the appearance of the landscape and enhance the recreational opportunities of the area;

(ii) to assess what effect, if any, this method will have on farmers' attitudes towards recreation and landscape.'

The success of the works undertaken during the Lake District experiment can be measured largely from the fact that, for a number of years, most of the other National Park Authorities have been operating schemes of similar scope.

Other experimental and research projects which illustrate the interest shown in finding solutions to complex rural problems, and which all show some signs of a logical analytical process, include the Urban Fringe experiments (Hall 1975), and the various Heritage Coast Reports (e.g. Glamorgan) (Countryside Commission 1975). Overall, it seems fair to say that it is rare for a complicated iterative planning process to be used for such studies, but they do acknowledge the complexities of the issues studied, and do set out priorities for action. The best examples remain flexible to changing circumstances, and are still relevant today.

The new pragmatism

In recent years there has been a steady increase in interest in landscape management, and a continuing shortage of resources for capital projects. In some respects this shortage can be seen as the biggest constraint on the achievement of lasting improvements in the quality of the environment, and yet it is also an opportunity. There is, for example, a considerable source of volunteer labour which is being activated by local and national conservation groups. There is also the emphasis on local, low-cost solutions to problems rather than capital-intensive projects. Of particular note is the growing concern for the management of open spaces in towns and cities. Teagle's *The Endless Village* (1978) opened many eyes to the wildlife resource already present in the West Midlands. Newly created conservation bodies such as the Avon Wildlife Trust, which concentrates its activities in and around Bristol and Bath, and of the Ecology Section at the Greater London Council, have proved just how much concern there is about the management of the urban landscape as wildlife habitat.

In the countryside, two generations of management plans have now been produced for the ten National Parks. On a smaller scale the acquisition of land by public bodies is frequently followed by the preparation of detailed management plans. Section 39 of the 1981 Wildlife and Countryside Act gave Local Authorities an unequivocal and much-needed power to negotiate Management Agreements with landowners in order to safeguard and enhance the conservation interest of land, if necessary including the payment of compensation. All such agreements are closely tied to management plans which set out in detail the approach, the objectives and the particular prescription for management.

The greatest value of these documents is that they seek to set out in the clearest possible terms what is being proposed for a particular area of land, and why. The *ad hoc*, sectoral approach to land management is being replaced by something rather more organized and explicit. Progress is of course not even; there are bad management plans and incomplete understanding is a constant problem, yet the progress is real.

MANAGEMENT OBJECTIVES

Having made the case for a systematic, logical approach to landscape management it is now necessary to look at objectives in more detail.

Hall (1975) has commented in detail on the use of objectives in the context of urban and regional planning, but the principles apply equally well to landscape management. 'They are defined in terms of actual programmes capable of being carried into action. . . . They also require the expenditure of resources (using that word in its widest sense. . . .) so that they imply an element of competition for scarce resources.'

The physical context for the definition of objectives is whatever unit of land is being considered. This can vary from a whole National Park to a single field, and from a Country Park to a housing estate. The first step is to define the overall aims of management. These are 'essentially general and highly abstract' (Hall 1975). Objectives amplify the aims. They must:

1   be explicit enough to amplify the major, underlying assumptions;
2   be general enough to permit flexibility of approach as resources permit or other circumstances change;
3   be precise enough for their achievement to be measured and assessed;
4   relate to the resource itself and to the people who have an interest in the land, whether they live, work or enjoy themselves there;
5   be capable of implementation, bearing in mind the available resources of time, finance and manpower;
6   make sense in the overall legislative and political context;
7   be capable of being ranked in order of priority.

If all these guiding principles are taken into account, then constraints will not need to be considered separately. They will automatically be taken into account in the formulation of a list of objectives which are explicit, flexible, precise, relevant, feasible, realistic and ranked.

Objectives would, ideally, be listed at the very beginning of a project, after the aims have been set, but before any survey work is undertaken. Used in this way they then direct the attention of the manager towards those areas of concern where insufficient information is to hand. Naturally, if other relevant matters come to light during the course of the survey, the objectives can be altered once the new information has been analysed.

Once objectives have been defined, refined and agreed, management policies and programmes of work should follow on logically. Of course, in the real world, courses of action will not always follow directly from this logical process. Nevertheless, all ideas for action should be tested against the objectives. If they do not fit, they must either be discarded or the objectives must be reassessed. It must be stressed that no list of objectives, although it should be the best that can be defined at any one time, must ever be regarded as final. The process is essentially an iterative one. Clearly, reviews of the objectives are usually best done in an organized way following a pre-determined cycle, but there is of course nothing to prevent a review at any time.

PROFESSIONAL ISSUES

Any manager must of necessity take his or her own values and viewpoints into account when undertaking an exercise in management planning, but these should not be over-emphasized. The objectives must be those of the project and not those of the professional who is responsible for the project. Landscape architects and ecologists can easily fall prey to a kind of professional arrogance. Designers can rely too much on the importance of intuitive solutions to problems. Scientists can be blinkered by their specialist knowledge and should take particular care to widen the scope of their investigations to encompass all relevant issues. It is useful to try to list all the attitudes and assumptions which the individual takes to a project: it may not be possible to discount them, in fact many may be relevant to the process, but making them explicit greatly reduces the likelihood of misunderstanding and misinterpretation at a later date.

CONCLUSIONS

The pressures on land are great and are increasing, and land management professionals must be able to analyse these pressures and produce management prescriptions which are balanced and realistic. The most valid approach to landscape management is that which accepts the many facets of interest of any one piece of land, and which attempts to balance these in the face of known pressures and constraints.

In recent years land management in Britain has moved from a simple, sectoral approach towards one which is based on a systematic, analytical and iterative process. The definition of management objectives is an essential stage of such a process. Objectives should be explicit, flexible, precise, relevant, feasible, realistic and ranked. The consideration of constraints is necessary to the derivation of objectives.

relied upon to provide a sanctuary for wildlife and spiritual refreshment for man. A similar change has occurred also in forestry practice.

It has been suggested (Grime 1972; Bradshaw 1977) that an appropriate response to these developments is to indulge in 'creative' ecology and conservation. The objective in this approach to land-use and management is to compensate for shrinkage in the area of attractive countryside by the introduction of more sophisticated techniques which, with low management inputs, will transform 'new' or derelict land into varied, species-rich landscape accessible to the public.

The creation and maintenance of extensive areas of diverse species-rich landscape calls not only for an understanding of the habitat requirements of individual species or populations of plants and animals, but also for sound prediction and control of the species interactions which determine the structure and dynamics of complex, perennial communities over extended periods of time. For such manipulation of processes a greater penetration of ecological concepts into landscape planning and practice seems essential. Here there are communication problems which deserve careful consideration.

COMMUNICATION PROBLEMS

In Fig. 11.1, the communication interface between plant ecologists and those who manage vegetation is represented (a) in an ideal world, (b) as it exists at present and (c) as it might operate in future. Model (b) asserts that whilst the theories and research methods of many ecologists remain strongly influenced by agriculture, forestry and horticulture, efforts to transmit ecological information have encountered strong resistance. This is hardly surprising

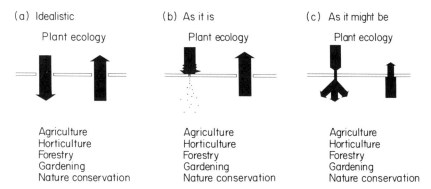

FIG. 11.1. Models representing communication between plant ecologists and those who manage vegetation (a) in an ideal world, (b) as it exists at present and (c) as it might operate in future.

since ecology is a new science and is seeking to influence traditions of land-use which, despite recent transformations, have been established over many centuries. In this situation the main responsibility for the failure in communication must rest with the ecologist. Little is achieved where the information provided by the ecologist is detailed and fragmentary. As suggested in model (c) in Fig. 11.1, greater success is likely where the evidence and ideas are distilled into a broad conceptual framework, the essentials of which are easily communicated by the ecologist and subsequently adapted to particular systems and objectives by those directly engaged in design and management.

Translation of ecological insight and expertise into guidelines for management is hindered further by problems of communication between ecologists. The rapid expansion of ecological research over the last decade has been associated with increasing specialization into schools exploring different approaches such as systems analysis, demography and comparative physiology. Whilst there is no reason to deny the essential contributions of these various fields of activity, it is clear that excessive devotion to particular methodologies (e.g. Gould & Lewontin 1979; Harper 1982; Woolhouse 1981) is limiting communication between specialists and causing delay in the development of ecology as a generalizing and predictive science.

DISTILLATION AND USE OF ECOLOGICAL INFORMATION

During the initial phase of its development, plant ecology has been concerned with the description and classification of vegetation, and much of this research has been conducted by taxonomists who have brought to the subject a characteristic appetite for accuracy and detail. Without denying for a moment the scientific bedrock which this work provides, it would be foolish not to recognize that many of the resulting data are too detailed and specific to be applied directly to management. A similar conclusion must be drawn with respect to the wealth of information which is now becoming available from investigations of the demography, phenology, reproductive biology and physiology of individual species and populations.

It seems imperative, therefore, that means are found by which effectively to distil ecological information into guidelines for vegetation management. As a tentative step towards this objective, the concept of 'plant strategies' has been developed in an attempt to describe, in broad terms, various aspects of the way in which environment and management determine the characteristics of vegetation.

THE CONCEPT OF PLANT STRATEGIES

Before it is possible to predict the consequences of management it is necessary to achieve some general understanding of the processes which cause the structure and species composition of vegetation to vary from place to place and with the passage of time. One approach to this problem is to attempt to recognize the major types of ecological specialization (strategies) which have evolved in plants, and to analyse the role of these strategies in the processes which control the structure, dynamics and species composition of vegetation.

The concept of strategies is of value to vegetation management because it provides a compact framework in which to bring together many disparate threads of ecological information. In particular, it allows a synthesis of data from the work of population biologists and physiological ecologists. Adoption of the term strategy by plant ecologists has attracted some criticism, mainly from 'molecular chauvinists' unwilling to recognize the value of generalizations formulated at the level of the whole plant, rather than its constituent parts. With its teleological implications, the term is not ideal. However, provided that a proper definition is applied, the word can be used with precision. Here a strategy is defined as a grouping of similar or analogous genetic characteristics which recurs widely among species or populations and causes them to show similarities in ecology. Use of the word strategy is also a mark of respect for the pioneer ecologists who first used the term; their achievement was to recognize that organisms do not consist of random assemblages of characteristics but exhibit sets of co-adapted traits which are predictably related to their ecology.

THE THREE PRIMARY PLANT STRATEGIES

In recent years, evidence has been growing that there are, in fact, three primary strategies in plants. Moreover, there is a growing suspicion that analogous strategies occur in algae (Dring 1982), fungi (Pugh 1980; Cooke & Rayner 1984) and animals (Greenslade 1983). This is to suggest that Darwin's 'struggle for existence' can be dissected by recognizing three distinct threats to existence (severe stress, disturbance and competitive exclusion), each occurring under particular types of environmental conditions and each resulting in a different type of evolutionary specialization. The three major threats to plant existence and the strategies which they have evoked will now be described in turn.

Severe stress may be an inherent characteristic of an impoverished environment (e.g. the conditions experienced by lichens on a bare rock), or may be induced or intensified by activities of the vegetation such as dense shade and sequestration of mineral nutrients in biomass litter or humus (e.g. the conditions experienced by ferns and tree seedlings in a mature beech wood). In contrast with competitive exclusion (see below), we are concerned here with circumstances in which one or more stresses are operating almost continuously throughout the year and constrain the growth of all of the species present in the environment. In consequence, there is little opportunity for the morphology or seasonal pattern of growth of the plant to provide mechanisms of stress avoidance. Under these conditions the successful strategy is that of the *stress-tolerators* which are slow-growing, long-lived evergreens which rely upon the conservative utilization of captured resources and their capacity to survive for long periods during which there may be little growth and reproduction (Fig. 11.2). As a result of their slow growth, stress-tolerators are sensitive to damage; many appear to have evolved highly-effective mechanisms of defence which reduce their palatability to herbivores.

Disturbance is a potent threat to existence in circumstances where there is frequent and severe destruction of the vegetation by herbivores, pathogens or various management procedures (e.g. ploughing, trampling), wind damage, fire or from abrupt changes in climate (frost, drought or flood). Where severe disturbance coincides with continuous stress the habitat is untenable. However, the effect of frequent disturbance in more fertile environments is to promote *ruderals* with rapid growth rates, abbreviated life-spans and prolific reproduction, all of which allow the intervals between disturbances to be effectively exploited (Fig. 11.3).

Competitive exclusion is characteristic of environments which contain an abundance of resources and experience a low intensity of disturbance. Such conditions lead inevitably to occupation by a dense cover of large, rapidly-growing perennial plants capable of high rates of resource capture. This results, over the course of each growing season, in the development of expanding zones of resource depletion above and below the ground surface. In such vegetation, high mortalities occur in those individuals which are outgrown and have their leaves and roots confined to the depleted zones. In these circumstances, the successful strategy is that of the *competitor* which has evolved mechanisms of escape from the depleted zones by constant replacement of the effective leaf and root surfaces during the growing season. In this essentially 'capitalistic' situation, sustained high rates of resource capture and survival depend on high rates of reinvestment of captured resources in the construction of new leaves and roots (Fig. 11.4). A marked contrast may be observed between this system of resource capture and that associated with the stress-tolerator (see above) in which growth is slow and the leaves and roots are relatively immobile, long-lived structures which rely upon resource

Fig. 11.2. A community of 'stress-tolerators' on bare limestone and occupying a shallow crevice of calcareous nutrient-deficient soil: 1, lichens; 2, bryophytes; 3, common rockrose (*Helianthemum nummularium*); 4, meadow oat (*Avenula pratense*); 5, sheep's fescue (*Festuca ovina*).

FIG. 11.3. 'Ruderals' in a frequently disturbed garden plot: 1, groundsel (*Senecio vulgaris*); 2, pearlwort (*Sagina procumbens*); 3, sowthistle (*Sonchus oleraceus*); 4, American willow-herb (*Epilobium adenocaulon*); 5, annual meadow-grass (*Poa annua*); 6, hairy bitter-cress (*Cardamine hirsuta*); 7, large field speedwell (*Veronica persica*).

FIG. 11.4. Two 'competitors' forming a tall dense stand on a stream terrace: 1, reed-grass (*Phalaris arudinacea*); 2, meadow-sweet (*Filipendula ulmaria*).

capture during the brief periods in which there is a relaxation of the severe stresses characteristic of their environments.

The extreme conditions favouring either competitors, stress-tolerators or ruderals form only part of the range of environments exploited by plants. The full spectrum of habitat conditions and associated strategies can be represented in the form of an equilateral triangle (Fig. 11.5) in which the relative importance of competition, stress and disturbance is represented by three sets of contours. This model allows recognition of not only the three extremes of plant specialization described above, but also a range of intermediate strategies associated with less extreme equilibria between stress, disturbance and competition.

A review of the full implications of the model has been presented elsewhere (Grime 1979) and is beyond the scope of this paper. However, reference to the descriptions of the primary strategies (Table 11.1) provides an insight into the wide range of plant characteristics (e.g. life-span, potential growth rate, palatability) which may be expected to change in a predictable manner as we move from one set of environmental conditions to another. The model also

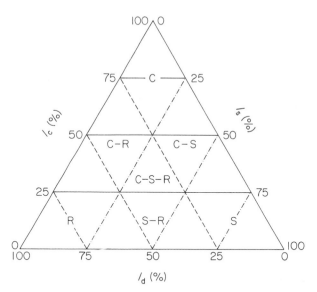

FIG. 11.5. Model describing the various equilibria between competition, stress and disturbance in vegetation and the location of primary and secondary strategies: C, competitor; S, stress-tolerator; R, ruderal; C–R, competitive–ruderal; S–R, stress-tolerant ruderal; C–S, stress-tolerant competitor; C–S–R, 'C–S–R strategist'. I_c, relative importance of competition (——); I_s, relative importance of stress (· — · —); I_d, relative importance of disturbance (— — —).

TABLE 11.1. Some characteristics of competitive, stress-tolerant and ruderal herbaceous plants

	Competitive	Stress-tolerant	Ruderal
1. Morphology	High dense canopy of leaves. Extensive lateral spread above and below ground	Extremely wide range of growth forms	Small stature, limited lateral spread
2 Leaf form	Robust, often mesomorphic	Often small or leathery, or needle-like	Various, often mesomorphic
3 Canopy structure	Dense rapidly-ascending monolayer	Often multilayered. If monolayer not dense or rapidly-ascending	Various
4 Life-span	Long or relatively short	Long–very long	Very short
5 Longevity of leaves and roots	Relatively short	Long	Short
6 Leaf phenology	Well-defined peaks of leaf production coinciding with periods of maximum potential productivity	Evergreens, with various patterns of leaf production	Short phase of leaf production in period of high potential productivity
7 Phenology of flowering	Flowers produced after (or, more rarely, before) periods of maximum potential productivity	No general relationship between time of flowering and season	Flowers produced early in the life-history
8 Frequency of flowering	Established plants usually flower each year	Intermittent flowering over a long life-history	High frequency of flowering
9 Proportion of annual production devoted to seeds	Small	Small	Large
10 Perennation	Dormant buds and seeds	Stress-tolerant leaves and roots	Dormant seeds
11 Maximum potential relative growth-rate	Rapid	Slow	Rapid
12 Photosynthesis and uptake of mineral nutrients	Strongly seasonal, coinciding with long continuous period of vegetative growth	Opportunistic, often uncoupled from vegetative growth	Opportunistic, coinciding with vegetative growth
13 Storage of photosynthate and mineral nutrients	Most photosynthate and mineral nutrients are rapidly incorporated into vegetative structure but a proportion is stored and forms the capital for expansion of growth in the following growing season	Storage systems in leaves, stems and/or roots	Confined to seeds
14 Defence against herbivory	Often ineffective	Usually effective	Often ineffective
15 Litter decomposition	Rapid	Slow	Rapid

provides a convenient basis on which to predict or explain the changes in species composition which result from alterations of the intensities of stress or disturbance inflicted on the vegetation. These include the changes which follow agricultural dereliction, eutrophication or increases in the frequency of grazing, mowing or trampling. An economic lesson to be drawn from the model is that no vegetation is capable of surviving the combined effect of severe stress and frequent disturbance. This suggests that in attempting to revegetate heavily-trampled but unproductive habitats (e.g. paths which are shaded, on mountains or on maritime cliff-tops), effort has been wasted in pursuit of an unrealistic objective.

USE OF STRATEGY CONCEPTS TO MANIPULATE SPECIES

Despite the growing number of local population studies, it seems inevitable that information relating to *species* will remain as the major currency in communications between ecologists and those engaged in vegetation management. Quite clearly, therefore, it will be helpful to classify species with respect to strategy. It is desirable, however, that any effort to link species with strategies should take account of the capacity of species to expand their ecological range through genetic and phenotypic variation with respect to strategy. By using current knowledge of species and their field distributions it is now possible to begin to estimate the strategic range of common species in Britain. In Fig. 11.6 a triangular ordination of 2008 vegetation samples drawn from all the major habitats of the Sheffield region has been used to examine the ecological amplitude of selected species. The contours in each diagram plot the frequency of occurrence of the species in a matrix of vegetation types, within which each m² vegetation sample has been located by reference to selected characteristics (life-history, morphology, phenology, reproduction, etc.) of the component species. A brief account of the ordination procedure and the strategy concepts upon which it is based is provided in Appendix 1. In order to assist interpretation of the contour diagrams, Table 11.1 lists some of the plant characteristics which may be expected to become increasingly prominent in populations, species and vegetation types as we approach the respective corners of the triangle.

When the distributions in Fig. 11.6 are examined, many differences are apparent in terms of both the centre and spread of the contours. The analyses which follow are taken from an earlier attempt (Grime 1984) to interpret the patterns exhibited by the eight species.

Veronica persica: there is a very compact distribution. Restriction to the left-hand corner of the model indicates that this ephemeral species exploits

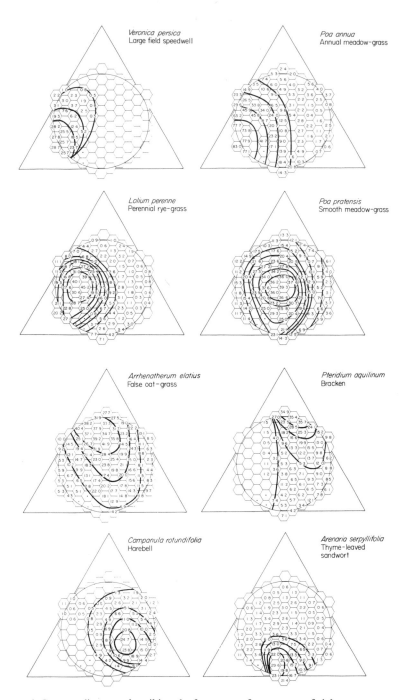

FIG. 11.6. Contour diagrams describing the frequency of occurrence of eight common herbaceous species in a matrix of functionally-defined vegetation types. The procedure used in the triangular ordination is described in Appendix 1. Values indicate the percentage of m² samples containing the species. Contours drawn by eye. *Insufficient data.

productive, heavily-disturbed vegetation, but is unable to colonize infertile habitats and is sensitive to competition from perennial species.

Poa annua: displays a similar concentration in the 'ruderal' corner, but expands into other parts of the model. This distribution is consistent with the results of experimental studies (Law, Bradshaw & Putwain 1977) which show that in addition to the ephemerals of disturbed ground there are perennial genotypes of *Poa annua* which are particularly common in productive pastures.

Lolium perenne: the contours again suggest ruderal characteristics, but they are centred rather higher, indicating sensitivity to heavy disturbance and some ability to persist in perennial communities. The frequency of occurrence of *L. perenne* falls away sharply towards the apex and right-hand side of the triangle, reflecting both the failure of the species to survive in competition with large perennials in productive undisturbed habitats and its exclusion from infertile sites.

Poa pratensis: the distribution is focused in the centre of the triangle, suggesting that the species is particularly associated with vegetation experiencing moderate intensities of stress, disturbance and competition, conditions likely to obtain in the pasture habitats exploited by the species. The contours indicate that, to a remarkable extent, *P. pratensis* is able to extend its range into widely different vegetation types. In our present state of knowledge it is uncertain to what extent this wide amplitude is the result of genotypic variation.

Arrhenatherum elatius: a wide strategic range is also suggested. The highest occurrence coincides with fertile relatively undisturbed conditions (road verges, derelict land, etc.), but the contours also extend towards the base of the triangle, a pattern which may be explained, at least in part, by the occurrence of prostrate genotypes of *A. elatius* shown to be unusually resistant to defoliation (Mahmoud, Grime & Furness 1975).

Pteridium aquilinum: shows a high frequency of occurrence towards the apex of the triangle, a distribution which suggests that the species is a strong competitor with the potential to monopolize plant communities of fertile, relatively-undisturbed sites. The complete absence of the species from the left-hand side of the matrix reflects the failure of *P. aquilinum* to exploit frequently-disturbed habitats. From the descending contours on the right there is evidence of the ability to persist in some unproductive communities such as those associated with heathlands, and heavily-shaded and/or highly acidic woodland herb layers.

Campanula rotundifolia: the contour pattern indicates a well-defined ecology. The species appears to be intolerant of high intensities of disturbance and competition and is restricted to relatively unproductive vegetation.

Arenaria serpyllifolia: concentrated in the area of the triangular model which corresponds to conditions of moderately severe intensities of both stress and disturbance. This accurately portrays the ecology of the species, which is a small winter-annual associated with localities where infertile soils are subjected to disturbance by drought, solifluction or animal activity.

These examples illustrate the potential of strategy concepts to convey in a simple diagram some of the main features of the ecology of a species. It seems reasonable to suggest that information in this form could guide the selection of species for introductions to specific sites and could be used in attempts to manipulate the abundance of particular species. Perhaps most important is the possibility of predicting the fate of species during succession or in response to changes in soil fertility and management regime.

It would be misleading to suggest that all the information required for effective manipulation of a species is contained in the contour diagram. Despite the large number of attributes which vary in association with the axes of the triangular model (Table 11.1), reference to other types of information may be necessary to expose additional species characteristics which are responsible for 'fine-tuning' of their ecology. Data on the frequency of occurrence of species on soils of different pH classes allow predictions of edaphic tolerance and field or laboratory indices of drought and shade tolerance are also desirable. Recently it has been suggested (Grime & Mowforth 1982) that measurements of nuclear DNA content may be used to predict the onset of spring growth, a feature of considerable importance in relation to the timing of cutting, grazing and fertilizer applications.

The most important omissions from the contour diagrams, however, relate to the regenerative biology of the species. In common with other organisms, herbaceous plants may adopt quite different strategies as mature individuals and juveniles, and any summary of the ecology of a species must include consideration of regenerative strategies. Seedlings and vegetative offspring of herbaceous plants are comparatively small and they are exposed to hazards which may be quite different and much more severe than those experienced by established plants. Successful regeneration in many species depends upon exploitation of local and unusually favourable sites, and plants appear to have evolved a variety of regenerative mechanisms which vary in efficiency according to habitat. Five major types of regenerative strategies are distinguished in Table 11.2, which contains also a brief description of the conditions to which each strategy appears to be adapted.

A detailed account of the regenerative strategies and their relevance to management has been provided elsewhere (Grime 1979) and will not be re-stated here. We may conclude, however, that no effort to manipulate a plant species should be attempted without reference to its regenerative biology, a

TABLE 11.2. Five regenerative strategies of widespread occurrence in terrestrial vegetation

Strategy	Habitat conditions to which strategy appears to be adapted
1. Vegetative expansion (V)	Productive or unproductive habitats subject to low intensities of disturbance
2 Seasonal regeneration in vegetation gaps (S)	Habitats subjected to seasonally predictable disturbance by climate or biotic factors
3 Regeneration involving persistent seed or spore bank (B_s)	Habitats subjected to spatially predictable but temporally unpredictable disturbance
4 Regeneration involving numerous widely-dispersed seeds or spores (W)	Habitats relatively inaccessible (cliffs, tree trunks, etc.) or subjected to spatially unpredictable disturbance
5 Regeneration involving persistent juveniles (B_j)	Unproductive habitats subjected to low intensities of disturbance

point already stressed by Grubb (Chapter 6). This is particularly important in projects designed to 'insert' additional species into established vegetation. Failure to provide suitable conditions for establishment is undoubtedly the most common cause of failure in attempts to introduce native species into habitats.

Consideration of regenerative strategies is also highly relevant to efforts to control the abundance of species. In many of the most widely-successful plants (e.g. *Chamaenerion angustifolium*, *Epilobium hirsutum*, *Poa annua*, *Holcus lanatus*) the same genotypes exhibit several regenerative strategies. It seems likely that this versatility explains, at least in part, the ability of these species to invade and persist within a wide range of plant communities. As a corollary, we may suspect that the relative scarcity of certain species, e.g. *Orchis apifera*, *Ophioglossum vulgare*, is related to dependence upon a single regenerative strategy.

MANIPULATION OF PLANT COMMUNITIES

As autecological information accumulates and experience is gained in controlling the distribution and abundance of individual species it seems inevitable that attention will turn to the manipulation of plant communities. Here a particular objective is likely to be the creation of species-rich herbaceous communities, many of which have declined in response to recent changes in land management. Figure 11.7 illustrates a very tentative exploration of the use of strategy concepts to analyse plant communities and devise guidelines for manipulation. Each diagram refers to a m² sample of

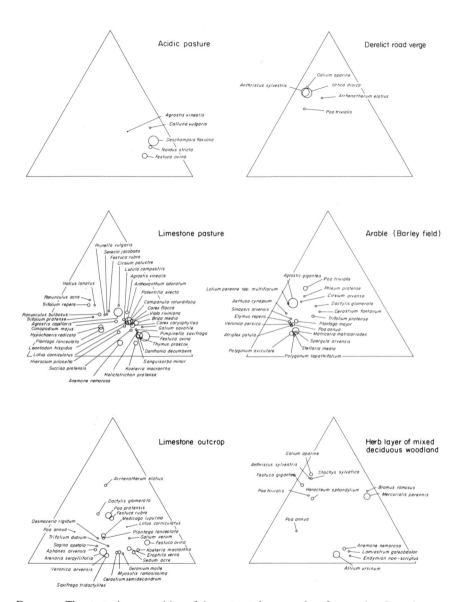

FIG. 11.7. The strategic composition of six contrasted m² samples of vegetation. In each figure the position of each species corresponds to the centre of its contour diagram (see Fig. 11.6) and the size of the circle indicates the abundance of the species within the vegetation.

vegetation for which the abundance of each species is indicated by the size of a circle, the location of which corresponds to the centre of a contour diagram based upon the data and procedure used in Fig. 11.6.

A rather narrow strategic composition is evident in the acidic and calcareous pastures, the arable field and the derelict road verge. A more diverse array of strategies occurs on the limestone outcrop and the woodland herb layer. It may be argued that the range of strategies within these communities has been exaggerated by failure to take account of the particular genotypes and phenotypes represented in each sample. A more likely explanation for the patterns in Fig. 11.7 is that co-existence between different strategies is a widespread phenomenon in plant communities and arises from spatial and temporal variation in the equilibrium between stress, disturbance and competition. In the particular example of the grazed limestone outcrop, spatial niches occupied by contrasted species would be expected to result from effects of variation in soil depth upon mineral nutrient availability and severity of summer drought. Co-existence between plants of very different strategy in the woodland herb layer may be related, at least in part, to temporal niche-differentiation arising from seasonal changes in the quality of growing conditions beneath a deciduous tree canopy.

This first attempt to examine the structure of plant communities by reference to component strategies is both speculative and incomplete; in particular it is necessary to examine also the regenerative strategies which are known to vary considerably within communities (Thompson & Grime 1978). Despite these limitations, however, it is possible to recognize ways in which the diagrams in Fig. 11.7 could be used to initiate or predict changes in the species composition of communities. In the specific case of the grazed limestone outcrop, for example, we may predict that removal of grazing would result in an expansion of the more competitive species *Arrhenatherum elatius*, *Dactylis glomerata* and *Festuca rubra*, and contraction or even loss of many of the small winter annuals situated towards the base of the triangular model. Conversely, an increase in grazing pressure might be expected to shift the balance in favour of the annual species.

Field tests of the usefulness of strategy concepts as a basis for explanation, prediction and control of community processes are now required. Some interesting results are already available from studies in which changes in strategic composition have been recorded in the course of succession (Shepherd 1981), or following changes in management (Buttenschon & Buttenschon 1982), or during fluctuations in climate (Leps *et al.* 1982). There is how an urgent need to set all landscape management on a properly scientific basis; plant strategy concepts will be helpful in this task.

ACKNOWLEDGMENTS

I wish to thank my colleagues, Dr J.G. Hodgson, Mr S.R. Band, Miss J.M.L. Shacklock and Mrs J. Fletcher for their assistance in the collection and analysis of data used in this paper. This research was supported by the Natural Environment Research Council.

REFERENCES

Bradshaw, A.D. (1977). Conservation problems in the future. In *Scientific Aspects of Nature Conservation in Great Britain. Proceedings of the Royal Society of London B*, **197**, 77–96.

Buttenschon, J. & Buttenschon, R.M. (1982). Grazing experiments with cattle and sheep on nutrient poor, acidic grassland and heath. I. Vegetation development. *Natura Jutlandica*, **21**, 1–48.

Cooke, R.C. & Rayner, A.D.M. (1984). *Ecology of Saprotrophic Fungi*. Longman, London.

Dring, M.J. (1982). *The Biology of Marine Plants*. Edward Arnold, London.

Gould, S.J. & Lewontin, R.C. (1979). The spandrels of San Marco and the Panglossian paradigm: a critique of the adaptationist programme. *Proceedings of the Royal Society of London B*, **205**, 581–598.

Greenslade, P.J.M. (1983). Adversity selection and the habitat templet. *The American Naturalist*, **122**, 352–365.

Grime, J.P. (1972). The creative approach to nature conservation. In *The Future of Man* (Ed. by F.J. Ebling & G.W. Heath), pp. 47–54. Academic Press, London.

Grime, J.P. (1979). *Plant Strategies and Vegetation Processes*. John Wiley & Sons Ltd., Chichester.

Grime, J.P. (1984). The ecology of species, families and communities of the contemporary British Flora. *New Phytologist*, **98**, 15–33.

Grime, J.P. & Mowforth, M.A. (1982). Variation in genome size—an ecological interpretation. *Nature*, **299**, 151–153.

Harper, J.L. (1982). After description. In *The Plant Community as a Working Mechanism* (Ed. by E.I. Newman), pp. 11–25. Special Publication No. 1, British Ecological Society, Blackwell Scientific Publications, Oxford.

Law, R., Bradshaw, A.D. & Putwain, P.D. (1977). Life history variation in *Poa annua*. *Evolution*, **31**, 233–246.

Leps, J.J., Osbornova-Kosinova & Rejmanek, K. (1982). Community stability, complexity and species life-history strategies. *Vegetatio*, **50**, 53–63.

Mahmoud, A., Grime, J.P. & Furness, S.B. (1975). Polymorphism in *Arrhenatherum elatius* (L.) Beauv. ex J. and C. Presl. *New Phytologist*, **75**, 269–276.

Pugh, G.J.F. (1980). Strategies in fungal ecology. *Transactions of the British Mycological Society*, **75**, 1–14.

Shepherd, S.A. (1981). Ecological strategies in a deep water red algal community. *Botanica Marina*, **XXIV**, 457–463.

Thompson, K. & Grime, J.P. (1978). Seasonal variation in herbaceous seed banks. *Journal of Ecology*, **67**, 893–921.

Woolhouse, H.W. (1981). Aspects of the carbon and energy requirements of photosynthesis considered in relation to environmental constraints. In *Physiological Ecology: An Evolutionary Approach to Resource Use* (Ed. by C.R. Townsend & P. Calow), pp. 51–85. Blackwell Scientific Publications, Oxford.

APPENDIX 1

Brief description of the triangular ordination used in Figs 11.5 and 11.6

The initial step in the ordination was to attempt to classify with respect to strategy the herbaceous species of the Sheffield region. The method involved the use of a dichotomous key (Fig. 11.8) based upon characteristics of life-

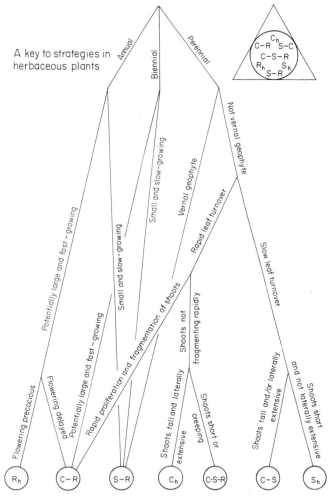

Fig. 11.8. A dichotomous key for the provisional classification of herbaceous plants with respect to strategy. C_h, herbaceous competitor; S_h, herbaceous stress-tolerator; R_h, herbaceous ruderal; C–R, competitive–ruderal; S–R, stress-tolerant ruderal; C–S, stress-tolerant competitor; C–S–R, 'C–S–R strategist'.

history, morphology and phenology and allowed species to be assigned to one of seven positions (Fig. 11.8 inset) corresponding to a characteristic pair of stress and disturbance coordinates and situated within a circular area of the triangular model conforming to the strategic range proposed for herbaceous plants (Grime 1979, Figure 18, page 73). It should be emphasized that this was merely an approximate and provisional classification which could be applied with certainty to only a restricted number of species. No attempt was made to classify species for which critical data were lacking and species known to exhibit major variation in life-history and morphology were also omitted. This procedure allowed the classification of 204 species (henceforward described as marker species) including many of the commoner herbaceous plants of the region.

The next step was to ordinate vegetation samples drawn from the range of habitats represented in the Sheffield region. Each 1-m^2 sample was located in the triangle by reference to the positions and frequencies of the component marker species. This was achieved by calculating mean stress and disturbance coordinates in which the contribution of each marker species was weighted according to its frequency in the vegetation sample. By this procedure, 2008 vegetation samples were ordinated and found to be distributed fairly evenly within the central circular area of the triangle. This allowed a calculation of the percentage occurrence of each species within each of 91 hexagonal zones of the circle and this in turn permitted contours to be drawn describing the distribution of each species.

12. CONTROLLING ECOSYSTEMS FOR AMENITY

B. H. GREEN

Wye College, University of London, Ashford, Kent

SUMMARY

The objective of management of many amenity lands is the maintenance of high-diversity, low-production ecosystems. Disturbance and depletion of nutrients to arrest natural successions and favour earlier seral stages is thus an important part of amenity land management. Unfortunately such management is difficult to implement, for both practical and doctrinal reasons, in landscapes where high-production, low-diversity ecosystems are the overwhelming objective of maximum sustained yield resource management. In such situations prescribed burning is both the easiest and most effective means of managing amenity lands.

INTRODUCTION

In the UK there are nearly half a million hectares of land managed primarily for amenity purposes, i.e. for the protection of wildlife and landscape and for access into the countryside for informal recreation. Over half this land (277 697 ha) is administered by voluntary conservation organizations, namely the National Trusts, County Naturalists Trusts and the Royal Society for the Protection of Birds. The remainder (203 876 ha) is administered by the state conservation agencies and local authorities as national nature reserves, local nature reserves, country parks, picnic sites and other public open spaces and statutory access areas. There is, in addition, a very substantially larger area of land where amenity purposes are statutorily a secondary objective of land management. This applies to nearly 4000 Sites of Special Scientific Interest (1 438 404 ha) and to the ten National Parks of England and Wales (1 360 000 ha) where landowners must now, under the 1981 Wildlife and Countryside Act, manage their land in consultation with the conservation agencies. There are also large areas of land where some amenity objectives are normally, but not statutorily, integrated with the primary uses. These include commons (607 050 ha), Ministry of Defence lands (282 679 ha), road and rail verges (130 000 ha) and unknown, but considerable areas on airfields, golf courses and marginal farmland (Green 1984; O'Riordan 1983).

Whilst there are considerable overlaps between many of these kinds of land designations, it seems that amenity is an important objective of management for as much as four million hectares of land, or 17% of the surface of the UK.

Most, if not all, of the ecosystems on this land, whether woods, heaths, moors, downs, dunes, peatbogs, fens and marshes, are the products of past management practices such as coppicing, outfield/infield grazing, and turf, furze, bracken and reed cutting which were vital to the economy of a rural industrial society. In most parts of Britain—apart from the uplands, where hill grazing is heavily subsidized to sustain rural communities, and some exceptional areas like the New Forest—these traditional kinds of land management are now almost invariably unprofitable and therefore rarely practised. In some ecosystems rabbit grazing masked the effects of this change in rural land use for a long time until myxomatosis almost eliminated the rabbit population in the late 1950s. As a result, many of these ecosystems are now subject to succession; large areas of down, heath, dune and marsh have scrubbed over, passing eventually to secondary woodland (Wells 1969; Williams 1983). For example, heathlands on the Greensand and Tertiary sands and gravels in Southern England, once dominated by heather (*Calluna vulgaris*), are disappearing under birch (*Betula pubescens* and *B. pendula*), pine (*Pinus sylvestris*) and oak (*Quercus robur* and *Q. petraea*). All the chart heathlands in Kent, including those described by Tansley (1939) are now oak/birch woods with heather confined to the rides (Fig. 12.1). Since secondary woods are not, in general, scarce ecosystems whilst open heathland and downland are, amenity management has concentrated mainly on maintaining the open seral, arrested, or deflected 'climax' communities.

Such management is difficult and the extent of scrub invasion, even on nature reserves, bears testimony to this. The problem is not lack of knowledge of how to maintain different kinds of communities. Numerous trials have shown the efficacy of a great variety of grazing, mowing, burning and chemical treatments (Gimingham 1972; Duffey *et al.* 1974). It is the translation of this research into field-scale management systems which needs to be more widely implemented. There are both practical and ideological barriers to be overcome before this can be achieved. The practical problems of managing such ecosystems, and possible solutions to them, have been widely discussed (Thiele-Wittig 1974; Lefeuvre 1980; Wells 1980; Green 1983). The ideological problems are more intransigent, for they relate directly to the very basis of conservation philosophy.

CONSERVATION IDEOLOGY AND LAND MANAGEMENT

Two quite separate concepts have characterized the modern conservation movement since its inception in the US towards the end of the last century.

FIG. 12.1. Bracken and birch colonization of Hotfield Common, Kent. Open heather swards are declining rapidly in lowland Britain due to lack of management by grazing or burning.

Gifford Pinchot, head of the developing US Forestry Service, was instrumental in developing the idea of *efficient use of resources*. The concept of exploitation at the maximum sustainable yield has determined the policies and practice of all forestry, fishery, game and other agencies concerned with renewable resources ever since. High production of utilizable material is the main aim of this kind of land management. Other early conservationists—such as the naturalist and founder of the Sierra Club, John Muir, and the landscape architect Frederick Olmsted—were much more concerned with *environmental protection* for cultural reasons. They wanted to protect Nature for itself, and for the knowledge and pleasure, or amenity, to be derived from it; whether through scientific or historical inquiry, spiritual renewal or just the contemplation of its beauties. Maintaining a diversity of species and landscapes was thus a prime objective.

The conservation movement in Britain developed almost exclusively within this second paradigm, which is a more difficult basis on which to justify and practise land management: 'When I am reporting on the merits of a proposed nature reserve, after describing the scientific importance of its flora and fauna, I often find it hard to resist bringing in the scenic beauty of the landscape or the attractiveness of the vegetation, though my allusions to these tend to take on an almost apologetic tone. It is as if I were trying to say "And of course the place really is beautiful as well, though perhaps I ought not to mention the fact" ' (Tansley 1953). This revealing insight into the thinking of one of the prime architects of conservation in Britain illustrates the striving for respectability and cogency which has always bedevilled that part of conservation concerned with cultural or aesthetic, rather than material, objectives.

This need for persuasive arguments to promote environmental protection, together with the realization that only relatively small, and perhaps inadequate, areas are ever likely to be able to be specifically managed primarily for amenity conservation, has led its proponents to embrace the concept of rational resource exploitation into a modern, integrated conservation ideology. Thus, the World Conservation Strategy (IUCN 1980) is based upon the ideas that (i) conservation and development are interdependent, and (ii) adverse environmental impacts are incidental rather than intrinsic to resource exploitation and can, therefore, be avoided or remedied by appropriate levels of exploitation and measures such as environmental impact analysis. This policy is clearly the one most likely to succeed in protecting the environment in a world in which continued massive resource exploitation is inevitable. In its promulgation, however, two important ecological considerations must not be overlooked. First, ecosystems which can bear continuous heavy exploitation of utilizable material, without artificial inputs, are rather scarce in nature and commonly poor in species. Second, many ecosystems considered desirable on amenity grounds are extremely unproductive.

EFFECTS OF CROPPING ON ECOSYSTEM PROPERTIES

Harvesting material from ecosystems, whether by grazing, mowing or fire, has two main sets of effects: disturbance and nutrient impoverishment.

Disturbance

Disturbance caused by the physical removal of species creates new opportunities for others. Resources such as light, soil nutrients and water are left unused which which can be exploited by new individuals of the species lost, or by new species. The colonization of gaps thus created, particularly by ruderal and other early successional species, is an important means by which community diversity of species is maintained (Jones 1945; Grubb 1977; Ricklefs 1977). In forests a very substantial proportion of the total flora and fauna are absolutely dependent on natural disturbance caused by windthrow, avalanche, flood, earth movements and volcanic activity (Gilbert 1980). It is thus not surprising that harvested woodlands, such as those cut for coppice (Stuttard & Williamson 1971), or selectively logged (Johns 1983), are so rich in many species compared with those less regularly disturbed. But species will be eliminated from the community if lethal disturbances, e.g. fire, recur at intervals shorter than the species needs to successfully propagate (Moll *et al.* 1980). Some places are thus rarely colonized by the potential dominant for the site conditions (Noble & Slatyer 1980). Such considerations may explain the relative scarcity of shade-bearing, superior competitors such as the small-leaved lime (*Tilia cordata*) and beech (*Fagus sylvatica*) in British woodlands. With their long history of selective exploitation they are widely dominated by ruderal colonist species, particularly the oaks (*Quercus petraea* and *Q. robur*), despite the inability of these species to regenerate well under shade (Shaw 1974).

As our knowledge of the effects of natural disturbances grows, the concept of the steady-state climax ecosystems becomes more and more questionable (Bormann & Likens 1979; Whitmore 1982). In those apparently rather rare situations where neither natural nor human disturbance are prevalent, it seems possible that the self-regenerating ecosystems of unperturbed dominants may be rather species-poor (like beech woodlands in Britain?). It is the control of competitive dominants which allows other species to survive and generate diverse ecosystems (Connell & Slatyer 1977). For this reason, ecosystems managed for amenity, where maximizing species richness is a main objective, will rarely need to be completely protected from disturbance, although protection against intervention was a view once prevalent amongst conservationists (Streeter 1974). On the contrary, where protected areas are too small or isolated to benefit from natural perturbations, deliberate disturbance must be part of their management (Pickett & Thompson 1978; Foster 1980). The short cycle, large patch disturbance of much commercial exploitation of ecosystems aimed at maximizing yields from productive early successional stages commonly, however, reduces diversity. Species richness is probably maximized at intermediate disturbance frequencies and community composition depends on the scale, pattern and phasing of disturbance in relation to the autecology of individual species (Miller 1982). Juniper, for example, seems to need ground bared by fire or overgrazing in which to germinate, but then several years free of grazing or fire to become established (Miles & Kinnaird 1979; Gilbert 1980; Ward 1981).

Nutrient impoverishment

Abundant, free, inorganic supplies of the major nutrients (nitrogen, phosphorus and potassium) in their forms needed for plant growth are rather scarce in nature. They occur mainly in sites flushed with the products of erosive processes, e.g. river valleys, in animal dunging areas and where organically-bound or otherwise unavailable forms are released through fire, upwellings of hypolimnion waters or other disturbances. It is no surprise that until recently such situations were those most commonly cropped by Man. Sedentary agriculture evolved on alluvial terraces where nutrients removed in crops were replaced in the silt of annual floods. Away from river valleys, settled agriculture was dependent on nutrient replacement, or fertilization, from animal manure and its concentration of nutrients scavenged from a wide area. Otherwise cropping could only be of the intermittent slash-and-burn type of shifting cultivation or nomadic pastoralism dependent on the accumulation of nutrients which takes place with succession in fallow periods. Until the widespread introduction of synthetic fertilizer in this century these facts determined the nature of the rural economy and the countryside: 'so essential were sheep for the growing of corn that Hugh Latimer, the famous bishop, took it for granted in one of his sermons that "A ploughland must have sheep; yea, they must have sheep to help fat the ground; for if they have no sheep to help fat the ground they shall have but bare corn and thin" . . this . . . was one of the great assumptions on which English people based their lives' (Kerridge 1973).

Sheep and corn husbandry was the mainstay of agriculture over much of the sand and limestone lands of England from the medieval period to the early part of this century. It was an outfield/infield system which relied entirely on the transference of fertility from extensive outlying pastures, where the sheep were grazed in the day, to the arable, where they were folded to dung at night. Old chalk country sheep 'were bred specifically for folding on arable. The sheep masters paid little or no attention to improving the carcass for the butcher and looked only for the hardiness the sheep needed to pick up a living on a close-fed down, to walk two or three miles for its keep and then to carry its dung back the same distance to the fold on the tillage' (Kerridge 1973). Corn production was the overriding objective of this farming system and there was little concern for the outfield 'wastes'. Grazing tenancies laid down the numbers of sheep that must be folded largely irrespective of where they might find their keep (Armstrong 1975; Chadwick 1982).

There were attempts to extend the arable by 'denshiring' or 'downsharing'. This was a slash-and-burn system of rotational agriculture, whereby the turf was pared and the sods burnt to provide fertility. The land was then folded and used for growing a succession of arable crops such as turnips and barley, before finally undersowing with clover and sainfoin for return to sheepwalk on a 7–10 year cycle. There was a lively debate at the beginning of the nineteenth century as to whether this was a productive or destructive use of the land (Boys 1805). There is no doubt that the harvesting of furze (*Ulex* spp.) for fodder, bracken (*Pteridium aquilinum*), cut green for bedding stock, and of peat and turf for fuel were destructive to the fertility of these areas (Helsper *et al.* 1983). Firing them was also widely practised to provide fresh bite for stock. All these processes, like folding, can extract large amounts of nutrients which take a long time to be replaced in rain (Green 1972). Since there seems to have been little attempt other than downsharing to restore the nutrients of the downs and heaths thus cropped or pumped into the arable by the sheep, overall this system was degrading in the extreme to the fertility and production of these outfield lands. They were overgrazed and eroded to deserts of bare chalk and shifting inland dune systems. Defoe (1724) described Bagshot Heath in Surrey as '... a sandy desert, and one may frequently be put in mind here of Arabia Desert, where the winds raise the sands, so as to overwhelm whole caravans of travellers, cattle and people together: for in passing this heath on a windy day, I was so far in danger of smothering with the clouds of sand, which were raised by the storm, that I could neither keep it out of my mouth, nose or eyes: and when the wind was over, the sand appear'd spread over the adjacent fields of the forest some miles distant, so that it ruins the very soil'. There are similar contemporary accounts of the heaths of Breckland and of Jutland. The later destructive effects of rabbits, which concentrated nutrients in their warrens, on downland can be seen in photographs in Tansley (1939). The fertility of woodlands was also run down by grazing and the cropping of timber and underwood (Rackhan 1980).

LOW FERTILITY ECOSYSTEMS

Intensive agriculture now occupies most of the potentially fertile land in Britain. Amenity land is largely the exhausted and impoverished residue of the pastoral outlands which were plundered of their fertility to sustain the croplands before synthetic fertilizer broke the shackles of the old system. The amenity value of downs, dunes, heaths and moors is a direct result of their infertility. Numerous fertilizer trials, notably the Park Grass Trial at Rothamsted (Brenchley 1958) show a reduction in vascular plant species richness with increasing fertility (Fig. 12.2). On extremely infertile soils there may be a small increase in species with low levels of fertilizer application, but thereafter any increase in fertility favours a few vigorous competitive species which exclude most of the rest (Grime 1973; and Chapter 11).

Natural succession towards 'climax' ecosystems involves an accumulation of nutrients and their conversion from the inorganic to the organic form (Stevens & Walker 1971; Gorham *et al.* 1979). Reichle *et al.* (1975) suggest that

FIG. 12.2. The Park Grass Experiment of Rothamsted. Fertilized plots are much more productive but less species-rich than the unfertilized controls.

'... given adequate water and energy resources, the level of maximum persistent biomass in an ecosystem will be determined by the supply of essential elements'. It is largely the balance between loss and storage of resources which controls succession. Any surplus of resources invites colonization of the next seral community, any deficit a repression (Dansereau 1974). Leguminous shrubs, or other nitrogen-fixing bushes, are the key species in many seres which provide the excess of nitrogen necessary for progression to communities of greater biomass. The land manager who has this objective will want to encourage them (Roberts *et al.* 1981). The amenity land manager who wishes to keep open downland or heathland must do the opposite and somehow remove the accumulating nutrients (Green 1972).

Unfortunately, the impoverishment of ecosystems is far more difficult to achieve today than their enrichment. The old infertile lands that are left are subject to eutrophication both through direct fertilization for more productive grassland or woodland production, or, if abandoned, through natural succession. Nutrient accumulation in abandoned land, and managed amenity land, is boosted by unintended artificial inputs of nutrients from aerial fertilizer application, nitrate-rich groundwaters and nutrient-enriched rain and dust containing nitrogen and phosphorus compounds released from crop residues by stubble burning. Modern agriculture, with its profligate use of fertilizer, is enriching the total environment.

MANAGEMENT TO PERTURB, IMPOVERISH AND DEGRADE

The old pattern of rural land use is thus now completely reversed. Consistent with the modern philosophy of maximum sustainable yield exploitation, nutrients are pumped in the opposite direction and cropping carried out in a steady, regular way. Extraneous disturbances caused by flood, wind, fire or pest outbreak are controlled as far as is possible. This, in many ways, also accords with the minimal intervention management policies of many of those concerned with the environmental protection who still adhere to Clementsian ideas of the steady-state climax ecosystem. But in reality the very opposite is required to maintain amenity ecosystems: they need to be degraded by disturbance and impoverishment as they were in the past. Naveh (1982) has recently argued for such management of Mediterranean ecosystems. Thomas (1983) has even suggested the use of explosives to create suitable habitats within chalk downland for the Adonis blue butterfly (*Lysandra bellargus*).

For a destructive force one need look no further than the armed services, and it is no coincidence that Ministry of Defence lands now bear some of the best downland and heathland in the country (Wells *et al* 1976). Elsewhere such management is scarce. The traditional type of outfield/infield grazing management is, for example, difficult to operate on fragments of outfield left within the modern agricultural system. Whole farms, even parishes, were once designed primarily to accommodate it: 'The sheep principally manure the land. This is to be done by folding; and, to fold, you must have a flock. Every farm has its portion of down, arable and meadow;...' (Cobbett 1830). Specialization in agriculture today means that every farm does not have these components, many have no stock at all. Even where it is possible to graze downland or heathland, the stock can rarely be folded in the old way. Stock permanently or seasonally enclosed in paddocks may be effective in the short term in controlling scrub, but, unless the paddock is large enough to allow some internal outfield/infield system to develop by zonation of distinct dunging areas, in the long term, nutrient accumulation may gradually change the sward. Rotation, whereby areas are heavily overgrazed and then left to recover for a few years, is also highly desirable for maximizing species richness, particularly of invertebrates (Morris 1971); but is also difficult to implement where the needs of the stock are an important consideration.

Fire management

In the US, amenity area management has recently become much more interventionist. The National Park Service has changed its fire suppression policy to one of fire management. The change in attitude dates back to the *Report on Wildlife Management in the National Parks* where the disastrous ecological consequences of *laissez-faire* policies were identified and much more positive management advocated: 'Of the various methods of manipulating vegetation, the controlled use of fire is the most "natural" and much the cheapest and easiest to apply.... Other situations may call for the use of the bulldozer, the disc harrow, or the spring tooth harrow to initiate desirable changes in plant succession' (Leopold *et al.* 1983). Those few European conservationists who have experimented with fire have invariably come to the same conclusions as to its undoubted value as a most useful management tool (Lloyd 1968; Ward 1972; Webb 1980; Green 1983) (Fig. 12.3).

The views of the majority of conservationists on the use of fire are however, at best, ambivalent (Tubbs 1974; Bibby 1978). Although fire is widely and effectively employed for managing mountain grazings and grouse moors, it is very rarely used by amenity land managers. There are three main reasons for this: (i) it is felt to be dangerous and difficult to control; (ii) it is believed to encourage succession, particularly the invasion of 'undesirable'

FIG. 12.3. Spring burning to control tor-grass (*Brachypodium pinnatum*) on Wye Downs, Kent.

species such as purple moor grass (*Molinia caerulea*), bracken and birch; (iii) it is seen as a destructive agency, particularly to invertebrates and species such as sand lizards (*Lacerta agilis*) and Dartford warblers (*Sylvia undata*). All these reservations are justified, particularly that regarding safety. Proper precautions must always be observed when burning. They fail, however, to appreciate the versatility of fire and its quite natural role in ecosystems. Accidental summer fires in heathland are, for example, quite different in their effects from controlled burns which are carried out in the spring. Hot summer fires which burn off the soil humus layers and their content of heather seeds and roots may preclude heather regeneration and encourage bracken and birch invasion. But spring or autumn fires taking place when the humus is damp do not have this effect. Indeed, by removing highly flammable accumulations of litter, from patches burnt in rotation, they are the only sure way of protecting against the disastrous summer fires such as those of 1976 which swept completely over many lowland heaths, doing great damage to wildlife (Surrey Naturalists' Trust 1976). Controlled spring or autumn burns outside the breeding season, when many species are inactive below ground, are far less damaging to wildlife. Species of fire climaxes such as steppe and prairie, including animals, have evolved with fire and are well equipped, not merely to survive it, but to exploit the new resources it makes available (Force 1981). It is even suggested that some plants have evolved flammable properties, such as the accumulation of volatile oils and deep litter, encouraging fires which may prevent their invasion and replacement by later successional species (Mutch 1970).

The pattern of litter accumulation and breakdown of British grasslands dominated by tor (*Brachypodium pinnatum*), upright brome (*Zerna erecta*) and red fescue (*Festuca rubra*) is very like that of some North American fire-climax prairies (Gay & Green 1983). All these species, when left ungrazed, accumulate deep litter which can resist most scrub invasion for periods of up to 20 years before its breakdown allows rapid scrubbing over. In the past, when such grasslands were much more extensive and continuously linked together, it seems likely that they might have been naturally burned at normal temperate grassland fire intervals from lightning strikes which are every 2–8 years (Rundel 1981). Their fragmentation and isolation now makes the probability of lightning fires very low; Wells (1970) has shown the importance of uneven topography, rivers and other discontinuities in giving fire protection and enabling tree growth to take place on the prairies. Protection from natural fires, and from the deliberate burns which were part of past grazing management, is probably more 'unnatural' or, at least, atypical, of these ecosystems than their firing on a cycle of less than 20 years. Systematic burning is certainly very effective (Fig. 12.4). The fear of fire, and its

FIG. 12.4. Trials to assess the effects of different burning frequencies and mowing regimes on tor-grass dominated swards on Wye Downs.

reputation as an unpredictable and destructive force in ecosystems, derives from foresters and other resource managers concerned with maximizing yields. It is a degrading agency. Large amounts of the nutrients—up to 100% of the nitrogen and 46% of the phosphorus—in the standing crop are lost in fires (Rundel 1981). To a forester this is disastrous, but it is precisely what is required for the management of amenity lands.

CONCLUSIONS

To maintain most kinds of ecosystem now used for amenity purposes, amenity land managers must forget many of the principles of forestry and agronomy on which their management is now often based. Traditional methods of husbandry, very different from modern practice, must be re-introduced or new techniques, especially controlled burning, developed. In the meantime, if the amenity land manager is ever in doubt as to his best course of action, he has merely to think of what a modern farmer or forester would do, and do the opposite. His objective is to make one blade of grass grow where two grew before.

REFERENCES

Armstrong, P. (1975). *The Changing Landscape: the History and Ecology of Man's Impact on the Face of East Anglia.* Dalton, Lavenham.

Bibby, C.J. (1978). Conservation of the Dartford Warbler on English lowland heaths: a review. *Biological Conservation*, **13**, 299–307.

Bormann, F.H. & Likens, G.E. (1979). Catastrophic disturbance and the steady state in northern hardwood forests. *American Scientist*, **67**, 660–9.

Boys, J. (1805). *General View of the Agriculture of the County of Kent.* P. Norbury, Brentford.

Brenchley, W.E. (1958). *The Park Grass Plots at Rothamsted.* Rothamsted Experimental Station, Harpenden.

Chadwick, L. (1982). *In Search of Heathland.* Dennis Dobson, London and Durham.

Cobbett, W. (1830). *Rural Rides.* Penguin, Harmondsworth (1967).

Connell, J.H. & Slatyer, R.O. (1977). Mechanisms of succession in natural communities and their role in community stability and organisation. *American Naturalist*, **111**, 1119–44.

Dansereau, P. (1974). Types of succession. *Vegetation Dynamics. Handbook of Vegetation Science Part VIII* (Ed. by R. Knapp), pp. 123–5. Junk, The Hague.

Defoe, D. (1724). *A Tour Through the Whole Island of Great Britain.* Strahan, London.

Duffey, E., Morris, M.G., Sheail, J., Ward, L.K., Wells, D.A. & Wells, T.C.E. (1974). *Grassland Ecology and Wildlife Management.* Chapman & Hall, London.

Force, D.C. (1981). Postfire insect succession in Southern California chaparral. *American Naturalist*, **117**, 575–82.

Foster, R.B. (1980). Heterogeneity and disturbance in tropical vegetation. In *Conservation Biology* (Ed. by M.E. Soulé & B.A. Wilcox), pp. 75–92. Sinauer, Sunderland, Massachusetts.

Gay, P.A. & Green, B.H. (1983). The effects of cutting on the growth and litter accumulation of coarse grass swards on Chalk soils. *South East Soils Discussion Group*, **1**, 140–52.

Gilbert, L.E. (1980). Food web organisation and conservation of neotropical diversity. In *Conservation Biology* (Ed. by M.E. Soulé & B.A. Wilcox), pp. 11–33. Sinauer, Sunderland, Massachusetts.

Gilbert, O.L. (1980). Juniper in Upper Teesdale. *Journal of Ecology*, **68**, 1013–24.

Gimingham, C.H. (1972). *The Ecology of Heathlands.* Chapman & Hall, London.

Gorham, E., Vitousek, P.M. & Reiners, W.A. (1979). The regulation of chemical budgets over the course of terrestrial ecosystem succession. *Annual Review of Ecology and Systematics*, **10**, 53–84.

Green, B.H. (1972). The relevance of seral eutrophication and plant competition to the management of successional communities. *Biological Conservation*, **4**, 378–84.

Green, B.H. (1983). The management of herbaceous vegetation for wildlife conservation. In *Management of Natural and Semi-natural Vegetation* (Ed. by M.J. Way), pp. 99–114. British Crop Protection Council, London.

Green, B.H. (1984). The impacts of agriculture and forestry on wildlife, landscape and access in the countryside. In *Planning and Ecology* (Ed. by R.D. Roberts & T.M. Roberts), pp. 156–64. Chapman & Hall, London.

Grime, J.P. (1973). Control of species density in herbaceous vegetation. *Journal of Environmental Management*, **1**, 151–67.

Grubb, P.J. (1977). The maintenance of species richness in plant communities: the importance of the regeneration niche. *Biological Reviews*, **52**, 107–45.

Helsper, H.P.G., Glen-Lewin, D. & Werger, M.J.A. (1983). Early regeneration of *Calluna* heathland under various fertilization treatments. *Oecologia*, **58**, 208–14.

International Union for the Conservation of Nature (1980). *World Conservation Strategy.* IUCN, Morges, Switzerland.

Johns, A. (1983). Wildlife can live with logging. *New Scientist*, **99**, 206–11.

Jones, E.W. (1945). The structure and reproduction of the virgin forest of the North Temperate Zone. *New Phytologist*, **44**, 130–48.

Kerridge, E. (1973). *The Farmers of Old England.* Allen & Unwin, London.

Lefeuvre, J.C. (1980). Possibilités d'élevage de moutons de race rustique dans les landes des Monts d'Arrée. I. Considérations générales. *Bulletin d'Ecologie*, **11**, 765–73.

Leopold, A.S., Cain, S.A., Cottam, C.H., Gabrielson, I.N. & Kimball, T.L. (1983). Wildlife management in the National Parks. *American Forests*, **69**, 32–5, 61–3.

Lloyd, P.S. (1968). The ecological significance of fire in limestone grassland communities of the Derbyshire Dales. *Journal of Ecology*, **56**, 811–26.

Miles, J. & Kinnaird, J.W. (1979). The establishment and regeneration of birch, juniper and scots pine in the Scottish Highlands. *Scottish Forestry*, **33**, 102–19.

Miller, T.E. (1982). Community diversity and interactions between the size and frequency of disturbance. *American Naturalist*, **120**, 533–7.

Moll, E.J., McKenzie, B. & McLachlan, D. (1980). A possible explanation for the lack of trees in the fynbos, Cape Province, South Africa. *Biological Conservation*, **17**, 221–8.

Morris, M.G. (1971). The management of grassland for the conservation of invertebrate animals. In *The Scientific Management of Animal and Plant Communities for Conservation* (Ed. by E. Duffey & A.S. Watt), pp. 527–52. Blackwell Scientific Publications, Oxford.

Mutch, R.W. (1970). Wildland fires and ecosystems: a hypothesis. *Ecology*, **51**, 1046–51.

Naveh, Z. (1982). Mediterranean landscape evolution and degradation as multivariate biofunctions: theoretical and practical implications. *Landscape Planning*, **9**, 125–48.

Noble, I.R. & Slatyer, R.E. (1980). The use of vital attributes to predict successional changes in plant communities subject to recurrent disturbances. *Vegetatio*, **43**, 5–21.

O'Riordan, T. (1983). Putting trust in the countryside. In *The Conservation and Development Programme for the UK.*, pp. 171–260. Kogan Page, London.

Pickett, S.T.A. & Thompson, J.N. (1978). Patch dynamics and the design of nature reserves. *Biological Conservation*, **13**, 27–37.

Rackham, O. (1980). *Ancient Woodland: its History, Vegetation and Uses in England*. Edward Arnold, London.

Reichle, D.E., O'Neill, R.V. & Harris, W.F. (1975). Principles of energy and material exchange in ecosystems. *Unifying Concepts in Ecology* (Ed. by W.H. van Dobben & R.H. Lowe-McConnell), pp. 27–43. Junk, The Hague.

Ricklefs, R.E. (1977). Environmental heterogeneity and plant species diversity: a hypothesis. *American Naturalist*, **111**, 376–81.

Roberts, R.D., Marrs, R.H., Skeffington, R.A. & Bradshaw, A.D. (1981). Ecosystem development on naturally colonised china clay wastes. I. Vegetation changes and overall accumulation of organic matter and nutrients. *Journal of Ecology*, **69**, 153–61.

Rundel, P.W. (1981). Fire as an ecological factor. In *Encyclopaedia of Plant Physiology*, **12A** (Ed. by O.L. Lange, P.S. Nobel, C.B. Osmond & H. Ziegler), pp. 501–38. Springer, Berlin.

Shaw, M.W. (1974). The reproductive characteristics of oak. In *The British Oak* (Ed. by M.G. Morris & F.H. Perring), pp. 162–81. Classey, Faringdon.

Stevens, P.R. & Walker, T.W. (1971). The chronosequence concept and soil formation. *Quarterly Review of Biology*, **45**, 333–50.

Streeter, D.T. (1974). Ecological aspects of oak woodland conservation. In *The British Oak* (Ed. by M.G. Morris & F.H. Perring), pp. 341–54. Classey, Faringdon.

Stuttard, P. & Williamson, K. (1971). Habitat requirements of the nightingale. *Bird Study*, **18**, 9–14.

Surrey Naturalists' Trust (1976). *Heath Fires in Surrey 1976*. Godalming.

Tansley, A.G. (1939). *The British Islands and their Vegetation*. Cambridge University Press, Cambridge.

Tansley, A. (1953). The conservation of British vegetation and species. In *The Changing Flora of Britain* (Ed. by J.E. Lousley), pp. 188–96. Buncle, Arbroath.

Thiele-Wittig H. Chr. (1974). Maintenance of previously cultivated land not now used for agriculture. *Agriculture and Environment*, **1**, 129–37.

Thomas, J.A. (1983). The ecology and conservation of *Lysandra bellargus* (Lepidoptera: Lycaenidae) in Britain. *Journal of Applied Ecology*, **20**, 59–83.

Tubbs, C.R. (1974). Heathland management in the New Forest, Hampshire, England. *Biological Conservation*, **6**, 303–6.

Ward, S.D. (1972). The controlled burning of heather, grass and gorse. *Nature in Wales*, **13**, 24–32.

Ward, L.K. (1981). The demography, fauna and conservation of *Juniperus communis* in Britain. In *The Biological Aspects of Rare Plant Conservation* (Ed. by H. Synge), pp. 319–29. John Wiley, Chichester.

Webb, N.R. (1980). Amenagement et conservation des landes: synthèse. *Bulletin d'Ecologie*, **11**, 655–8.

Wells, P.V. (1970). Postglacial vegetation history of the Great Plains. *Science*, **167**, 1574–82.

Wells, T.C.E. (1969). Botanical aspects of conservation management of chalk grasslands. *Biological Conservation*, **2**, 36–44.

Wells, T.C.E., Sheail, J., Ball, D.F. & Ward, L.K. (1976). Ecological studies on the Porton Ranges: relationships between vegetation, soils and land-use history. *Journal of Ecology*, **64**, 589–626.

Wells, T.C.E. (1980). Management options for lowland grassland. In *Amenity Grassland: an Ecological Perspective* (Ed. by I.H. Rorison & R. Hunt), pp. 175–95. John Wiley, Chichester.

Whitmore, T.C. (1982). On pattern and process in forests. In *The Plant Community as a Working Mechanism* (Ed. by E.I. Newman), pp. 45–59. Blackwell Scientific Publications, Oxford.

Williams, J.H. (1983). Species composition and above ground phytomass in chalk grassland with different management. *Vegetatio*, **62**, 171–80.

13. LOW COST SYSTEMS OF MANAGEMENT

J. C. PARKER
Estates and Valuation Department, Kent County Council

SUMMARY

Because of recent pressures for cost savings, there has been increased mechanization of open space management and simplification of ground layouts. Under these developing regimes much of the variation in vegetation has been lost, and ecological design and management has been suggested as a way of reversing the trend. These newer concepts are not necessarily more economic than the existing simple low cost systems and, if they are to be successful, they involve significant changes in the operating methods and technical skills of the work force.

Some of the changes can make the work more attractive for volunteer groups, who could enhance the basic maintenance provided by many local authorities. However, such enhancement can only be achieved if means can be established for volunteers to work in harmony with paid employees.

EFFECTS OF COST SAVING

Low cost or economy in the use of resources is a desirable attribute in any enterprise and activity. It can, however, so dominate an organization that it influences the end-product or goal, particularly where that end-product is difficult to define. In commerce or industry outputs can be measured with relative ease, in terms of product numbers, sales statistics or percentage profits, and these parameters form a clear yardstick for assessing the effects of any cost-saving exercises. Thus, in simple terms, if quality reductions create an unacceptable product it will fail in the market place and corrective action will have to be taken if the enterprise is to remain in operation.

In the management of public open spaces and amenity land the input costs are easily defined, but the output values can only be assessed in rather vague subjective terms. Beauty or enjoyment are very much in the eye of the beholder and very rarely are there any effective measures of public acceptance comparable to sales performance. As an example, the cost of a road planting scheme is very easy to assess but the public benefit, the output measure, in terms of travellers' enjoyment or safety is almost impossible to define. This in turn means that alternative schemes of varying costs can only be evaluated on the basis of careful subjective judgements at the best. Furthermore, these judgements may be made on a whole range of different criteria ranging from nature conservation to road safety, many of which may be almost impossible to reconcile without complicated analysis (Berry 1983).

Such difficulties apply to the management of nearly all open spaces, and it is not surprising that cost reductions have dominated the general policies, often to such an extent that the whole 'product' has been transformed, almost beyond recognition. Not all these changes are necessarily for the worse, but in the search for cost reduction, at least over the last 20 years, a number of common characteristics have become apparent.

  (a) Operations that need a high input of labour have been much reduced, e.g. summer bedding, herbaceous borders, formal hedge cutting, etc.
  (b) Mechanization has been introduced wherever possible and, particularly as far as grass mowing is concerned, the layouts have been modified to accommodate high capacity machinery.
  (c) Herbicides have largely replaced mechanical weed control.
  (d) Ground cover plants have been widely planted, often to replace annual bedding or herbaceous borders.

Coupled with these trends, there have been significant changes in the skills and organization of the work force. Mechanical skills and the ability to carry out routine tasks quickly have become predominant at the expense of plant knowledge and the horticultural skills that probably reached their peak in Victorian times. The numbers of staff have been reduced dramatically in concert with the changes in layout.

Thus, in summary, the typical open space, if such a thing exists, is generally neat and tidy. The site is likely to be level or gently sloping with a simple layout that is easily mown. What is not mown is usually flowering shrubs or ground cover. The smooth mown spaces are ideal for dog-walking, kite-flying, organized sport, etc. and the tidiness discourages dumping. It is, however, easy to see that many open spaces are dull. The scent, colour and marked seasonal change of, for instance, the traditional cottage garden are lacking, and the botanical variation and associated wildlife is severely limited by the dominating mowing regime. Utility and low cost maintenance are probably the most important attributes.

THE 'ECOLOGICAL' ALTERNATIVE

The relative drabness of the modern open space has been sensed by many (e.g. Ruff & Tregay 1982) and in particular the ecologists who have suggested that costs might be further reduced by adopting a vegetation pattern which is more natural and will need less intervention management to maintain it (Bradshaw

& Handley 1982; Chapter 2, this volume). In addition, it is hoped that such an approach would bring back a greater variation and interest in the vegetation, and perhaps do something to counter the limited scope for varied wildlife habitats in the modern agricultural landscape.

There is no doubt that this message has been well received by a number of landscape managers, and changes in attitude have taken place. For instance, the grass cutting of road verges is now much reduced in rural areas and almost completely stopped on motorway verges. On many open spaces there has been a gradual tendency to leave grass a little longer and set aside certain areas for natural growth.

In my own authority there has been a deliberate policy to use a 'meadow culture' technique for playing field perimeters and other non-operational pieces of land. I would like to analyse our experiences in a little detail to illustrate the effectiveness or otherwise of these experiments.

The policy was first introduced in 1980 as a means of overcoming a budget cut. It was, therefore, designed first and foremost to save labour at the peak time of the year. However, it was quite clearly hoped that the lower mowing frequency would encourage a much more interesting grassland flora that would be an attraction as well as an educational facility.

I will deal first of all with the economics. The cost of maintaining almost any amenity open space depends mainly on the labour demands during the peak growing season: usually from mid-April till about mid-June (Fig. 13.1). The number and type of mowing machines is also dictated by this peak demand when most lawns will need cutting at least once a week. The meadow culture technique involves not cutting the grass until July/August at the earliest, i.e. when the peak of grass growth has passed. This means

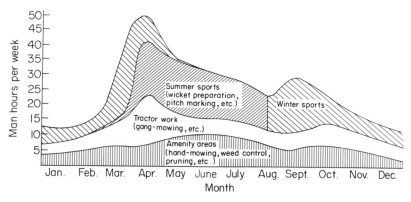

FIG. 13.1. Seasonal work loads on a secondary school site (playing fields and amenity grounds).

immediately that there is a reduced mowing work load in the early summer (in my experience up to about 5%), and manning and machinery levels can be reduced accordingly. However, the time savings on each site are likely to be very small. For instance, the leaving of a 4-m width around the perimeter of a 4-ha field would probably only save about 10 minutes of the gang-mower driver's time. Multiplied up to cover the whole season of, say, 20 mowings, this is only equivalent to about 3·5 hours. This small saving can be very quickly swallowed up in the cost of bringing in a separate machine that will cope with the accumulated growth in the autumn, particularly if there is a considerable distance to be travelled from the depot and the arisings have to be raked up and/or carted off site. Thus, the savings in total labour and machines hours are very small, if they exist at all. But it is possible to reduce the peak labour demand which in turn reduces costs, if the tractor rounds can be reorganized to make use of the individual savings.

So much for the costs; now for the benefits. In the first season or so, and in the early months, the uncut swards did make an attractive and welcome alternative to the former smooth cutting. However, the effect is relatively short-lived and by the end of June the dead seed heads—possibly flattened by wind and rain—lose a great deal of their visual appeal. Then in subsequent years there appears to be a gradual domination by the coarser growing species such as *Anthriscus sylvestris* (cow parsley), *Cirsium* sp. (thistles), *Rumex* sp. (docks), *Urtica dioica* (stinging nettle) and various tall grasses.

Even in the early honeymoon stages, it seems that the average public response was very much against the whole idea. Complaints of untidiness, fire risk and weed banks were extremely common and some claimed extra hay-fever suffering and harbouring of snakes, vermin and mosquitoes. These emphasized the undoubted problems of mixing nature and urban man (Manning 1982). In a few favoured places, generally where fertility and growth were poor, there developed more interesting stands of wild plants which appeared to be better received by the public. They were, alas, the exception rather than the rule, and only in a very few cases did we achieve what had been hoped for—a more varied and attractive flora at reduced cost.

As many informed observers have commented (e.g. Green 1981 and Chapter 12), the problem lies in either inherent fertility of the soil or the gradual build up of the biomass and fertility once the mowing regime is removed. This experience illustrates all too well the advantages of not putting back topsoil on site perimeters and banks, which must be considered for all new layouts. For existing ones we seem to need a machine or a method which will easily take away or crop the herbage so that the mulching and manuring effect of flail mowing can be countered. The forage harvester seems the best machine available at the moment but it is much too large and cumbersome for small areas, particularly between trees and other obstacles. Forage harvesters have been used successfully on large areas of rough open grassland, provided the slopes are satisfactory, but controlled burning has been equally effective where it is well away from public highways or habitations. Railway embankments were subjected to regular burning, accidentally or intentionally, but the technique is very unlikely to be acceptable in urban situations.

In my experience, it seems that most of the public reaction against a meadow culture technique is based on the appearance of neglect when grass is left uncut. In the extreme this sometimes leads to rubbish dumping in places where it would be very rare if the sward were regularly mown. It is, therefore, important that meadow culture should be seen as a deliberate policy and not merely one of default, and should be accompanied by:

(a) careful design and choice of areas;
(b) regular perimeter mowing where meadow culture abuts footpaths, roadways, etc;
(c) public education, by whatever means are available, of the aims and objectives of the exercise;
(d) careful education and instruction of the workforce, who will have to bear the brunt of the changed working methods and in particular the labour of the late summer or autumn clearance.

All of these things are more easily said than done, but I believe it is essential to have a carefully planned campaign if the operation is to have overall success.

COST-SAVING THROUGH PLANTING

As an alternative to a continual commitment to sward management, field perimeters of banks can be converted to shelter belts or copses by planting with trees and shrubs. In the long term this should bring a significant saving in labour, but the initial capital inputs can be relatively high as indicated in Table 13.1 and elsewhere (Penistan & Laing-Brown 1979).

Tree and shrub planting has considerable advantages in shelter and visual screening but in cost-saving terms it is likely to take at least 20 years to recoup the initial capital outlay, if the whole of the tree-planting costs are an extra cost to the organization. In practice, such planting can often be done by existing staff so that the only extra costs are those of materials (say one-third of the total). Then the payback period may be less than 10 years, which obviously improves the economic advantages of small-scale planting. Another alternative for large areas is to use direct sowing of tree and shrub seeds.

Direct sowing is a much more sophisticated technique than planting, but experience and experimentation by my colleague Mr LaDell, in association

TABLE 13.1. Planting costs compared with regular mowing costs per 1000 m²

	Management	Cost
Tree/shrub planting	Transplants at 2 m spacing and including 3 years maintenance	£300 (say)
Grassland maintenance	Tractor flail 1–2 times per year	£5–10 per year
	Gang mowing 15–20 times per year	£10–15 per year

with Dr Humphries of Cambridge University, have shown how seed sowing can provide good stands of trees and shrubs. This success does, however, depend on a detailed and carefully applied knowledge of the storage and germination requirements of the different species. The resulting growth is often quite spectacular when compared with the planted equivalent, and the cost overall is probably about half that of conventional planting if the operation is on a sufficiently large scale for the necessary cultivations to be done with conventional agricultural equipment. For this reason the technique is particularly suitable for large-scale woodland planting where the normal costs of establishment would be prohibitive (LaDell 1983).

Apart from the need for a high level of technical knowledge the only other serious problem with direct sowing appears to be in dealing with the weed competition in the emergent crop. Some success is being achieved with selective applications of herbicides such as propyzamide. If this proves reliable in the future I anticipate that direct sowing will become a much more popular method.

Woodland strips and copses can greatly improve the character of a site, but they do require careful, if infrequent, maintenance. Much of that maintenance is best done in the winter months, so that a mixture of woodland and grassland has the immediate advantage of providing a continuity of employment throughout the year. The practical difficulties in this are that woodland skills are somewhat different to those of routine grassland maintenance, and the rate and quality of work can sometimes be unsatisfactory if the same workmen are employed to do both.

THE VOLUNTARY SECTOR

Although a change to a more natural management may not necessarily reduce costs it does alter the nature of the maintenance tasks. This in turn can increase the opportunities for making use of more voluntary help and community involvement. The main differences in the work tasks are shown in

Table 13.2. An examination of these differences clearly indicates that very few people can be involved in formal low cost maintenance. It can have very little attraction to volunteers who are seeking the satisfaction of 'making a show' and the companionship of a work team. In simple terms, a few days mowing is much less fun than clearing scrub and having a good bonfire.

The involvement of volunteers and the adoption of naturalistic management programmes is very different from the routine maintenance by direct labour as it is practised in most local authority open space departments. These departments have often developed sophisticated and carefully controlled systems that rely on a well motivated staff concentrating on a relatively narrow range of routine maintenance tasks (Ellison 1982). As I have mentioned earlier, the costs generally are low and the biological expertise tends to be concentrated in a very small number of managers or technical assistants.

Naturalistic maintenance requires the technical officer to be much closer to the physical tasks, and probably a greater proportion of 'white' to 'blue' collar workers than is currently the case. This is unlikely to be popular with cost-conscious local authority employers, or with many of the professional staff themselves, who sometimes seem to perceive their status as being inversely proportional to their wearing of wellington boots. In addition, the manual workers, or at least their unions, obviously fear replacement of paid employees by volunteers. For all these reasons it is not surprising that most volunteer work on open spaces has been centred around special restoration projects or on countryside conservation, rather than the long-term care of public open space (Blencowe & Wood 1983).

In an age of increasing leisure and/or unemployment this could mean that opportunities will be, and are being, missed for making open spaces more attractive by topping up the very basic maintenance regimes that are provided

TABLE 13.2. Comparison of the maintenance tasks involved in formal and naturalistic managements

Formal	Naturalistic
Frequent and regular	Infrequent and irregular
Highly mechanized	Lower mechanization
Small use of hand tools	High use of hand tools
Repetitive	Little repetition
Few people involved	Large work teams possible
Low biological knowledge	Higher biological knowledge
Routine use of herbicides	Selective use of herbicides

by the local authorities. To alter the situation there has to be an initial rethink on the design and layout so that more volunteer work can be accommodated in a productive way. Secondly, the managers must alter their ways of working and organization; some have already shown examples of what can be achieved. Finally, the volunteers have to be seen as aids rather than supplanters of the existing paid labour. This perhaps is the most difficult problem of all. It needs a great deal of persuasion and trust between the staff and managers and can only be seriously attempted where there is already a background of good industrial relations.

REFERENCES

Berry, P.M. (1983). The landscape, ecological and recreational evaluation of woodland. *Arboricultural Journal*, 7, 191–200.

Blencowe, J. & Wood, D. (1983). The implementation of landscape schemes using Manpower Services Commission sponsored teams. *Landscape Design*, 141, 41–42.

Bradshaw, A.D. & Handley, J.F. (1982). An ecological approach to landscape design: principles and problems. *Landscape Design*, 138, 30–34.

Ellison, S. (1982). Push-button technology gives Bromley an accurate maintenance programme. *Turf Management*, July 1982, 33–34.

Green, B.H. (1981). *Countryside Conservation*. Allen & Unwin, London.

LaDell, T. (1983). An introduction to tree and shrub seeding. *Landscape Design*, 144, 27–31.

Manning, O.D. (1982). Designing for man and nature. *Landscape Design*, 140, 30–32.

Penistan, M.J. & Laing-Brown, J.R. (1979). Copse and spinney. *Landscape Design*, 127, 26–29.

Ruff, A.R. & Tregay, R. (1982). *An ecological approach to urban landscape design*. Occasional paper No. 8. Department of Town and Country Planning, University of Manchester.

14. DESIGNING FOR WILDLIFE

D. A. GOODE AND P. J. SMART

*Greater London Ecology Unit, Room 435, County Hall,
London, SE1 7PB*

SUMMARY

Although naturalistic planting and the creation of new wildlife habitats are now an accepted part of landscape design, there is a danger, especially in urban settings, that schemes may produce little more than a green veneer. It is relatively easy to create an impression of nature which in reality lacks much of the diversity which could be developed. For success in developing new wildlife habitats it is necessary to consider the detailed needs of a range of species. Such requirements are considered with particular reference to birds to illustrate some of the problems and solutions in accommodating a greater variety of species within an urban setting. In particular, it is argued that the right habitat is not sufficient in itself and that consideration must also be given to the management of people.

INTRODUCTION

Traditionally the protection of nature, and particularly the promotion of nature reserves, has been concerned with maintaining fragments of the natural world within the countryside. The widely established 'key-site' concept now forms the basis for selection of nature reserves in many countries. The relative importance of individual areas of semi-natural habitat is measured by means of criteria such as naturalness, rarity and diversity (Ratcliffe, 1977). These criteria reflect the intrinsic biological features of the areas concerned and also take account of the degree to which individual sites or habitats are threatened by man's activities. But this system has one clear disadvantage in that it pays little regard to the value placed upon such areas by society, i.e. to the needs and aspirations of people.

It is in this respect that a most significant change has occurred in recent years. The past 10 years, particularly, have seen the development in Britain of a new set of values in nature conservation. Apart from the moral arguments that we have a responsibility towards the natural world, or towards future generations of mankind, there are strong arguments for nature conservation for the simple enjoyment of nature (Goode, 1984). This rapidly developing movement seems to stem from a deep desire amongst many people for greater contact with nature through a more natural environment. As Mabey (1980) points out, the preservation of a series of Sites of Special Scientific Interest (SSSIs), and nature reserves, does little to provide for the continued existence of ordinary wildlife for people to enjoy in their own local surroundings. In this new approach, the emphasis is increasingly on the needs of people and their local environment, rather than on the 'scientific' justification for conservation of specific examples of the natural world.

The distinction is perhaps most pronounced in the case of the urban environment where the importance of habitats depends very largely on their value to local people. In cities many artificial sites and areas with potential for creative conservation are now highly valued, even though some may have little intrinsic wildlife value at present (Mostyn, 1979). 'Greening' has become an accepted term for this move to establish more vegetation of a natural kind in cities.

The development of an additional set of nature conservation criteria based on values for people has been acknowledged by the landscape profession in Britain. As Tregay (1982) points out, the nature conservation movement is no longer a fringe concern relevant only to wildlife watchers in special and remote landscapes. It now has an enormous public following, and we can conclude that many people wish to live in and experience a more natural world rather than be divorced from it. Manning (1982) writes of the need to design landscapes in which man and nature closely co-exist and where optimum values for both are achieved in a balanced relationship, reflecting a wide range of human needs; landscapes in which urban man himself is seen as occupying the central ecological role. Other socially orientated justifications for the so-called 'ecological approach' to landscape design are discussed by Tartaglia-Kershaw (1982), who stresses that the argument is for a new type of informal 'common land', useful to the community for a wide range of recreational pursuits.

In Holland, the 'ecological approach' to landscape design was under way in the 1930s when the Amsterdam Bos was created. This 400-ha planted native woodland was designed specifically for the urban public. The people's wish to bring nature within reach has resulted in urban wildflower gardens, school allotments and the laying out of haemparks (homeparks), where a range of created habitats comprise carefully designed natural landscapes (Ruff, 1979; Londo, 1983). Large nature parks have also been created and are managed as nature reserves. These habitat creation schemes are designed to *work with nature* by establishing plant communities suited to local conditions. In Sweden and Germany too, the ecological approach to landscape design has been refined by paying specific attention to the plant sociology approach, which emphasizes finding the right mixture of species for the soil and local climate (Gustavsson, 1982).

We now have the knowledge and technology to create many wildlife habitats from scratch (Harrison, 1974; Kelcey, 1977, 1978; Baines & Smart, 1984). Backed by research into reclamation techniques (Bradshaw & Chadwick, 1980; Tregay & Moffat, 1980) and research into the creation of species-rich swards (Wells, Bell & Frost, 1981), techniques continue to be refined (Cole, 1983). In responding to the desire for a more natural environment, landscape designers and managers have made use of habitat creation methods. However, the question remains, to what extent does this 'ecological approach' to the design and management of the green urban landscape actually cater for wildlife as well as people? It is possible that created naturalistic urban landscapes may result in only an impression of nature: a 'green veneer'. It is certainly acknowledged that an ecological basis for the design of the external urban environment is for the benefit of both indigenous flora and fauna and the city-dwelling population (Ruff, 1979; Corder & Brooker, 1981). But, as Bradshaw & Handley (1982) point out, landscape architects do not usually think of purposely providing habitats which will encourage animals. They suggest that a full 'ecological approach' should involve many different ecological principles, all the outcome of understanding ecosystems and natural processes, preferring to call the planting of indigenous plant species in natural situations a 'naturalistic approach' to landscape design.

Obviously, a great deal of progress has been made in establishing particular vegetation types, and it is possible to create a range of different communities in totally artificial conditions. In a sense this is the easy part of catering for wildlife. Attractive plant communities, like flower beds, can be created and managed with relative ease. The difficulties arise when one sets one's sights a little higher with the aim of creating and maintaining a more complete wildlife habitat. Success will not automatically depend on getting the vegetation right. A whole lot of other factors will determine what is possible on a given area, including the degree of disturbance, its size and the way that it is managed. The problems are exacerbated in urban settings where the predominant factor is the all-pervading influence of people. Yet, with a little care and attention to detail it is possible to create areas which are surprisingly rich in wildlife and can still accommodate large numbers of people.

If we are interested in increasing the amount of visible wildlife in an urban area then birds and plants must rank as major components. There is, of course, always the opportunity to encourage a greater variety of attractive insects such as butterflies and dragonflies. But these are closely wedded to the plant species or to the provision of appropriate habitats, such as a small pond.

It is possible to accommodate a range of different butterflies, even on a relatively small scale, so long as proper attention is given to the necessary management of appropriate food plants. It is the birds which might present the greatest problems, because their detailed habitat requirements may not be so easily understood or incorporated within a design scheme. All too often, inadequate attention is paid to effective areas of habitat and especially to the question of disturbance.

This chapter, therefore, considers how the existing bird life of a large city such as London throws light on ways in which a greater diversity of species might be accommodated within new landscape designs for the urban setting.

BIRD DISTRIBUTION IN LONDON

Different categories of species

Some indications of the suitability of different areas for birds can be gained by looking at their distribution in a city such as London (Montier 1977) (Fig. 14.1). Of the 115 species breeding in the London area (856 tetrads) only 15 breed in more than 500 tetrads (a tetrad being 2 × 2 km). These are the common and widespread species (group A; Table 14.1). Apart from the willow warbler, they are all easy to see in parks and gardens throughout suburbia. A further 41 species have been recorded breeding in the range 100–500 tetrads.

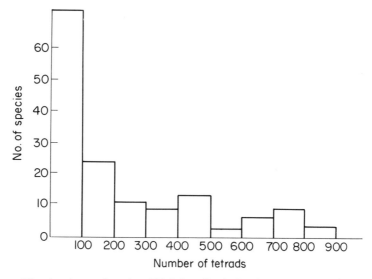

FIG. 14.1. The abundance of species of birds breeding in London represented by the number of tetrads (2 × 2 km) they occupy.

Table 14.1. Breeding birds in the London area which occur in more than 100 tetrads

Group A (breeding in more than 500 tetrads)		Group B (breeding in 100–500 tetrads)	
Blackbird	Wren	House martin	Pied wagtail
Blue tit	Blackcap	Jackdaw	Redpoll
Carrion crow	Bullfinch	Jay	Reed bunting
Chaffinch	Chiffchaff	Kestrel	Rook
Dunnock	Coal tit	Lesser whitethroat	Skylark
Great tit	Collared dove	Linnet	Spotted flycatcher
Greenfinch	Coot	Long-tailed tit	Swallow
House sparrow	Cuckoo	Magpie	Swift
Mistle thrush	Feral pigeon	Mallard	Tawny owl
Robin	Garden warbler	Moorhen	Tree creeper
Song thrush	Goldcrest	Mute swan	Tree sparrow
Starling	Goldfinch	Nuthatch	Turtle dove
Willow warbler	Greater spotted woodpecker	Partridge	Whitethroat
Woodpigeon	Green woodpecker	Pheasant	Yellowhammer

These (group B) include mallard, kestrel, moorhen, coot, jackdaw, magpie, jay, pied wagtail, spotted flycatcher, blackcap, nuthatch and long-tailed tit. Many of these are familiar birds within a suburban environment, but they depend on particular requirements being met. As many as 57 species of birds breeding in the London area occur in fewer than 100 tetrads (group C) and, of these, 27 species are restricted to 10 tetrads or less. Group C includes species which have very specific habitat requirements and many are nationally uncommon or rare. A large number are also directly associated with particular semi-natural vegetation types which are now scarce in the London area. Group C includes most ducks and waders, also heron, dabchick, great crested grebe, kingfisher, nightingale, sedge and reed warbler, tree pipit, grey and yellow wagtails and redstart. So half the species of birds breeding in London are very restricted (group C), others may be fairly widespread but have rather specific requirements (group B), and only 15 are ubiquitous (group A).

For London as a whole, it is generally true to say that the number of species breeding in a given unit area increases with distance from the centre. Thus, in Inner London (totalling 24 tetrads) there is an average of only 22 species per tetrad, and indeed within the heavily built up areas of the City there are only 12 species. Inner suburbs (200 tetrads) average 37 species, and outer suburbs 44 species per tetrad. In the more rural areas of Green Belt land around the periphery of London (372 tetrads) there is an average of 52 species per tetrad.

From the distribution maps of individual species in London (Montier

1977) it becomes apparent that a large number of group B species, and even several of group C, can breed in central areas of London so long as the appropriate conditions are maintained. This is particularly true of water birds such as mallard, coot, moorhen, grebes, tufted duck and Canada goose, all of which breed in the central parks. But a second group is equally significant. These are the birds of woodland and scrub which can survive in central areas as long as appropriate conditions exist. These include nuthatch, tree creeper, long-tailed tit, magpie, jay, blackcap and even spotted flycatcher, all of which are group B species.

Two questions, therefore, arise. What are the conditions necessary for these species, and why is it possible for them to survive in some places and not others? It is clear from the overall picture that, on the whole, species of groups B and C will be less likely to survive in the more central parts of the city. So the circumstances where they do exist might provide the landscape designer with important clues regarding the management of areas for wildlife, especially within urban areas.

Habitat factors and disturbance

The most significant factors influencing the distribution of birds are connected with habitat conditions, especially the nature and diversity of the habitat. The size, shape and pattern of distribution of habitats as well as the stage of development of any given area are also of paramount importance. Another factor, rarely considered in detail in the design of natural landscapes in urban areas, is the effect of disturbance on wildlife. This can be a critical factor which determines the species richness and number of birds. It is not possible to define disturbance exactly, as it is a quality subjective to the individual bird: what disturbs one bird might not disturb another individual or species (Hounsome, 1979). However, it can be considered broadly as 'people pressure' or 'human impact' (Davies, 1976). In urban areas it is likely that disturbance will be a very significant factor in determining the range of species which can exist, and it is worth exploring this further.

If one looks at the range of artificial habitats within London, one finds that some support a relatively large number of breeding species, even though they may be totally encapsulated within the urban or suburban environment. Some artificial habitats will inherently have a greater potential range of habitat conditions than others, especially those like disused sewage works and water works which contain a variety of wetland habitats. So the range of habitats will depend to some extent on the original conditions. But it will also depend on the extent to which any area has been allowed to 'go wild'. So much of the town environment is managed to keep it neat and tidy that the relatively wild patches which have escaped intensive management are a great contrast. On the whole, the heavily manicured town landscapes support few bird species and those that do occur tend to be of group A. As the degree of 'wildness' increases so species of groups B and C may occur.

Many of the places in London which have a more varied fauna of breeding birds have developed simply because they have been left alone or neglected over the years. Some have been totally undisturbed for many years. So there are two major factors which may interact: the gradual development of secondary vegetation and the absence of disturbance by people. Throughout the urban area there are numerous examples of disused industrial sites which have become colonized by vegetation, such as abandoned waterworks, old sewage farms, railway marshalling yards and even disused gas works. Many of these are actually inaccessible to the general public and are therefore not subject to much human pressure. They remain largely undisturbed and are comparatively wild areas. Many provide suitable conditions for a number of group B or C species such as redshank, lapwing, yellow wagtail, meadow pipit, sedge warbler, and even quite rare species such as the little ringed plover. All these species breed in derelict industrial sites in London, often in close proximity to working industrial plants or housing. Admittedly, many of the areas, like the sludge lagoons at Rainham, or Beddington Sewage Farm, are extensive open areas with little disturbance. But others are confined within the urban fabric, where the variety of breeding birds is something of a surprise.

The disused filter beds of a London waterworks are a case in point. The Middlesex filter beds adjacent to the River Lea in Hackney have been disused since 1969. Over the past 16 years they have been colonized by a range of wetland vegetation from open water with floating and emergent aquatic plants, to fringing reedswamp, and in places dense stands of willow carr (Fig. 14.2). The whole area is only 4·2 ha and is adjacent to a heavily built-up area of housing and industry within 7 km of St Paul's Cathedral. Yet this site is used by a wide variety of birds. A total of 23 species are known to breed including a number of aquatic and marshland species of groups B and C. Reed, sedge and willow warblers breed, and little ringed plovers have also attempted to nest. Herons, snipe and various duck use the area for feeding and it also provides an important feeding area for swifts and house martins.

This abandoned waterworks is a good illustration of what can happen as a result of natural succession over a quite limited period. But there are many other examples within the built-up parts of London. In the absence of active management, many Victorian cemeteries have gradually turned to woodland and some of these now have considerable value as nature reserves. They are the only sizable woodlands of any kind in the central parts of London. Tower Hamlets Cemetery, 4 km from the Tower of London, is largely composed of

FIG. 14.2. The disused Middlesex filter beds are very close to housing and industry yet, because they are undisturbed and well colonized by wild plants, contain 23 species of breeding birds.

sycamore woodland. Part has been deliberately left unmanaged for many years as a bird sanctuary within the cemetery, and it now has a remarkably wild appearance. Over 20 species of birds are known to breed, including a number typical of scrub woodland. A larger cemetery of about 25 ha at Nunhead, just south of central London, has a greater variety of breeding birds including chiffchaff and spotted flycatcher. Here there is a range of habitats from open grassy glades through scrub woodland to mature ash woodland (Fig. 14.3). Similarly, Abney Park Cemetery in Stoke Newington is now dominated by sycamore and ash woodland. It is a relatively small area (12 ha) but supports 27 breeding birds, many of which are typical of deciduous woodland, or woodland edge, including spotted flycatcher, blackcap, chiff-chaff, willow warbler, bullfinch, goldfinch and redpoll. There is a very considerable difference between the breeding birds of these neglected cemeteries and those of heavily manicured formal parks in the same parts of central London.

Precisely the same picture emerges when one looks at other habitats within the suburbs where natural succession has been allowed to proceed to scrub woodland. A disused railway line running west from Finsbury Park through the residential suburbs of Haringay to Highgate provides a good illustration. This is now managed as a parkland walk, and along the sides of the old railway track there is an extensive belt of scrub, much of it birch. Again, the birds are

FIG. 14.3. The overgrown cemetery at Nunhead has a wide variety of habitats and consequently a wide variety of breeding birds.

those of scrub and woodland edge, totalling at least 21 species. A secluded area of birch woodland which has colonized the disused station at Highgate, immediately adjacent to the busy Archway Road, supports breeding willow warblers and blackcap and elsewhere along the railway, the breeding birds include goldfinch, spotted flycatcher, bullfinch, jay and magpie. Another piece of unused railway land at Chiswick has similarly developed into birch and willow scrub. This is a small 2 ha plot, known as the Gunnersbury Triangle, which was left virtually undisturbed for about 40 years. In spite of its small size, 20 species of breeding birds have been recorded, including willow warbler, sedge warbler and redpoll. Again, this site lies within a predominantly residential area, and the absence of any form of management contrasts strongly with other open spaces in the vicinity.

But lack of formal 'park' management is only one factor in determining the range of breeding birds. Lack of direct day-to-day disturbance, through physical inacessibility, has also played an important part. For many years the Middlesex filter beds were totally inaccessible, though people frequently used a footpath along one side. So long as the filter beds remained undisturbed they were used regularly in winter by gadwall, pochard, heron and snipe. Recent attempts by little ringed plover to breed were prevented because of human disturbance, since the site was, by then, freely accessible to local people, many of whom used the area for walking dogs. It is also unlikely that other marshland species such as sedge and reed warblers will remain now that it is more generally accessible. The same may well apply to other habitats where the effects are less obvious. In the case of cemeteries and other areas of secondary woodland, lack of disturbance may be an important factor for certain species of group B such as blackcap and tree creeper, as well as for many of the more sensitive group C species. This will be particularly significant in the case of small undisturbed areas such as the Gunnersbury Triangle. It is unlikely that the breeding population of willow warblers would survive there if the woodland was continuously used by a large number of people.

Implications for management

If we are right in these assumptions, they have considerable implications for the management of urban parks and other open spaces. In central London, the Royal Parks support a relatively large number of breeding birds. The total of 29 species breeding in Hyde Park and Kensington Gardens is about double the number to be expected in the surrounding area of Inner London. This is primarily because these parks are relatively large and support a wide variety of habitats, but it is also a consequence of enlightened management. One element of this is the provision of refuges, in particular associated with park lakes. Although most of the Serpentine in Hyde Park is used for boating, and part even for swimming, the northern end of the lake in Kensington Gardens is designated as a sanctuary. One side is totally inaccessible to people, a situation which allows grebes and ducks to nest and other birds, such as herons, to feed in undisturbed conditions. At the same time they can, nevertheless, be viewed by the public from the opposite side. Even kingfishers occasionally feed along this side of the lake in the autumn and winter, in full view of people on the nearby bridge over the lake.

Despite the very large number of visitors, St James's Park is even more of a sanctuary for wildfowl. Here low fences separate the main paths from flower beds and grassy areas around the lake margins. This reduces substantially the amount of direct disturbance from people and allows coots, for example, to breed successfully in very close proximity to people. These refuges are also 'loafing areas' for a wide variety of ducks and geese. This would not be the case if people and their dogs had free access to the water's edge. The design of lake margins and paths is quite crucial to the success of the lake for wildlife. Islands in this lake also provide very significant sanctuary areas where more sensitive species such as dabchick can breed. On one island, adjacent to Horseguard's Parade, about 200 bird boxes have been provided for ducks to nest, and there is a substantial breeding population of tufted duck and pochard, yet the thousands of visitors to the park only see a tree-clad island in the lake. The resident wildfowl are joined in the winter by large numbers of diving ducks and gulls, providing a remarkable wildlife spectacle in the centre of London. It is a public park, but it could equally well be described as a nature reserve.

In other parts of London there are similar examples of sanctuary areas in parks or public open spaces which are equally effective. One of the ponds on Hampstead Heath is fenced off and has been allowed to develop more natural features than other ponds on the heath. In this small sanctuary reed buntings and willow warblers breed. More spectacular perhaps is a section of the Brent Reservoir in north London which is now managed as a nature reserve. It has up to 50 pairs of great crested grebes nesting, one of the largest colonies of this species in Britain. The nature reserve is within a very heavily built-up urban area and other parts of the reservoir are used for sailing and canoeing. Reservoirs in London are of considerable significance for wildfowl and gull populations, especially in winter (Oliver, 1982), but it is known that the more active forms of recreation, such as sailing and canoeing, may conflict with the minimal-disturbance requirements of wildlife. Observations at the Brent Reservoir have shown that despite recent intensification of sailing activities, large numbers of wildfowl and other water birds still use the reservoir because there is an adequate refuge area (Batten, 1977). An important feature of this refuge is the creation of deep-water barriers on the landward side to prevent people from entering the sanctuary.

The concept of refuges within woodland may be less easy to demonstrate, but there are examples which tend to give support to this notion. Despite being rather small and close to the centre of London, 24 bird species breed in Holland Park. Much of the area comprises mature woodland which has been well managed and it consequently caters for the relatively specialized requirements of several group B species (Brown, 1973). 'People pressure' has been minimized by effective channelling, so that there are still relatively undisturbed parts within this much-used park. The fact that 'trampling pressure' is limited allows the survival of some of the ground flora and understorey vegetation. This provides additional bird habitat and probably allows more birds to use the area than would otherwise do so. Ken Wood at Hampstead Heath provides a similar woodland refuge which is largely inaccessible except along fenced paths. The same is true of Lesnes Abbey Woods in Greenwich which also functions as a rather informal park. It is very noticeable in these woods that there is a substantially greater ground flora of woodland plants such as wood anemone, than in similar woods where people can wander at will. This is a crucial factor in the management of ancient woodlands in urban areas, where channelling of people is necessary to retain the full range of woodland conditions. In the same way that in a park people are expected to respect flower beds by walking round them, so it may be necessary to provide wildlife sanctuaries within a park or woodland which are equally free from disturbance.

NATURE RESERVE MANAGEMENT APPLIED TO URBAN OPEN SPACES

Ecological changes which result from high levels of recreational activity have long been of concern to those responsible for the management of semi-natural areas in the countryside (Goldsmith, 1983). 'Recreation ecology' is now a demanding interdisciplinary research subject which requires an understanding of ecosystem function, the requirements of component species and the attitudes and movement of visitors (Speight, 1973; Goldsmith, 1983). Indeed, it is now standard to give full consideration to the management of people in the preparation of nature reserve management plans.

Methods used to minimize the effects of high numbers of visitors range from the use of physical barriers such as fences, hedges of spiny shrubs, and ditches, to more subtle approaches involving signposting, nature trails and alternative routes. Pathways, including steps in steep areas, serve to attract and direct people. They can be designed to lead people away from sensitive areas and may comprise mown routes through grasslands or boardwalks and bridges over wetlands. Devices to concentrate people in one place (the 'honeypot' approach) include car-parks and refreshment huts. Careful positioning of hides also serves to attract people to less sensitive areas whilst still allowing, or even enhancing, enjoyment of the area. The RSPB make use of high-level viewing platforms to minimize disturbance to breeding birds. Islands in wetlands provide refuges for wildlife. In the same way, moats are excavated to make some areas completely inaccessible to visitors. Refuges may be made even more effective by proper screening. At the Sevenoaks Gravel Pit reserve and several RSPB reserves, reed bed screens are used as a subtle means of separating people from wildlife (Harrison, 1974; Batten, 1977).

All means of manipulating visitors allow the use of an area to be zoned in both space and time. Areas are divided into zones with different intensities of use, allowing the manager to direct visitors to the least vulnerable areas at any given time. The sensitivity of the zones may vary according to the time of year, e.g. areas with nesting birds are particularly vulnerable during the breeding season. On some reserves a 'recreation carrying capacity' has been calculated (Countryside Recreation Research Advisory Group 1970). This index serves as the upper limit to the number of visitors admitted to the reserve at any one time, although in actuality there is no single threshold because the value depends on the pressure exerted by individual visitors. Nevertheless, it is a useful means of limiting disturbance. It allows the maximum number of visitors to enjoy the wildlife of an area without undue detriment to the wildlife itself.

Management of public open space in cities tends on the whole to be geared to intensive use by people. Whether for a formal park or football pitch, the primary objective is very often incompatible with the maintenance of a variety of wildlife. But this need not be so. If the management of urban greenspace was viewed from a different perspective, it would be possible to accommodate a great deal more wildlife. Application of the techniques developed for management of people in nature reserves could result in a very significant change in the case of urban parks. The design of St James's Park illustrates that it can be done extremely effectively without detracting from enjoyment of the park. Indeed, in most cases it is likely to result in positive enhancement. Such techniques could equally well be applied to town squares, parks, informal open spaces; indeed, any areas of urban greenspace where a greater variety of wildlife is desired. The methods can be applied at any scale to suit the circumstances, without necessarily affecting the existing primary uses of particular places. Sanctuaries may be quite small but still effective. Equally, if the concept of sanctuary areas is not included in the design, then the opportunity for a greater variety of wildlife may be entirely lost.

TUMP 53: AN EXAMPLE OF URBAN DESIGN FOR WILDLIFE

In the centre of new housing developments in Thamesmead, there is a network of canals and islands which are a legacy of the old Woolwich Arsenal. One of the many moated islands which were originally used to store explosives has been retained within the new town development. This island, known as Tump 53, is enclosed by brick walls and earth banks and bounded by a canal and a moat with reed beds (Fig. 14.4). The Tump was earmarked for conservation by the GLC when plans were made for this part of the town. Existing reed beds were protected by installing weirs and pumps to maintain water levels in the moat. More reeds were encouraged to grow along the canal to provide habitat for birds such as moorhen. Now, with the aid of a grant from the Sainsbury Trust, the GLC has developed the Tump as a nature study centre. The Council commissioned a study by Land Use Consultants (LUC) to reconcile the use of the site as a community resource and teaching area with the protection of the existing habitats, which are highly sensitive to disturbance.

The final design of Tump 53 is shown in Fig. 14.4. A number of design features have been incorporated to achieve a balance between nature

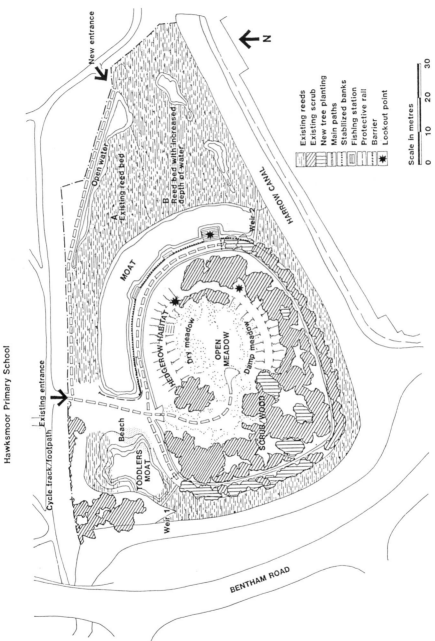

FIG. 14.4. Tump 53, a relic of the old Woolwich Arsenal, has been developed for wildlife conservation and teaching, using a carefully planned zoning to allow access by people and protection for the wildlife.

conservation and the use of the site by people. Water, rather then fencing, is being used as a means of controlling access and protecting the sensitive reed beds. Clearly, the layout of the Tump lends itself to this type of zoning, with the canal and moats preventing access from the road and footpaths. To refine this control the reed bed at the eastern end of the Tump has been deepened. The alternative of erecting a high security boundary fence might tend to alienate the general public and could act as a challenge rather than a deterrent to access (LUC 1984). The open water moat at the north-east access point also serves to direct people onto the adjacent path and away from the reed beds. The fact that mute swans have bred successfully in the reed beds to the west of the Tump (in full view of the road) may partly reflect the success of this approach to access control. It is hoped that the design will encourage waterfowl and other marshland species to breed. Reed buntings and reed warblers have nested and dabchicks can be seen feeding in a canal immediately adjacent to the houses.

Access to the two weirs (necessary both to protect the reed beds and for safety reasons) is prevented by physical barriers. The newly planted vegetation behind the barriers will eventually provide a more subtle means of directing visitors away from these areas. Fishing activities have been confined to one bank of the large moat where platforms have been erected. No fishing is allowed from the northern bank of the moat, again to minimize disturbance and damage to the reed beds. The pathways direct and attract people to the centre of the Tump where the most intense use is to occur. Here, an open meadow is being created and small plots are being provided for adoption by school groups. The path passes the smaller moat which is to be developed as a primary school/toddlers' pool. Pond-dipping activities are confined to a 'walk in' shingle beach.

The design of Tump 53 nature study area has also taken the eventual management and maintenance input into account. One of the major aims of the scheme is that the site should be resilient to public use and disturbance and should have a low maintenance requirement. This is important because it will not always be possible to have a warden on site. Tump 53 illustrates the point that even a very small site can accommodate a variety of species, and withstand heavy use, providing care is taken in the design phase to direct visitors away from sensitive areas.

CONCLUSION

Although wildlife conservation is becoming progressively more accepted as part of landscape design, there is still a great deal to be done. The urban environment has plenty of examples which show that effective wildlife

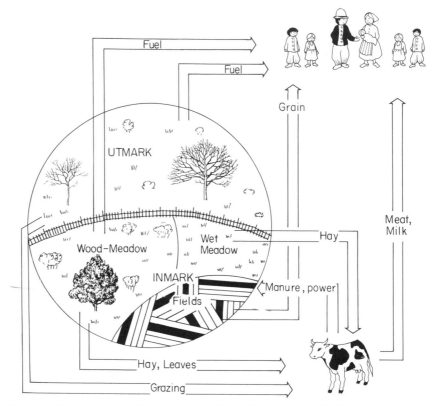

FIG. 15.2. The circulation of products in a self-sustaining traditional village of central Scania before the land reforms of the nineteenth century.

manure the fields could be farmed every year. On the other hand, in the south Scanian village, with its grain-dominated economy, very little 'utmark' existed. Therefore, one-third of the 'inmark' was left every year as grazed fallows.

The systems were often not in ecological balance. Descriptions from the eighteenth century often speak of environmental disasters. The 'utmark' was overgrazed and the woods cut down. In sandy areas the fallow periods were shortened, resulting in sand drift. The meadows were transformed into arable fields, resulting in a shortage of hay. Less animals could now be fed, and the reduced supply of dung in turn resulted in diminishing production from the arable fields.

As a consequence of the ecological crises and of general demands to increase the effectiveness of the agriculture, the land reorganization of 1803, already mentioned, took place. The result was a drastic alteration of the landscape during the nineteenth century. Many types of earlier management were abandoned and large areas brought under cultivation. The 'utmark', formerly a savanna-like landscape, was, if not brought under cultivation, afforested. The wood-meadows were no longer needed due to the availability of artificial fertilizers and the new practice of growing legumes. Long-term fallows could be abandoned for the same reason; huge drainage programmes destroyed most of the wet meadows.

TRADITIONAL VEGETATIONAL MANAGEMENT IN SCANIA

The 'utmark'

Most of the 'utmark' of the eighteenth century had, in early medieval time, been beech, oak or alder woodland. As the population grew, the 'utmark' was gradually opened up by the cutting of firewood and grazing of domestic animals. There were many intermediate vegetation types between the dense woodlands and the totally open grazing areas, some of which can still be found as relics. It is these intermediate savanna-like vegetation types that are now of considerable interest to the planning of present-day parks and similar areas.

A common 'utmark' type, the 'utmark' beech wood, was composed of trees, mainly beech but often also oaks, with wide, multi-stemmed canopies (Fig. 15.3). These canopies were the result of cutting the trees 2–4 m above ground. The reason behind this management was twofold: (i) it enabled the peasants to overcome a royal law against felling of beech, and (ii) it resulted in more food for the pigs. Canopies of this type are believed to have produced more acorns than canopies of dense woods.

Some relic areas still exist where regeneration has not taken place, due either to the dense canopies or grazing, and their conservation value is very large. Besides their great historical value, they are important as a genetic reservoir for the indigenous Scanian beech and, from a zoological view, are especially useful as sanctuaries for some birds and insects.

In grazed stands the vegetation is rich in grasses and various dicotyledons. Ungrazed woods are, depending on soil status, dominated by *Deschampsia flexuosa*, *Oxalis acetosella* or *Lamium galeobdolon*. In the planning of new urban green areas this may not be the most coveted type but new stands can quite easily be constructed by cutting and thinning of existing woodlands. Where the woods were more open, grazing led to either a 'juniper shrub-land' or a 'hawthorn shrub-land'. The former was typical on poorer soil, the latter on more calcareous soils.

FIG. 15.3. 'Utmark' beech wood north of Ållskog in the parish Baldringe, southern Scania.
(Photo: O. Nordell.)

Juniper shrub-land

Up to the 1950s there were still large areas of juniper shrub-land (Fig. 15.4). In
recent years, however, areas with this vegetation type have decreased
markedly. Many areas have totally disappeared due to forestry plantations.
Moreover, in many places where the juniper can still be found, the field layer is
today totally different from the field layer of the traditionally managed shrub-
land. Modern grazing practice has brought with it regular artificial fertilizing
of the grass sward so that these areas are very similar to any other type of
modern managed pasture, except for the presence of bushes.

Traditionally managed juniper shrub-lands had very low productivity.
The field layer of drier areas was dominated by *Calluna vulgaris* and
Deschampsia flexuosa, with *Vaccinium* species, *Empetrum nigrum* or *Nardus
stricta* as co-dominants. The wetter areas were dominated by *Erica tetralix*,
Myrica gale and *Molinia caerulea*. These vegetation-types are described in
more detail by Malmer (1965). On more fertile soils grass- and herb-
dominated juniper shrub-lands could be found. This type can today be found
on some coastal localities.

In north-western Scania, and especially in the adjacent province of
Halland, recurrent burning was traditionally used (Nilsson 1970). This

FIG. 15.4. Unfertilized, still traditionally managed juniper shrub-land at Kjugekull in the parish Kiaby, north-eastern Scania. (Photo: A. Schmitz. From the picture collection of the Scanian Nature Conservancy Society.)

practice, or a heavy grazing pressure in combination with mechanical clearance of shrubs, as in southern Scania, prevented extensive growth of juniper. Prostrate stems were used for fencing and building. However, in period of war or cattle disease, the juniper bushes could sometimes expand and totally cover 'utmark' areas. After such an encroachment period it was difficult or even impossible, due to lack of man-power, to restore the pastures.

Due to changing farming practices since the 1930s, many of the existing shrub-lands have gradually become overgrown by juniper. Some of the resulting successional stages have been regarded by conservationists as very attractive vegetation types. Nature reserves were, therefore, often established in such areas. Great difficulties have been met in developing suitable management to keep the succession in status quo. Such management is expensive and, from a historical point of view, is also questionable. An alternative, more in keeping with the traditional management, would be to clear some areas totally and thereby start new succession.

Hawthorn shrub-land

The term hawthorn shrub-land is here used as a collective name for a lot of quite diverse shrub-land types found on richer soils. Such soils were the main target for the land reorganization of the nineteenth century. Only a very tiny fraction of this shrub-land type is therefore left today. Of this tiny fraction, only very small parts have not been artificially fertilized.

The hawthorn shrub-lands, unlike the juniper shrub-lands, contained a lot of trees (Fig. 15.5). The tree and shrub communities were often very diverse, as the traditional economy relied on products from many different species. Most of the indigenous tree species occurred, and the most common shrubs were *Crataegus* spp., *Rosa* spp., *Rubus* spp. and *Prunus spinosa*. The field layer was also often very diverse with an attractive assembly of colourful herbs, including many orchids. Individual treatment of the shrubs was more common here than in the juniper shrub-lands, which were often extensively cut.

As an alternative to modern parks with alternating lawns and shrub plantations, new juniper and hawthorn shrub-lands could be used. In Scania the hawthorn type is preferred in and around most urban areas, where there are usually better soils. The juniper type is of more interest in northern Scania where relic sites still exist.

When constructing a hawthorn shrub-land the best method is aggregated planting of the shrubs. These aggregates can later be opened out. The major problem is probably the field layer since suitable sites for this kind of park usually have very fertile soil in their upper layers. This means that when planting the natural flora of the vegetation type, aggressive weeds are likely to invade leading to an uninteresting vegetation which will persist for a long time. To overcome this problem the upper layers of soil must be depleted as already discussed by Green (Chapter 12). This could be done by burning and/or by growing demanding crops, such as wheat, on the site for some years. Later, after sowing the preferred mixture of grasses and herbs, mowing could take place.

When the fertility of the sward has been lowered, management of the area by grazing can be confined to shorter periods of the year than is traditional. By

FIG. 15.5. Hawthorn shrub-land at 'Oxhagen' in the parish Baldringe, southern Scania. Note the relic tree of the former 'utmark' beech wood, which preceded the hawthorn shrub-land. In the background a modern managed beech forest. (Photo: L.G. Blomkvist. From the picture collection of the Scanian Nature Conservancy Society.)

dividing the area into folds and having quite intense grazing pressure in each fold, most of the area could be held free from livestock for much of the year. If the fertility of the site could be drastically reduced, it may even be possible not to graze the area every year. The alternative of sowing a normal grass lawn would require a lot of management work, as well as being uninteresting from both a historical and a biological point of view.

Wooded meadows

The most important use of the coppice woods of Britain and western Europe was the production of underwood and timber (Rackham 1980). The underwood was used for fuel, fencing, 'wattle-and-daub', etc. In Denmark and southern-central Scania this kind of coppice woodland was only found occasionally. Instead the stools were scattered and hay was produced between them. Normally the hay was the most important product.

The regular cutting of the underwood seems to have been the crucial condition for a high production of hay. The trees and shrubs collect nutrients from deeper layers of the ground and fertilize the upper layers by their leaf litter production. This process and the fertility produced by the degenerating root-systems of the cut coppices gave this system its excellent ability to produce hay for a very long time. The persistence of the production is well documented (Sjörs 1954). Details of the ecology of the system are, however, not yet fully understood, and the causal connection between the ecological processes outlined and the persistence of the production requires further investigation.

This type of coppice-meadow was, as already mentioned, the common type on drier ground in Denmark and southern-central Scania. Further to the north-east, in Sweden, the principles of management of the wooded meadows were the same, but pollards substituted for coppice stools. A far greater part of the fodder production in the pollarded meadows came from the foliage harvested from the pollards. The total fodder production was higher but the production of woody material less. Another difference was that fruit-bearing trees (*Malus, Prunus* spp., *Sorbus* spp. etc.) were an important component, while in Denmark and Scania these species were found more often in the 'utmark', especially in the hawthorn shrub-land.

The British coppices, the Danish-Scanian coppice-meadows and the pollard-meadows of south-eastern Sweden represent three types on a continuous gradient. At one extreme, in the west there has long been a lack of woody material but no problems of finding winter-grazing; at the other, in central Scandinavia, there have always been vast forests but a large demand for hay for livestock during the winter (Bergendorff & Emanuelsson 1982).

In Scania, the coppice-meadows were concentrated in the cattle-raising zone of the central part of the province mentioned above, which is also where we find most of the remaining examples. In some districts the coppice-meadows in the eighteenth century covered more than 30% of the total land area. Today there are no traditionally managed coppice-meadows in Scania, but the authors have recently started a project to restore this type of vegetation. Nevertheless, we estimate that there are about 100 ha of degenerated coppice-meadows distributed in 10–12 localities where the traditional management ceased so recently that the general structure of a coppice-meadow still persists. Additionally, there are approximately 100 localities, covering some 5–800 ha, where traces of former coppice-meadow management are still to be found.

The coppice-meadow management of Scania does not seem to have been as regulated as, for example, in Britain. The underwoods were cut when hay-production ceased, which normally occurred after 20–30 years. A division of the wood into parcels as described for Britain (Rackham 1980) did not normally occur, but due to fragmented land holdings the coppice-meadows were often cut in a similar way. On richer soils, coppice-meadows with *Tilia cordata*, *Fraxinus excelsior* and *Corylus avellana* occurred. More common, however, were combinations of *Betula* spp., *Salix* spp., *Alnus glutinosa* and *Corylus avellana*. *Quercus robur* often occurred as standards.

The decline of the coppice-meadows began in the early eighteenth century, when many areas were transformed into grain-producing fields. The final death-blow was the introduction of artificial fertilizers, which made the coppice-meadows superfluous by the end of the nineteenth century.

Many people regard the pollard-meadows of south-eastern Sweden as very beautiful, so a number of these have been restored and are now managed as nature reserves. However, because there had been considerable degeneration of the Scanian coppice-meadows by the time of the present increased interest in nature conservation, such areas have unfortunately not been preserved. They must, however, have been very attractive with great public appeal due to: (i) a diverse flora with many colourful species, including many rare species especially orchids, (ii) a rich bird and insect fauna, and (iii) vegetation suitable for walks. Their similarity to a well-arranged garden also appeals to many people.

The coppice-meadow and the pollard-meadow should not be too un-familiar to the ordinary park planner. Many of our parks today have features resembling these ancient cultural landscape types. However, exotic trees and shrubs are now mixed with indigenous species, fertilized lawns are common, and the shrubs are not allowed to mix with the grass sward.

Considerable management advantages could be gained if conventional parks were at least partly replaced by wooded meadows. Whereas lawns have to be cut regularly, it is sufficient to manage the wooded meadow by mowing only once or twice a year. Mowing has to be carried out only in the first 5–10 years of the turnover period of 20–30 years. In the later stages the shrubs will form a continuous canopy, shading the field layer and resulting in a lowered production. The wood-pasture will, thus, be accessible for walks all year round although, in the earlier stages, walking should be prohibited during May–July. With a well-planned division of the park into parcels, two-thirds of the park will be accessible even in May–July, while the remaining one-third would constitute a very attractive area with a rich variety of flowers.

Management of trees and shrubs would be more rational than that of a conventional park, since each parcel would only be cut once during the rotation period.

Until recently there was a very degenerated old coppice-meadow at the experimental site at Hörjel. This area is today partly restored with a modest input of manpower and already gives an impression of a true managed coppice-meadow (Fig. 15.6). Most woody areas could probably be converted quite easily into coppice or pollard-meadows.

Wet meadows

Wet-lands of many different sorts have for a long time been used for hay-making in Sweden (Larsson 1976; Regnéll 1976; Tyler 1981), e.g. mires in areas with archaean rocks have been mowed even though the production was low. Very high production was obtained in seasonally inundated meadows along streams in areas of calcareous moraines (Fig. 15.7). The economic

FIG. 15.6. Partly restored, newly mowed, coppice-meadow at Hörjel in the parish Vanstad, southern Scania. In the foreground, stools of *Ulmus glabra* and *Fraxinus excelsior* 1.5 years after cutting. In the background an unrestored parcel with a mixture of multi- and single-stemmed trees not cut for approximately 30 years. (Photo: O. Nordell.)

FIG. 15.7. Winter inundation of the small river Görslövsån in the parish Jonstorp, north-western Scania. (Photo: B. Petersson. From the picture collection of the Scanian Nature Conservancy Society.)

importance of the flood meadows in Scania was considerable until the twentieth century. Up to the early nineteenth century coppiced alders commonly grew scattered in many of these meadows. To provide regular irrigation, systems of canals were built in many areas during the nineteenth century. A great majority of these meadows are now drained. The few remaining are mostly grazed and not mown, which has had a radical effect on both flora and fauna. The mowing was particularly important for many birds, which have now become very rare, e.g. black-tailed godwit, ruff, corncrake, the southern race of the dunlin.

In Scania, outside the archaean area, about 10% of the total land area was wet meadow in the early nineteenth century (Emanuelsson & Bergendorff, in press). Wet meadows which are still mown now cover only about 50 ha (0·010%). Grazed wet meadows, including sea-shore meadows, cover some few thousand hectares (0·5%) (Emanuelsson & Kjellén, in press). We do not think it necessary to construct wet meadows close to urban areas, especially since there are partly destroyed wet meadows in the vicinity of most urban areas in Scania. A much more important task is to save these meadows from total destruction and to reinstate traditional management so that their species diversity is restored.

Finally, the calcareous fens, once common in the most fertile parts of Scania (Tyler & Waldheim 1983), should be considered. These fens were situated around wells and they were very rich floristically. It has been shown that such fens can develop if ground water is forced to the surface, for example by construction works (Englesson 1972). This very attractive type of fen can accordingly be created deliberately.

Neither the wet meadows nor the calcareous fens are suitable for unrestricted walking because of their range of beautiful plant species, but they can form very attractive spots in urban areas, and are certainly alternatives to bird ponds and flower-beds. The public can enjoy them without causing damage if paths and board-walks are provided.

Fallows

In the early nineteenth century fallows were a normal feature in nearly all types of agricultural systems occurring in Scania. It is, however, only in the sandy areas with their long fallow periods that the fallow and its associated flora has persisted up to the twentieth century (Fig. 15.8). In many of the old fallow areas in Scania, other factors besides cultivation, causing disturbance, have recently replaced the traditional farmer, e.g. rabbits and even military tracked vehicles.

On acid sand a *Corynephorus canescens* association develops (Andersson 1950; Mattiasson 1974; Olsson 1974), while on more calcareous soils it is replaced by a *Koeleria glauca* association. Both these very species-rich associations will, however, disappear if the disturbance of the sand ceases and a heather-dominated association develops. This transition leads to a much lower diversity, floristically as well as faunistically.

Both these fallow associations have decreased considerably during the twentieth century, but the *Corynephorus canescens* association including degenerated forms, still covers some 10–15 km^2. The *Koeleria glauca* association is, however, very rare today, covering only 50 ha (Mattiasson 1974). These vegetation types are perhaps not important when discussing urban parks, but preservation of areas where they are still to be found, albeit in a somewhat degenerated form, must be considered an important task in nature conservation.

Although the activities of rabbits and military tracked vehicles have saved the vegetation in certain spots, they are not sufficiently effective to preserve enough of the fallow areas. To restore and preserve larger areas, it is necessary that they are ploughed regularly but not too often. In some places, when pine

FIG. 15.8. Area where some traces of traditional fallow management were still to be seen in the 1940s. The '*Briza*-bracken' in the parish Brösarp, south-eastern Scania. (Photo: O. Andersson, 1945. From the picture collection of the Scanian Nature Conservancy Society.)

plantations have been felled to restore the open undulating landscape, ploughing has been used in an effort to restore the fallow vegetation.

CONCLUSIONS

In the present paper we have discussed the history, ecology and possible future use of traditional vegetational management. We would argue strongly that the historical aspect of the management of vegetation is neglected in many nature reserves and in the common landscape, as well as in urban parks where it is totally ignored by planning authorities.

During the twentieth century a gradual degradation of the common landscape has occurred due to modern forestry and modern farming practice, the draining of wetlands, etc. Nature conservation resources have been spent on many small, selected sites. The aims of these reserves are often vague and contradictory, due to compromises. As a result, the present management produces areas where neither biological, historical nor social aspects are fully considered. The historical aspect, in particular, is neglected, which means that vegetation succession may be overlooked. There is, of course, a great need for more resources in nature conservation work, but an alternative landscape policy, which could be executed even under present conditions, would be the transfer of resources from the management of nature reserves to use in the management of the common rural landscape. Subsidies and laws could be used to preserve larger areas of particular interest. Economic support could, for example, be given for every cow grazing in unfertilized pasture. Laws similar to the Swedish Beech Forest Law (1974) could be introduced. This law forces the land-owner to replant beech after beech forest has been felled. Subsidies and laws of this kind must, however, always be adapted to regional management traditions.

In the future the resources left for nature reserves should be concentrated on fewer sites for which the aims of conservation must be very clear. Such reserves could be divided into two categories.

1   Nature reserves which are very important for the preservation of particular endangered species or outstanding geological features.

2   Nature reserves where the management is carried out in strict accordance with old traditions.

We believe that it is preferable to keep a few reserves of the second type managed in full accordance with old traditions, than to preserve several reserves in only a semi-historical state. Such semi-historical management is common today in Scanian shrub-land reserves set aside for the preservation of the old cultural landscape. Due to the use of artificial fertilizer in such reserves, the sward is gradually being transformed into a very uniform and uninteresting plant community.

In recent years, there have been discussions in Sweden about ways of making urban parks more diverse and 'natural' (Gustavsson 1979; Gerell 1982). In planning what should be done it is most important that the historical aspect is considered. In this paper we have put forward reasons for using traditional vegetational management as a guideline in alternative urban park management. These proposals are likely to meet opposition from horticulturalists and others; opposition which may sometimes be well founded; for example, the possibility of hawthorns transmitting *Malus* diseases. There may also be some practical problems—obtaining indigenous plant material, etc. We argue, however, that the opposition is mostly based on a lack of knowledge of the history and biology of the traditional cultural landscape.

ACKNOWLEDGMENTS

We thank B. Nihlgård for valuable comments on the typescript, M. Varga for drawing the figures, K. Richter for typing the manuscript and the Scanian Nature Conservancy Society for giving us the opportunity to use their Hörjel farm for experimental work and also for providing illustrations for this paper.

REFERENCES

Andersson, O. (1950). The Scanian sand vegetation: a survey. *Botaniska Notiser*, **103**(2), 145–172.

Bergendorff, C. & Emanuelsson, U. (1982). Skottskogen: en försummad del av vårt kulturlandskap. *Svensk Botanisk Tidskrift*, **76**, 91–100.

Burenhult, G. (1982). *Arkeologi i Sverige. I. Fångstfolk och herdar.* Förlags AB Wiken, Höganäs.

Campbell, Å. (1927). *Skånska Bygder under Förra Hälften av 1700-talet. Etnografisk studie över den skånska allmogens äldre odlingar, hägnader och byggnader.* Ph.D. Thesis, University of Uppsala.

Dahl, S. (1942). *Torna och Bara. Studier i Skånes bebyggelse och näringsgeografi före 1860.* Ph.D. Thesis, University of Lund.

Englesson, N. (1972). *Norra Nöbbelövs mosse.* Generalplanekommittén i Lund. (General planning committee of the town of Lund.)

Gerell, R. (1982). Faunavård i Stadsmiljö. *Statens Naturvårdsverk PM 1622.* Solna.

Gustavsson, R. (1979). Samutnyttjande av tätortsnära jordbruksmark. *Landskap*, **52,**

Hannerberg, D. (1971). *Svenskt Agrarsamhälle under 1200 år.* Läromedelsförlagen, Stockholm.

Helmfrid, S. (1961). The Storskifte, Enskifte and Laga Skifte in Sweden. General features. *Geografiska Annaler*, 1961, **1–2.**

Larsson, A. (1976). Den sydsvenska fuktängen. Vegetation, dynamik och skötsel. *Meddn Avd. Ekol. Bot., Lunds Univ.*, **31,** 1–107.

Lindquist, B. (1931). Den skandinaviska bokskogens biologi (The ecology of the Scandinavian beech-woods). *Svenska Skogsvårdsföreningens Tidskrift*, **3,** 179–532. (Summary in English.)

Malmer, N. (1965). The south-western dwarf shrub heaths. *Acta Phytogeographica Suecica*, **50,** 123–130.

Mattiasson, G. (1974). Sandstäpp. Vegetation, dynamik och skötsel. *Meddn Avd. Ekol. Bot., Lunds Univ.*, **2:4,** 1–83.

Nilsson, J. (1970). Ljunghedar och deras skötsel. *Meddelanden från Forskargruppen för Skötsel av Naturreservat, Avd. Ekol. Bot., Lunds Univ.*, **3,** 1–38.

Olsson, H. (1974). Studies on South Swedish sand vegetation. *Acta Phytogeographica Suecica*, **60,** 1–170.

Rackham, O. (1980). *Ancient Woodland, Its History, Vegetation and Uses in England.* Edward Arnold, London.

Regnéll, G. (1976). Den sydsvenska kalkfuktängen i litteraturen. *Meddn Avd. Ekol. Bot., Lunds Univ.*, **4:3,** 1–40.

Sjöbeck, M. (1927). Bondskogar, deras vård och utnyttjande. *Skånska Folkminnen* 1927. Reprinted as: Vång och utmark i Skånes skogsbygd. *Sveriges Natur*, 1966, 149–182.

Sjörs, H. (1954). Slåtterängar i Grangärde finnmark. *Acta Phytogeographica Suecica*, **34,** 1–135.

Sjörs, H. (1963). Amphi-Atlantic zonation, Nemoral to Arctic. In *North Atlantic Biota and Their History* (Ed. by A. & D. Löve), pp. 102–25. Oxford.

Tyler, C. (1981). Sydsvenska kalkkärr. Hävd i gången tid och skötselförslag för framtiden. *Meddn Växtekol. Inst., Lunds Univ.*, **47,** 1–115.

Tyler, C. & Waldheim, S. (1983). Kalkmyrar och fuktängar i 1940-talets Skåne. *Meddn Växtekol. Inst., Lunds Univ.*, **51,** 1–181.

IV
MANAGEMENT PRACTICE

INTRODUCTION

Translating principles into practice is a difficult task in any sphere, none more so than landscape management where it is essential to have an understanding of the habits and limitations of people. The papers that follow vary considerably, both in the scale of subject and direction of approach. But certain features are common, the major one being that implementation of any management plan requires a change in the thoughts and habits of users, owners and maintenance staff.

The problem begins in urban areas. The management of urban landscapes involves balancing conservation against other uses. If new ideas arc to be explored, it is clear that careful consultation with, and consideration of the views of, the public and staff is required. Then, in new towns the desire to create more natural urban landscapes has resulted in new techniques of design and construction. Here there is a need for a new breed of creative landscape manager who cannot only understand the characteristics but also develop the potential of these particular landscapes.

This may sometimes seem difficult, but progress is possible. Even in a habitat so pressured on all sides as wetland, consideration of the ecology of waterplants and water's-edge vegetation suggests that there are many simple and effective ways of creating and managing aquatic vegetation, which need not fall foul of the many, often conflicting, uses of the water. Motorways similarly provide us with a good example where progress is possible. Pressure to reduce maintenance costs is now paramount, with a need to introduce low cost techniques to the maintenance of motorway landscapes. But, properly thought out, these can produce both a more attractive result and more valuable reserves for wildlife. We have to realize what can be done and take action.

In the light of these examples it is perhaps remarkable that the greatest determination to improve management, and capitalize on what nature can create with little assistance, should be in West Germany. Berlin's positive steps to conserve and develop its urban wild life should be a salutary lesson to the rest of us, who let opportunities slide by apathy and lethargy.

It is clear that we must improve our management practice. Awareness of nature and the environment has resulted in nature conservation becoming both a political and strategic planning issue in many parts of the world. The desire to conserve wilderness has resulted in the creation of many national parks. But the termination of traditional economic land uses, coupled with

259

enormous development of tourism, have brought about unwanted changes. These demonstrate how essential it is to relate planning and management to the dynamic balance of the natural ecology as well as to people's desires.

We have the understanding to put into practice the ecological principles which are so important, but we must also consider communication of those principles. Hard experience suggests that people have their own vision of the ideal landscape depending on, for example, whether they are designers, users or maintenance staff. At the same time many have no vision at all. To change these different visions requires skill in communication and simplicity of message.

How does one convince a farmer, farm labourer, highway maintenance crew, or neighbour, that an untidy, overgrown hedge or verge is desirable? Not with a 30-page document. What will persuade a holiday maker not to speed his motor boat, light fires, or feed bears? Penalties may alienate users, persuasion is time-consuming, and over-regulation destroys the very sense of freedom that is essential for relaxation.

Accepting new management proposals may conflict with economic interests, not merely in the obvious sense of, e.g. loss of yield from the land, but also even in terms of maintenance operatives' pay packets, which often depend on carrying out particular tasks. To reduce people's income while destroying their vision of a good job, is not the formula of success.

To achieve practical results in management it is essential to be able to express the objective of the plan in the various languages of all those who will use it, describing objectives which they will not only understand, but accept, creating images which remain in the mind.

16. MANAGEMENT OF URBAN PARKS

A. L. WINNING

Department of Recreation and Amenities, City of Sheffield

SUMMARY

Management of urban parks places the responsibility on the politician and the professional to maintain a balance between the conservation of the natural resource of parks and, at the same time, to maximize safe use of the facilities by the general public, as conservation without management could create a hazardous habitat.

Urban parks present a wide range of habitats, ranging from established Victorian parks and amenity woodlands to recently constructed functional recreation grounds or large sports complexes, playing fields or golf courses. Management problems related to daily operations are increasing, due to pressures associated with increased use, anti-social behaviour and to social policies related to the present recession. Constraints in the availability of finance and the consequences of local and national political direction can condition the standards of maintenance and the quality of the environment, and the nature of recreational experience to be gained from visiting the urban park.

The manager's role is diverse; apart from providing a service to the community within these identifiable constraints, he has an educative role, and an important part to play in public relations and in consultation with the users of park facilities.

INTRODUCTION

Urban parks must be considered as a long-term land investment for active and passive leisure pursuits. The standard of landscape treatment and maintenance will depend on the origins of the park, the subsequent treatment and the current management policies of the local authority.

In many areas the ecology of the park is under stress from the pressure of the users. Brian Clouston, as past President of the Landscape Institute, at an Institute of Leisure and Amenity Management Conference in Edinburgh in 1983 (and Chapter 1) argues that urban parks are in crisis, and that this has been conditioned by the effects of obsolescent design and less than efficient management. Whilst one would not immediately agree with these emotive

261

conclusions, there are undoubtedly variable maintenance standards; these are often related to the continuing experience of financial stringency in local authority operation. Considerable emphasis in the last decade has been towards the development of sophisticated recreational facilities. The much-needed capital for refurbishing and redesigning the urban park, to ensure maximum use at minimal maintenance cost, is not forthcoming. Resources tend to be concentrated on the provision of new physical facilities providing intensive use, e.g. floodlit synthetic play surfaces.

The general public use parks for a wide range of formal and informal activities, and the majority are totally oblivious of any ecological complexities or implications resulting from their visits and activities. Some types of sports and activities can be extremely damaging to the actual habitat and in some cases long-lasting effects can arise.

CLASSIFICATION OF URBAN PARKS

The origins and evolution of most areas now used as urban parks can be very diverse. The earlier breed of Victorian parks was largely modified from designs of private estates and, although many of these still provide fine landscape features, they have limitations in terms of use in relation to changing needs. The landscape content, especially the trees, is reaching a mature condition and the extent of successional planting has not always been adequate. Smaller urban parks or recreation grounds have developed, where active sporting facilities are provided for local communities, e.g. the conventional bowls and tennis areas with fringe landscape treatment and minimal social facilities. The expansion of urban areas has resulted in many former private estates being incorporated into the expanding towns. Many of these parks have provided excellent opportunities for a range of urban recreational facilities, e.g. the development of major golf courses, animal parks and even zoos. On the other hand, limitations of the initial design have not always been overcome, and the present needs for outdoor promotions in parks are quite sophisticated. Large areas for car parking are necessary; activities have to be held in a manner which does not contravene the various legal controls on noise and air pollution; environmental health regulations lay down stringent requirements in terms of temporary catering and the provision of toilets. All this requires sites more to the standards of the Milton Keynes Bowl. The ultimate development in parks arrived in Liverpool's International Garden Festival, opening in 1984. Ideas which have already been used in Europe for some time are being adapted and developed for Liverpool, Stoke-on-Trent and Glasgow.

In the last decade, the renewal of inner city areas, which previously suffered from considerable environmental and social deprivation, has sprouted a new generation of urban sports parks. There have been opportunities to provide specialized facilities—especially with floodlit sport areas, synthetic surfaces, and children's playgrounds—for the local communities within a total landscape concept. In some quarters, however, adverse comments are made on the grounds that these so called environmental improvements provide a very inferior type of landscape provision. These developments are part of the overall environment, with housing and industrial development being integrated into a parkland setting (Fig. 16.1), which leads to integrated maintenance techniques being applied to larger land areas, and conservation activities in a real and proper site instead of by artificial provision in parks and schools. The final product, a complex of many different facilities and areas, can be seen in any city, for instance Sheffield (Sheffield, undated) (Fig. 16.2).

SOCIAL AND POLITICAL OBJECTIVES

A definition of management suggested by the American Society of Mechanical Engineers seems appropriate in relation to this discussion. The definition

FIG. 16.1. The new landscape: a redeveloped amenity surrounding car-parking in an industrial area of Sheffield.

FIG. 16.2. The park areas of Sheffield. The total parkland of a present-day city is very complex, with many different areas catering for very different needs and uses.

reads 'The art and science of organising human effort applied to control the forces and to utilise the materials of nature for the benefit of man'. Applying this to parks management, the aim of a responsible and sympathetic manager is to balance the conservation of the natural resource and the degree of human usage to which the site is subject. This re-echoes what has already been argued by Kirby (Chapter 10).

The manager has firstly to deal with political objectives set by the local authority. The political dimension has become much more acute in recent decades, with interpretations ranging from the monetarist right to the extreme socialist left. The manager is also required to apply business acumen and professional skills in delivering a service to the community which has a wide variety of attitudes and needs. Recent legislation, as in the Health and Safety at Work Act, imposes general duties of care, qualified by the phrase 'so far as is reasonably practicable'; it requires all urban parks to be managed in a safe and functional manner, to have well documented operational systems, and to be monitored to ensure safety. Much more emphasis is being given to such matters, e.g. in children's playgrounds the provision of rubber shock-absorbent safety tiling or 'pour on' surfaces. The use of low lead paints is a current area of concern in the routine maintenance of heavy duty play equipment.

The attitudes and patterns of behaviour of park users are equally changeable. Society is tending to have a decreasing respect for law and order and for conventional standards of control (Fig. 16.3). Recession has polarized the situation, and there tend to be different approaches from young persons in employment with a purposeful approach to life compared with many unfortunates who, at the present time, are without work and tend towards anti-social behaviour. Urban parks have always been the gathering point for a limited number of social outcasts, and the gathering of the drop-outs, drug addicts, meths drinkers, perverts and gangs of mindless hooligans have to be controlled by the combination of urban park security services, private organizations and the police force.

The presence of such social outcasts can deter the public from using certain parks and requires careful handling. Nowadays bye-laws, rules of management, etc. are less respected and local authorities are often unwilling to undertake expensive court proceedings unless the situation is very serious. The park keeper with pointed stick has been superseded by mobile dog patrols and in some cases surveillance organizations covering all properties owned by the local authority.

The political dimension requires explanation as the park manager often has little say in the policies by which the urban park is to be managed. Some examples of recent policies introduced in Sheffield show how managers are

FIG. 16.3. A decreasing respect for law and order: rubbish dumped in a well managed amenity area in Sheffield.

given some guidelines to the political aims. Unfortunately this is not necessarily accompanied by the appropriate finance.

1   To develop modern purpose-built play sports areas to complement the existing series of Victorian parks and to consider the varying needs of different age groups.

2   To develop a policy for unemployed sports recreation coordination and to attempt to encourage unemployed people of all ages to participate in group sports and recreation activities.

3   To provide space where particularly noisy recreation can take place without disturbing local residents.

4   To develop a leisure plan for the city, a key to which will be the provision of all-weather high quality athletic provision.

FINANCIAL CONTROLS

The present battle between local and central government is over the right to establish the level of local government expenditure. The controversial views of central government in rate-capping are deplored by some large authorities; they consider the present policies to be a threat to local democracy. Whatever the political situation, urban parks can be considerably influenced by periods of financial cut-back. As a result, standards of maintenance and the range of facilities provided are, all too often, cut. This situation is all the more ironical, as it is generally accepted in recessions that urban parks absorb a great number of the people who are unable to afford holidays outside the cities and that, as a social resource, the urban park becomes much more important in periods of recession than in periods of comparative prosperity.

The process of managing cuts is one in which the politicians, professionals and trade unions have to attempt to control expenditure between many options. They have to consider what services are essential or mandatory; what activities can be stopped or reduced; what functions can be performed by others or done more effectively; what operations or promotions can become self-financing or show minimal loss. In some cases it may be possible to introduce additional sources of revenue.

MANAGING THE NATURAL RESOURCE

The average town dweller has little or no appreciation of the components of an urban park in terms of their role in an ecosystem, and only a few are likely to understand that pressure of use can have considerable influence on the ecosystem. The local conservation lobby discharges an invaluable role in raising issues and the professionals, including planners, landscape architects and parks officers, attempt to respond. There has been slow acceptance, other than in New Towns, of the specialism of landscape ecology, and the degree of resistance to change is not only related to shortage of resources for additional staff. The voluntary sector, including enthusiastic members of natural history societies, can provide information but, in general, the quality of organization of concerted effort is variable and documentation insufficient or patchy.

The parks manager wants (i) to know what natural resources require conservation and (ii) to predict the ecosystem sufficiently to prepare a management plan. Obviously the professional can now muster the aid of computers in prediction calculations, but such technology is not yet available for urban parks. At this stage a compromise is possible. Parks staff can prepare operation plans listing sites of interest previously recorded by the voluntary organizations; these sites can then be monitored in conjunction with the natural history societies and agreed maintenance undertaken. It is possible to identify a limited number of typical ecosystems in urban parks and associated open spaces which could provide a resource for the community. They could also provide opportunities for further studies and monitoring exercises, in conjunction with a range of interested organizations and educational establishments concerned with environmental studies.

It is traditional to accept that many parks will experience excessive wear in the process of routine use, e.g. in sports areas for team games, the worn-out areas of goalmouths or central areas in football, or the problems of divots from golf tees. In the process of routine greenkeeping, methods have been developed, and trained staff using specialized machinery can maintain these areas to reasonable standards. However, i.. recent times the pressures on the general park areas have been intensified, especially from sports such as BMX and trial motorcycles, which have resulted in appreciable physical damage. In the case of sloping ground, the erosion of the broken surface can cause considerable damage. The importance of having car-parking in urban parks is now well established but, on occasions of outdoor promotions, the use of large sections for car parking can consolidate the ground and, in inclement weather, can do appreciable damage.

In woodlands, individual damage to trees can result from soil compression due to the volume of visitors. Areas developed for children's play or adventure playground can have ropes, nails and timbers which are hardly beneficial, in particular to ageing trees. The use of amenity woodlands for orienteering, especially if large rallies are being held in bad weather conditions, can result in considerable consolidation and damage to the woodland carpet. Horses and ponies are acceptable when ridden on bridleways, but when taken indiscriminately into woodlands they can destroy the woodland carpet and cause excessive consolidation, leading to the death of trees.

The parks manager has to balance the positive recreational use of the urban park against the comparable loss of amenity due to the upset of the ecosystem resulting from the pattern of use. Some may wish to over-simplify the difficulty of this responsibility and may consider that more areas of urban park should be left natural. From a practical viewpoint this approach has many limitations; taking the example of long grass, this can (i) be an environmental hazard, (ii) encourage deposit of litter or major dumping depending on the length of grass, (iii) create a major health hazard for sufferers from hay fever, and (iv) constitute a major fire hazard in the spring and autumn, especially in dry seasons. In more wet seasons the long grass discourages use of the areas. In general, most people like long grass and have a wish to see it retained, but not at their back door.

The largest component of any urban park is grass and the regime for cutting requires to be carefully coordinated with the use of the area. Nowadays there is a wide range of grass-cutting equipment, ranging from cylinder, rotary and flail, and contractor-operated machinery, as well as a variety of motorized equipment for edging and trimming. The parks manager is required to produce a maintenance plan which sets out clearly the working plans of the areas and the rotas.

The manager may be pressed to consider wider implications, and there is a responsibility to consider such questions as the aims of the maintenance in relation to the landscape plan. This area is one of professional conflict between the landscape designer, with his concept of what the design should lead to, and the horticulturally-minded parks manager, who may be more concerned with day-to-day operations than with predicting whether the results of maintenance will produce the original vision of the landscape architect. There is an overall responsibility for the designer and the maintenance officer to work in close harmony. The professional institutes must continue to liaise and strive to have greater understanding of each other's philosophy; only when working together will they ensure that the design and maintenance aspects do not become rival camps but, rather, a common area for dialogue leading to better professional working relationships.

MANAGING THE PARK USER

Gone are the days when the average park user adopted a subservient attitude to the manager. The attitudes of the younger generation are all about public consultations: participation is the order of the day. Any wise and realistic parks manager will ensure that, along with his manual staff, he has close contacts with communities using the services of the park.

There are many advantages in involving the communities, in particular with summer activities, e.g. galas, picnics, children's events on the one hand and on matters concerned with the environment on the other. Many parents can adopt new attitudes to the park if their youngsters have been involved in nature study, or tree or bulb-planting and in some cases this may be done in conjunction with local schools. The involvement can include having a say on redevelopments within the park or in the provision or construction of facilities, e.g. in the siting of seats, or in the provision of a BMX track for youngsters. Another emotive area is concerned with the control of dogs. This is a problem in which much has been done in recent times, by exercising a greater control over dog owners and in educating the general public. The general public can also be involved as members of municipal sports clubs, in operating facilities within parks. Although privatization is not in universal favour and is very much resisted by trade unions, this form of practical community involvement is on the increase.

There are many aspects of the operation of urban parks in which specialist advisers, some professional and some amateur, can be involved and the local skills can be used to help to maintain the quality of the park. In Sheffield, the operation of the Amenity Woodlands Advisory Group over the last 10 years has provided management assistance in the preparation and monitoring of semi-natural amenity woodlands which are a feature of the Sheffield parks system. Representatives from local authority departments concerned with the environment, e.g. Planning, Estates, Recreation, Museums and members of voluntary organizations and Sheffield University staff have made a significant contribution. There have been experiments carried out, in conjunction with Sheffield University, by Dr Oliver Gilbert, and new glades have been created which are considered to be more resistant to recreational pressures than previous examples (Gilbert 1982). There have also been investigations with Peter Conlon into the establishment of trees and shrubs in peat blocks, and into the use of herbicides to control the ground flora, to maximize growth of woody stock and to maintain variety of flora.

The provision of an advisory service for the community, to assist ratepayers in general with advice on the management of their own gardens and trees, is another valuable service, which results in the public being more appreciative of the natural resources and becoming more responsible. A recent summary of facilities provided by local authorities appeared in two issues of *Gardening Which*, and indicated the range of educational support given by local authorities.

Public relations are another essential activity for the parks manager, who must cultivate sound relations with the media, local radio, television, etc. He must be able to respond to whatever situation arises, to explain developments. He may be involved in controversial issues and, in some cases, have to supply detailed background information for politicians to present to the media. Routine decisions of the local authority, to graze public moorlands with sheep or to fell trees, may result in considerable public reaction, and the manager must be competent to present the professional view in a responsible way. The involvement of local groups, specialist societies, and the provision of press releases on the matter, are all part of modern management. People will respond to, or at least attempt to understand, a decision if they know what is happening. Information, if presented correctly and appropriately, can defuse many difficult situations.

MANAGING THE WORKFORCE

Urban parks are by their very nature labour intensive. Average labour costs may contribute 50–65% of the total revenue budget for operations. The percentage, of course, can be varied; present-day trends are realistically reducing the previous excessive attention to seasonal floral effects. The staffing levels in general in urban parks in the early 1960s were high compared with the present, and it has to be admitted that overmanning existed but was not always recognized. The relatively low wages in local government manual service compared with industry, resulted in considerable dissatisfaction. As a consequence, introduction in the late 1960s of work study and bonus schemes resulted in increased performance and decreased manning. The threat of making additional payments in lieu of bonus unwisely forced the pace of introduction in early 1972, and resulted in too many open-ended schemes which are now receiving attention from the Audit Commission. Defects have been experienced in bonus schemes and few local authorities would now introduce such a system. Greater emphasis is now being given to rationalization, simplification, or alternative means of incorporating the productivity payments in a new grading structure for staff.

The beginning of the recession in the late 1970s resulted in local authorities cutting back on general expenditure in parks and playing fields, with many adopting policies of reduced hours of opening and reduced manning levels. At the same time, large numbers of unemployed, especially young people, have been involved in Job Creation Schemes, Youth Opportunities Programmes and now in Youth Training Schemes. The manager has had the difficult task of controlling and eventually maintaining new developments arising from these schemes while, on the other hand, attempting to cut back established facilities.

In the early 1980s the level of use of recreation facilities, with over 3 million unemployed, has accelerated to a stage where wear and tear on facilities has been considerable; municipal golf courses, for example, may now be absorbing 60–65 000 rounds per year compared with *c.* 40–45 000 a decade ago. Although cuts are the order of the day, in many cases extra staff are required to ensure average maintenance of the facilities. The additional complication of anti-social behaviour by the users does complicate the attitudes of the workforce. With the present power of trade union activity, withdrawal of labour under certain extreme circumstances is becoming a more common practice which can only be resolved by detailed consultation with trade unions, politicians and the community at large.

The concept of industrial democracy, in which the workers have a right to be involved to a greater extent in the day-to-day affairs of their work, is now accepted and is an essential way to get full cooperation of workers. This can be time-consuming, but undoubtedly many operatives have useful information for management, and team work in this respect is essential. Workers nowadays are being given every encouragement through formal Joint Consultative Committees to cooperate and sometimes to challenge traditional management attitudes. The role of the shop steward and, in large departments, of the convenor of shop stewards is an essential process of the development of industrial democracy.

TRAINING

In the past, opportunities for training of operatives have been minimal, but this has been greatly developed during the last two decades through the operation of the Local Government Training Board; there are also various modular schemes which enable the workers to be trained after induction, and to graduate through varying grades with increased responsibility and remuneration. Most large Parks Departments have a training section with Training Officer and Job Instructors, and more emphasis is being increasingly given to the training of staff on the job. Apprenticeships are now being replaced with placements within youth training schemes; the resources of the horticultural colleges provide very valuable opportunities for those entering the profession. Specialized courses are required to instruct in new technology and in Health and Safety matters. Possibly a more advanced training in principles of landscape ecology would be of long-term benefit. More advanced training in management beyond basic professional training is now being developed in most large authorities, for both junior and middle management, and in some cases corporate courses are provided for the whole authority. Local Government Training Boards are also encouraging management skills development programmes. In these, through a process of action learning, selected managers work through projects of concern and interest to the authority, and, at the same time, with the help of polytechnic lecturers, develop their management abilities.

OPERATIONAL MANAGEMENT

The operation of urban parks must be based on sound administrative processes with the assistance of modern equipment. The back-up of computers is now being applied to control general capital and revenue programmes, to monitor stores, and to keep records of nursery stock, transport, etc. Individual section managers should be trained and accountable, with responsibility for not only horticulture and arboriculture but also financial and manpower budgets of work programmes.

In most large authorities divisional staff work through line managers and have responsibility for a number of urban parks and possibly school playing fields and housing estates. The whole system of operation requires carefully coordinated deployment and operation of staff, transport and materials. The departmental back-up needs to be well organized for nursery production—either within modern glasshouse units or by purchase from the commercial nurseries—especially for trees, shrubs and ground cover. It also needs to be well organized in the provision of a thoroughly efficient workshop service to ensure that all machinery is in good working order. Although the present-day tasks of a park's manager are fraught with many pressures, there have been positive advances in recent years: modern technology can help to make his/her operational function more productive and positive than in the past; modern horticultural technology provides a wide range of mechanical equipment for maintenance; improved strains of ornamental plants and grasses provide material for landscape treatment; improved fertilizers and a range of chemicals are available for use in weed and pest control, subject to extreme care in application, dosage, etc.

CONCLUSION

Although there may be some justifiable claims that many urban parks are obsolescent and in a sense a neglected heritage, it must be accepted that parks continue to perform a vital function, although they may suffer from vandalism, some degree of neglect and, by their age, some degree of obsolescence. They still have a significant function to perform in providing recreational space and conserved environmental ecosystems which are essential in urban areas.

Due to changing attitudes to other people's property and persons, social considerations are becoming increasingly important to the parks manager. In his relations with the workers, joint consultation is the order of the day and a participative style of management is essential. The general community must be involved and, above all, there must be continuous and effective public relations exercised amongst the community using the urban park. Unless society can be encouraged to accept personal commitment and interest in the environment, there is little hope for the quality of life or for the continued useful functioning of urban parks.

REFERENCES

Gilbert, O.L. (1982). Management of urban woodlands in Sheffield. *Ecos*, **3**, 31–34.
Sheffield, City of (undated). *Recreation and Leisure in Sheffield*. British Publishing, Gloucester.

17. DESIGN AND ECOLOGY IN THE MANAGEMENT OF NATURE-LIKE PLANTATIONS

R. TREGAY

Warrington and Runcorn Development Corporation, New Town House, Buttermarket Street, Warrington, Cheshire WA1 2LF

SUMMARY

Methods for creating more natural landscapes in urban areas are being sought. The dynamic character of these landscapes implies that the design process continues into after-care. Inspiration can be found in nature and in established landscape design concepts. Ecology and related earth-sciences convey soundly-based working principles. The urban situation then imposes complex functional demands on the new landscapes. Drawing together creative, ecological and social factors will require a new breed of urban landscape manager. This is illustrated by reference to the creative development of vegetation structure in nature-like plantations.

INTRODUCTION

The last decade has seen the development of new ideas governing the extent and character of green space in urban areas. Nature is increasingly regarded as an essential component. There are several reasons for this (Tregay 1982) but two stand out. The first is that the great majority of people live in towns and cities and many of these urban areas are manifestly failing as a stimulating environment for habitation. A strong natural component in the outside environment can create diversity, richness and sheer delight, and for some people it can help satisfy a deep emotional need for contact with nature. Secondly, urban nature is a vital and proximate resource for environmental education. Through urban nature, people can begin to understand from first-hand experience some of the environmental issues of global concern.

New design solutions and redirected management policies can create more natural, usable, cost-effective landscapes as part of the overall provision of open spaces in a town which includes playing fields, gardens, and formal spaces. Diverse ecosystems can be introduced as an integral part of the urban fabric, bringing the special qualities of nature into people's daily lives—in parks, school grounds, residential areas and in landscapes associated with places of work.

A key element in the provision of these new landscapes is nature-like plantations of trees and shrubs with associated herbaceous plants which can form woodland and forests, as well as smaller copses, narrow shelterbelts and incidental areas of trees. Much experience has been gained in establishing commercial plantations and in managing existing natural woodlands, but far less is known about creating new plantations with a natural character, particularly in the difficult situations found in urban areas. Here, different priorities and principles are involved, leading to methods not normally associated with the typical Parks Department (Scott 1983).

Urban woodlands have been created before, of course: Vestkoven, west of Copenhagen, and the Amsterdam Bos are examples (Tregay 1979). The Gilles development in Delft set about creating a woodland in a housing court (Ruff 1980). Similar small-scale nature-like plantations are being created in Sweden (Gustavsson 1982a) and England (Tregay & Gustavsson 1982; Greenwood 1983). These and other examples depend for their success upon skilled management programmes to create ecological and spatial diversity. Vestkoven has been managed along rather traditional forestry principles and lacks diversity of habitat, structure, age and spatial character. Despite the vision of its planners, the end-result is sadly rather uninteresting. Much of the Amsterdam Bos suffers from the same lack of diversity in detail (Fig. 17.1). Similarly, the radical design concept at Gilles has not, arguably, been carried through into the management phase.

These examples provide clear pointers for the newer nature-like plantations such as those of Oakwood in Warrington (Fig. 17.2). Indeed, the landscape team at Warrington is aware that, even now, some bold management decisions are necessary to create more spatial variety and clarity within mixed-species plantations.

This paper discusses one technical aspect in the management of nature-like plantations—the creative development of vegetation structure—and will endeavour to illustrate the role of an ecological approach and the need for an on-going design process. To provide a context, fundamental differences between maintenance and management-oriented approaches to after-care are first examined.

AFTER-CARE: MAINTENANCE OR MANAGEMENT?

Maintenance is normally considered to consist of the complex of everyday operations involved in the care of a landscape. Management implies

FIG. 17.1. Many of the woodland plantations in the Amsterdam Bos are characterized by a lack of structural diversity.

operations governed by a long-term strategic view, in which fundamental changes in the structure, composition and character of the landscape take place. Current practices of landscape after-care often fail to recognize the need for management in general, and more specifically the need for it to exploit the dynamic qualities of vegetation processes and to use them to creative ends. There are several reasons for this, including stretched resources which only permit the carrying out of basic operations, simply to 'keep the place in order'. The problem of inadequate resources for landscape management is usually justified by financial stringency, although the real reasons are more deep-rooted and arguably related to a lack of political vision and professional leadership. The latter problem is itself rooted in the professional divisions between those who design and those who maintain the design, i.e. landscape architects and landscape (or parks) managers. This division is historical in Britain and it continues to help perpetuate a widely held assumption that a landscape design contains a germ that will lead inevitably to one sole end-result: the Grand Design. Given this premise, landscape after-care continues to be seen as a 'holding operation', maintaining, retaining, holding on (at best) to what the designer hands over. The system has perpetuated itself. Designers are reluctant to design dynamic landscapes if these are to be maintained along

FIG. 17.2. In this 6-year old plantation in Oakwood, Warrington, management through thinning and coppicing has created an open canopy and mixed height shrub understorey, a structure which provides shelter from wind for the houses behind and a dense habitat for wildlife. Through a creative management programme the objective is to achieve a variety of plantation structures throughout the area. The structure of each plantation will be related to functional and visual requirements.

traditional civic lines. Similarly, managers find it difficult to creatively develop landscapes which are designed to be static and inflexible. Nowhere has the political or professional will been strong enough at national level to break into the system and generate comprehensive change.

For these reasons, design and management tend to continue along separate roads in most British towns and cities. This is a major obstacle to change, to the introduction of design solutions which demand creative management expertise. In some minds there is, therefore, a strong case against nature-like landscapes, on the grounds that existing systems will not cope. This outlook, although seemingly quite realistic is, in fact, a negative one. An alternative viewpoint is that the well marketed introduction of nature-like landscapes will, of necessity, generate change in management procedures and ultimately bring about a union between the design and management professions. One can already detect hopeful, albeit isolated, signs pointing in this direction.

It is worth examining why nature-like landscapes in particular hold the key to this union, as well as creating a higher quality urban environment. All landscapes, including playing fields and gardens, require expert management and none will thrive when a maintenance regime is pursued irrespective of changing circumstances. Nature-like landscapes demand a special and creative kind of management expertise—even though they can demand little maintenance in terms of time and cost—because fundamental to them are the dynamic qualities of vegetation development and succession. Natural systems are ever-changing; a response to both external influences and internal developments. Landscape as a dynamic system is, therefore, a central concept for the designer. The design, although the single most important vehicle of creative intent, can also be viewed as a starting or change point in the continuous development of vegetation, soils and other natural components, and expressed only for convenience of implementation and illustration in static form—as a drawing. The landscape manager is concerned with the same dynamic qualities. Management in this sense is the design process put onto a time continuum: a kind of on-going design. Design and management are thus inextricably linked. The tools differ, the emphases differ, and the working day for those involved differs, but the basic principles are the same. For both the designer and the manager of nature-like landscapes, artistic creativity and a scientifically-based ecological approach become fused into a single way of thinking. This unity of landscape creation can best be illustrated if we narrow discussion to some detailed aspects of plantation management.

CREATIVE DEVELOPMENT OF VEGETATION STRUCTURE

Gustavsson's work (Gustavsson 1981, 1982b) has concentrated our minds on the concept of vegetation structure: the three-dimensional composition of trees and shrubs. His work is especially interesting because it proposes a simple language in which objectives for management can be defined. Most significantly, the language of vegetation structure is based on visual criteria, natural to the designer and easily communicated to the manager. It is a language which finds parallels in architecture, trunks and foliage being synonymous with walls and roofs. Scale, enclosure and complexity can be manipulated to create infinite design variation. It is through vegetation structure that the designer can convey, and the manager can achieve, creative intentions.

The design of a plantation relies also on an ecological understanding, which will enable the system to be set up in such a way as to encourage the development of desirable natural changes. It also ensures that the appropriate species are chosen for the site in question and that they will grow compatibly with one another. Design objectives in terms of species composition and succession require another language, that of the ecologist. Through this language, information is conveyed to the manager which concerns medium and long-term objectives for species associations and habitat creation.

The two languages have to be considered together, in practice. In particular, the objectives for plantation structure can only be achieved if the ecological relationships are appreciated. Here then, we have the basis of a creative approach to management in which the languages and objectives of the designer and ecologist can be integrated.

To date, the design and management of plantations of native trees and shrubs have taken an ecological slant. For example, a mixture of species, planted together, could be developed along successional lines with the pioneers, such as alder, gradually being thinned and replaced by climax species. The management objective might be, for instance, a canopy of oak and ash, with a hazel–holly shrub layer. This may be perpetuated on a group-selection system. Statements such as this are certainly useful to the manager, but they are nevertheless limited in respect of creative objectives for vegetation structure.

Developing structural variation through management is best illustrated by an example. In practice, vegetation structure can to a degree be anticipated at the design stage through the use of different plant mixes. However, for the sake of illustration, we shall assume as our starting point a single-species tree plantation. Of course, most plantations will involve species mixtures and, therefore, involve more complex ecological developments. Figure 17.3 illustrates how this simple plantation can be developed over time into a variety of quite different vegetation structures. In the first case (a), thinning is late and light, thereby leading to the development of straight, closely-spaced trunks with little foliage below the high canopy. At the other extreme, illustration (b) shows how an early and heavy thinning allows the lateral spread of single trees which develop a quite different form to type (a), and create large open spaces beneath a heavy, low and arching canopy. The thinned trees may regrow but will eventually be suppressed by shade. The third type of structure (c) is what Gustavsson would term a 'low woodland type' and is developed through a cyclical coppice system. Here, a dense mass of trunks at low level is created, with fairly intensive shading by a canopy still largely above eye-level. In type (d), heavy thinning creates a high canopy which lets in sufficient light to permit a one-level coppice understorey. The final category (e) is the most complex in structural terms, and is developed by a programme of thinning and coppicing, creating many layers of vegetation with great age diversity. The vigour of each layer depends largely on available light levels, which are controlled by the density of the canopy above, and on the shade tolerance of constituent species.

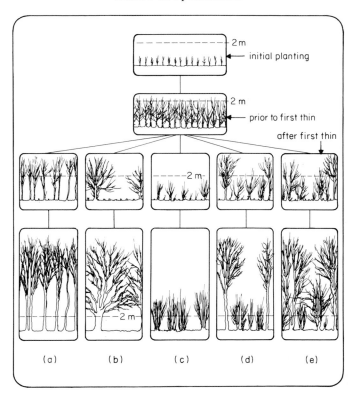

FIG. 17.3. Creative plantation management: through continuing the different thinning and coppicing regimes five structures (a)–(e) are achieved.

This example, although involving only one species, still illustrates the ecological basis of management. The plantation is perceived as a composite mass of trees, each individual competing for light and space, each interacting with its neighbour. In a mixed-species plantation, these interactions are far more complex, and a much deeper understanding is demanded of the landscape manager.

Figure 17.3 also illustrates the range of functional and creative possibilities presented through the practice of thinning. Type (a) structures allow views through shaded, closely-spaced trunks. Type (b) creates a forest 'room', and type (c) a low, dense mass with no high level forest screen. Type (d) can be used for screening or shelter from wind since a mass of foliage is retained at both low and high levels. Type (e) is the most dynamic structure. It is an open, high canopy with a dense, vigorous mass of vegetation at mid and lower levels, well suited to situations of heavy use where continuous cover is required.

Within a large plantation, all the different regimes identified in Fig. 17.3 can be combined, creating diversity, even if only one species is used. Along a footpath route a constantly changing pattern of light and shade and spatial qualities could be created. Perhaps the plantation is entered through a dense barrier, type (e), which controls access points and forms the 'walls' to a forest room, type (b). Further on, a screen is formed by type (d) and, behind this, is a complex pattern of open ground and low forest, type (c), formed by cyclical coppicing on a patchwork basis. Finally, structural type (a) can be used to dramatic effect for the transition between the shaded plantation and open ground beyond (Fig. 17.4).

Time, as well as felling and coppicing, will change vegetation structure and spatial qualities in a plantation. Simpler vegetation structures are especially ephemeral in this respect. This is well illustrated in a primarily commercial forest such as Delamere Forest in north Cheshire, which is managed by the Forestry Commission. Here, sensitive but decisive clear-felling on a patchwork basis, together with the retention of key tree groups creates an ever-changing landscape of great spatial variety (Fig. 17.4).

FIG. 17.4. Dramatic spatial qualities in rural Delamere Forest, the result of sensitive commercial management. Simple vegetation structures such as this have their place in urban parks and other green spaces, especially when used creatively in a planned relationship with other structural types.

This discussion, as most landscape architects will appreciate, is framed in a designer's language. This is not to deny the role of the silviculturalist or ecologist, but rather to provide another outlook which can help broaden our perception. Many other examples could be given, involving plantation management and the management of meadows, water edges and so on. Each would involve the same balance of ecological understanding, creative intent and functional appreciation.

To lay such emphasis on the dynamic qualities of plantations, and therefore on dynamic and creative management, does not admit a diminished role for the designer. Certainly, a plantation could be developed entirely through management, by relying on natural regeneration, instead of planting. Generally, however, the initial design will be an exercise of careful creative intentions for it is at the design stage that the main structure and composition of a new landscape is laid down. No one who sees, for example, the Dutch heemparks can believe that they are the work purely of a technician. In a similar way, the importance of the initial design and of creative intent has been stressed for the nature-like plantations in Warrington (Tregay & Gustavsson 1982).

DISCUSSION

The principles of plantation management discussed here include a number of detailed issues. Nevertheless, they are illustrative of the many new concepts and technical practices that need to be introduced if nature-like landscapes are to find their place within the fabric of green spaces in urban areas. This follows because the objectives and methods used in managing nature-like vegetation differ markedly from those currently employed in the management of most designed urban landscapes. The differences are so marked, in fact, that one can envisage within a typical Parks Department three separate teams dealing with different categories of landscape: 'natural and nature-like', 'garden and formal' and 'sport-related'. This would encourage specialism and the development of expertise, especially at landscape supervisor level.

New ideas need to be sold at all levels in local authority organizations. Senior managers need to be aware of the role of ecological approaches and nature-like landscapes within a broad overview, typically including indoor recreation centres and theatres, as well as parks and open spaces. Courses should be set up for those responsible for work on the ground, such as landscape supervisors and their manual staff. However, it is at the professional and technical level where the expertise is now most needed. It is, therefore, encouraging to hear that a number of university courses are endeavouring to establish landscape management options. Hopefully these will include concepts such as those discussed in this paper. If the enormous potential offered by natural design approaches is to be realized, and is to contribute to the real improvement of urban areas, we must now encourage a new breed of creative landscape manager.

REFERENCES

Greenwood, R.D. (1983). Gorse Covert, Warrington—creating a more natural landscape. *Landscape Design*, 143, 35–38.

Gustavsson, R. (1981). *Natur-lika grönytor i parker och bostadsområden.* Konsulentavdelningens rapporter, Landskap 58, Sveriges Lantbruksuniversitet, Alnarp.

Gustavsson, R. (1982a). Nature-like parks and open spaces in housing areas in Sweden. In *An Ecological Approach to Urban Landscape Design* (Ed. by A.R. Ruff & R. Tregay), University of Manchester.

Gustavsson, R. (1982b). Nature on our doorsteps. *Landscape Design*, 139, 21–23.

Ruff, A.R. (1980). *Holland and the Ecological Landscapes.* Publication available from the author, University of Manchester.

Scott, D. (1983). *Hand-over of New Town Landscapes to Local Authorities.* Paper given JCLI Conference, Liverpool.

Tregay, R. (1979). Urban woodlands. In *Nature in Cities* (Ed. by I.C. Laurie), pp. 267–296. Wiley, Chichester.

Tregay, R. (1982). Nature and an ecological approach to landscape design. In *An Ecological Approach to Urban Landscape Design* (Ed. by A.R. Ruff & R. Tregay), University of Manchester.

Tregay, R. & Gustavsson, R. (1982). *Oakwood's New Landscape: Designing for Nature in the Residential Environment.* Sveriges Lantbruksuniversitet and Warrington and Runcorn Development Corporation, Warrington.

18. WATERPLANT ECOLOGY IN LANDSCAPE DESIGN

J. W. EATON

Department of Botany, University of Liverpool, Liverpool L69 3BX

SUMMARY

Aquatic vegetation is an important landscape feature in many lakes, rivers and canals. The distinctive ecological characteristics of waterplants have to be considered in the design and management of such waterbodies for landscape purposes. Clear design objectives are required which take account of the ecological suitability and stability of the plant communities for intended function and also accommodate other hydraulic and amenity uses of the waterbody.

INTRODUCTION

'Water is the most interesting object in a landscape, and the happiest circumstance in a retired recess; captivates the eye at a distance, invites approach, and is delightful when near; . . . it may spread in a calm expanse to sooth the tranquility of a peaceful scene; or hurrying along a devious course, add splendour to a gay, and extravagance to a romantic, situation.'

In this way Thomas Whatley recognized the dominating influence of water in landscape in his *Observations on Modern Gardening*, 1771. This dominance has long been acknowledged by landscape designers, both in their careful treatment of natural waters and in their frequent construction of artificial ones (Jellicoe & Jellicoe 1971; Hackett 1976). They follow a very general, although almost unconscious, human attraction to water in landscape. From the obsession with brooding lakes in eighteenth century Picturesque, through the horrid Gothic torrents and placid Pre-Raphaelite waters of Victorian times to the endlessly sunlit scenes of twentieth century tourist brochures, there is an emphasis on water quite out of proportion with the areas which it occupies in the landscape.

The attention which focuses so strongly on water, inevitably also focuses upon its vegetation, giving waterplants and the water's edge fringe an influence in the landscape which far exceeds that of much greater surrounding areas of terrestrial plant cover. This influence can be positive, as for example in

285

graceful reedbeds fringing a lake, or in the delightful textural complexity of vegetation in small lowland rivers, or in the carefully chosen and precisely sited clump of lilies in a formal water-garden. It can also be negative when weed-infested or harshly eroding 'water features' ruin landscape developments and incur massive rehabilitation and maintenance costs.

This paper reviews some of the distinctive ecological characteristics of waterplants and water's edge vegetation, and suggests ways in which this information may be applied to the successful incorporation of aquatic vegetation into landscape design.

THE BOTANICAL RESOURCE TO BE MANAGED

There are more than 150 species of waterplant in the British flora (Haslam *et al.* 1975). Most are herbaceous perennials which die back to buried vegetative organs in autumn, leaving no visible structures during the winter months. Although taxonomically very heterogeneous, waterplants are divisible for present purposes into four forms.

    (a) Submerged: plant wholly underwater and usually comprises rootstock, stem and leaves.

    (b) Floating-leaved: buried rootstock produces stems or petioles which extend to water surface and bear floating leaves.

    (c) Free-floating: whole plant floats unanchored at the surface, with roots (if any) trailing freely in the water.

    (d) Emergent: buried rootstock with foliage mostly borne above the water surface.

Submerged species constitute about half the British waterplant flora and, although often inconspicuous in landscape terms, they are important as components in balanced aquatic ecosystems (Marshall & Westlake 1978), within which the more conspicuous waterplants are likely to be most successfully managed. In addition, they can be important for fisheries (Reynolds & Eaton 1983) and, along with other groups of waterplants, for nature conservation (Newbold 1981).

More immediately visible are the floating-leaved species. Waterlilies (Nymphaeaceae) have been highly regarded for their ornamental value from time immemorial. Indeed, these and other types of waterplants were incorporated into the highly sophisticated landscape management schemes of the Chinese Han Dynasty over 3000 years ago. In Britain only two species (*Nuphar lutea* and *Nymphaea alba*) are common, though various other families of rooted waterplants have species with floating foliage. In contrast, the tiny free-floating duckweeds (Lemnaceae) have little aesthetic appeal, and are generally considered objectionable when they multiply rapidly to form a continuous green carpet over the water surface, perhaps because their obscuring effect on what lies beneath them imparts a curiously sinister quality to the waterbody. In tropical waters, floating species form mats which disfigure water surfaces over very large areas; fortunately, this problem does not arise in temperate regions where short growing seasons and severe winters prevent build-up of mats (Sculthorpe 1967, p. 423).

Among emergent plants, narrow-leaved reed-like forms are important to landscape management, though the number of common species is again small. There is a taxonomically wider range of sedges, rushes and broad-leaved species, but these are usually confined to the extreme edge of the water, merging into the marsh and damp soil flora on river and lake margins.

In addition to the native flora, many species have been imported to supply aquaria and ornamental pools (Perry 1961). Proposals to introduce new species of unknown competitive abilities as part of landscaping projects should be considered with great caution for reasons of practical management as well as nature conservation. Most will probably be ill-adapted to their new conditions, there being a number of past introductions which survive in Britain only as very local rarities. However, there is always a risk of releasing a well adapted new species capable of explosive spread. A recent example is the spread of *Myriophyllum spicatum* through the Okanagan Basin in British Columbia, where the severe consequences have not proved controllable despite costly research and management efforts (Scales & Bryan 1979; Dove & Wallis 1981; Dearden 1982).

Although such species often decline after a period, they may continue to cause localized problems at any time, e.g. *Elodea canadensis* in Britain (Sculthorpe 1967, p. 360). The other group of plants which affects the appearance of waterbodies is the algae. Filamentous forms produce unsightly floating mats and phytoplankters occasionally achieve similar effects when they congregate at the surface as scums (Reynolds 1971), but more generally the algae may tint the water green or brown (Reynolds & Walsby 1975).

SOME ECOLOGICAL CONSIDERATIONS

Depth and water movement

Light is absorbed quite rapidly by water and beyond a certain depth illumination becomes insufficient for growth of submerged plants. In clear, uncoloured lakes this depth can be 8 m or more, but in peat-stained waters and in the plankton-rich lakes and silty rivers typical of lowland Britain it may be reduced to 2–3 m or less. Thus, submerged vegetation is commonly confined to a peripheral band (the photic zone) close to the shoreline. Floating-leaved and emergent rooted plants are limited by the length of stem or leaf-stalk which they can produce to reach the surface. For lilies this rarely exceeds 2 m. Common reed (*Phragmites australis*) sometimes extends to 2 m depth at sheltered sites, but other common emergent species such as *Sparganium erectum*, *Typha latifolia* and *Glyceria maxima* are limited to shallower waters. Clearly the width of the vegetation band depends upon the underwater depth gradient. It can be many metres in gently sloping areas, but may exist in highly compressed form on even quite steep gradients, provided these are physically stable. Thus, the diverse fringes sustained along some lowland waterways produce an attractive but very compact landscape feature (Fig. 18.1).

Superimposed on depth limitations are further restrictions due to water movement. Exposure to even quite moderate wave action reduces the maximum depth for *P. australis* to 1 m and more severe exposure prevents colonization altogether, maintaining a bare, often coarse substratum from well below the normal water level to some distance above it. The control of waterplant zonation in lakes by the interactions of species with depth, light penetration, water movement and the stability and particle size composition of the substratum is discussed by Spence (1982).

For rivers, Haslam (1978) describes how the scouring action of flow similarly restricts colonization even in shallow, clear waterways where depth and light penetration are not limiting. Emergent plants are nearly always confined to river margins and are absent altogether in lengths where eroding flows extend across the whole width of the channel, either continuously or during spates. Some submerged species are confined to areas of slow flow, others colonize into faster flows, but only where suitable substrata are available for firm attachment. As with exposed lakeshores, high proportions of bare substrata have to be recognized as a normal feature of some aquatic systems, in marked contrast to terrestrial systems where exposed soil is usually a sign of a degraded landscape.

Water and sediment quality

Natural waters and sediments vary widely in chemical composition. Studies reviewed by Hutchinson (1975, pp. 369–398) and Haslam (1978, pp. 120–149) have shown that adaptations to particular ranges of chemical composition are important influences on the species distribution.

In relatively undeveloped areas, especially upland ones, the qualities of both water and sediment can vary greatly between catchments, depending on geology, soil type and vegetation. Lowland waters are often naturally ion-rich, and agricultural and urban drainage invariably add further ions. These predominantly 'eutrophic' conditions favour eutraphent species (those requir-

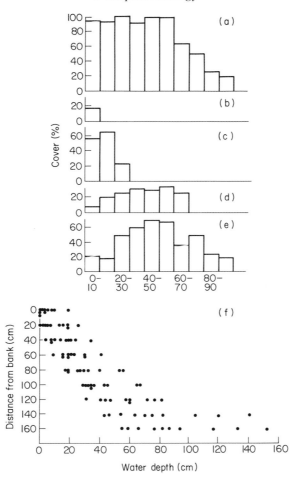

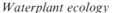

Fig. 18.1. A diverse and compact emergent plant fringe on a steep depth profile, Leeds & Liverpool Canal, Rufford, Lancashire. Surveyed September 1983 but known to have been stable for at least 5 years previously: (a) total cover; (b) *Iris pseudacorus* with small quantity of *Rumex hydrolapathum* and *Epilobium hirsutum*; (c) *Juncus effusus*; (d) *Sparganium erectum*; (e) *Typha latifolia*; (f) depth profiles from 10 transects.

ing high concentrations of ions) and eurytopic species (indifferent to ionic concentrations), and virtually exclude the 20–30 oligotraphent species in the British flora which are closely adapted to low-ionic, nutrient-poor (oligotrophic) waters. In extreme cases, treated sewage effluents or fertilizer-rich agricultural drainage may so increase nutrient loading that waterplants are replaced by massive blooms of phytoplankton and filamentous algae. Possible mechanisms of this replacement are discussed elsewhere (Bolas & Lund 1974;

Moss 1977; Phillips *et al.* 1978). It is sufficient to note here that this is an extremely undesirable trend for landscape purposes as well as for water supply, fisheries, conservation and other reasons.

Pollution is less widespread than formerly, but continues to exert local effects, usually reducing waterplant diversity and favouring tolerant species (Haslam 1978, pp. 350–367). Occasional serious pollution from accidental discharges is an inevitable risk in developed areas. Immediate effects may be devastating, but recovery usually occurs if the sediment escapes serious contamination and rootstocks therefore survive.

Change in aquatic vegetation

In water, as on land, vegetation is stable only so long as its environment remains stable. Alterations in depth, substratum, water quality or water movement usually cause changes in aquatic vegetation. This is most obvious in natural, unrestrained rivers where erosion and accretion processes constantly redistribute the areas available for colonization, and the position of vegetation may change markedly within a few years. In lakes, accretion exceeds erosion, causing them to be filled in by accumulating sediment. Inflows carry silt which settles mainly near their points of entry, creating marshy deltas whose rates of advance depend largely upon the siltiness of the supplying rivers. Less obviously, lakes accumulate sediment by precipitation of solutes from the water, principally calcium, iron and silica complexes, and by deposition of lake plankton and vegetation remains (Wetzel 1975, pp. 156, 170–171, 280–282; Davison *et al.* 1982; Reynolds *et al.* 1982). Accumulation from these processes occurs in deep waters and in sheltered inshore areas throughout the lake. Rates depend upon the chemistry and biological activity of the lake, but can be substantial in lowland waterbodies rich in solutes, plankton and vegetation. The resultant shallowing and shoreline advance is accompanied by migration of the vegetation zones. In some cases this is so slow that there is only a very long-term effect upon the landscape (Spence 1964). In others, such as small, shallow lowland lakes and very slowly-flowing channels fed by silty, solute-rich water, the vegetation may advance markedly within a few years and this strong trend towards change has to be taken into account in landscape management.

Most waterplant species spread vegetatively. Rhizomes or similar organs often achieve rapid linear advance and quite small vegetative fragments are capable of being transported by water movement and then rooting and establishing new stands. Seeds are also dispersed by water and accumulate in sediments where they may germinate in large numbers when conditions become suitable. These features provide effective survival and colonization mechanisms in eroding and accreting habitats. They also result in quick establishment of plantings for landscape purposes, such that a visually satisfactory appearance may be achieved within as little as one year.

Waterplant communities

Waterplant communities are therefore the products of interaction of species, climate, water and substratum quality with environmental gradients of depth, light and exposure to waves or flow. Although some sheltered waters support a classical series of vegetation zones, from submerged plants in deep areas through floating-leaved rooted ones to emergent plants at the water's edge, in many other waters the zonation may be so greatly modified and fragmented by local features that it becomes practically unrecognizable. This creates difficulties in the definition of communities, but some progress has been made using a variety of phytosociological techniques. Hutchinson (1975, pp. 486–508), Haslam (1978, pp. 204–308) and subsequent individual studies (e.g. Wiegleb 1981a, b) seek to define communities and in some cases suggest how they may form orderly successions over a period of time. The ecological processes distinguishing, sustaining and modifying communities are very imperfectly understood. Nevertheless, these sources are useful to the landscape manager for the interpretation of existing vegetation patterns, for planning planting programmes and for predicting likely changes in waterplant colonization.

SOME LANDSCAPE CONSIDERATIONS

The visual character of aquatic vegetation depends greatly upon the species present. For example, reedbeds of *Phragmites australis* are especially attractive because the individual stems are tall and graceful, giving stands a delicate visual texture. The shoots persist through the winter in a dead, yellow to light brown state, continuing the interest at a time when most other waterplants have disappeared. *Typha latifolia* and *T. angustifolia* have large, brown inflorescences which contribute a distinctive feature from midsummer to late autumn. The individual shoots are more robust than in *P. australis*, giving the stands a denser appearance in summer, and there is a gradual collapse and decay from late autumn onwards. *Glyceria maxima* is less attractive from a landscape viewpoint. Shoots are short and form very dense stands which collapse rather abruptly into an untidy, greyish-brown rotting mat in autumn.

Along the water's edge, species diversity and visual texture are more varied than in reedbeds. Clumps of broad-leaved species, such as *Rumex hydrola-*

pathum, contrast with the narrow foliage of 'reeds'. Marsh species contribute, when flowering, substantial patches of colours which are almost wholly absent from reedbeds, for example *Epilobium palustre* and *E. hirsutum* (pink-red) and *Myosotis scorpioides* (blue). Alder (*Alnus glutinosa*) and willows (*Salix* spp.) provide further contrast in size, shape and texture with the wholly herbaceous aquatic vegetation.

The appearance of a complete series of zones is therefore of an inshore progression from a featureless open water surface through uniform, finely-textured reedbeds to a diversity of texture and colour along the water's edge and on into the taller vegetation of bankside shrubs and woodland. The array of textural zones and the upwards curve of vegetation height are attractive features in the overall landscape. An observer walking near the water's edge is in a good position to appreciate these contrasts, especially where the reedbeds are not continuous and there are some direct views of open water and floating-leaved vegetation, for example where sheltered and exposed areas alternate.

Floating-leaved rooted species contribute most when present as small patches near enough to the shore to be observed fairly closely. Extensive, dense beds, by contrast, exert an oppressive effect similar to that of duckweed mats, by shutting out the open water surface.

A few species produce morphologically-different submerged, floating and emergent leaves and colonize a wide range of depths, although the different leaf types are not necessarily produced as direct responses to depth (Sculthorpe 1967, p. 224). An example is *Sagittaria sagittifolia* with ribbon-like submerged leaves in deep water, variously-shaped floating leaves and boldly arrow-shaped emergent leaves in shallower areas. Common in some rivers are batrachian *Ranunculus* spp. which have filiform submerged leaves and 'buttercup'-like emergent leaves. These types of plant can provide considerable visual interest along a water's edge.

Submerged species only make a positive contribution visually when observed closely from a water's edge vantage point. Then delicate leaf shapes (e.g. *Myriophyllum spicatum*) and brilliant green underwater patches (e.g. *Callitriche* spp.) can be attractive on a purely local scale.

MANAGEMENT

Rivers

The tendency of rivers to erode, accrete, shift position and flood has long been regarded as intolerable in developed lowland areas. Increasingly massive engineering works have been constructed to restrain flow and reduce flood hazard, creating straightened, dredged and embanked channels practically devoid of fringing vegetation. The visual effect is harsh and damage to the landscape is often very great. Even the visual softening of occasional bankside trees may not be permitted on the grounds that such trees obstruct access for maintenance and cause flood hazards. The landscape architect is presented with an intractable problem. Nature conservation and fisheries inevitably suffer as well through the ultra-uniform conditions and lack of vegetation cover.

In Europe, controlled planting methods have been developed for many decades to ameliorate these damaging side-effects of river management; in some cases simply to produce visual softening of the engineering works, but in others to actually replace the artificial materials with vegetation-based structures. Seibert (1968) provides an interesting and practical review, describing how different *Salix, Alnus* and hardwood associations are adapted to different functions. Fully-coherent softwood stands are considered to be good protectors of banks, even on the outside curves of strongly-flowing rivers where erosion is most severe. Various reed and sedge associations are specified for use where conditions are less demanding. Reinforced planting is used to protect the vegetation during its vulnerable establishment phase. Notable in this approach is the application of ecological information from the phyto-sociology and hydraulic attributes of the various plant species.

Binder & Gröbmaier (1978) and Pfeffer (1978) apply these methods to Bavarian rivers subject to flow regulation and channel re-modelling. Vegetative bank protection is used wherever possible to stop erosion. Where channels are straightened, meanders are retained as sidewaters. Precisely-deployed plantings create some of the ecological character of a more natural waterway. Pfeffer (1978) suggests that periodic vegetation maintenance is needed to preserve diversity. Otherwise with the natural erosion–accretion cycles suppressed, vegetation succession would proceed without interruption to woody climax communities of low diversity and reduced landscape and conservation interest.

Some similar schemes are now being tested in Britain. Their fisheries and wildlife conservation value is widely emphasized (Vivash 1982; Weeks 1982; Brookes *et al.* 1983; Newbold *et al.* 1983; Lewis & Williams 1984; Mason *et al.* 1984), but the potential for gains in landscape quality would also appear to be great (Purseglove 1983).

In urban areas, storm run-off is especially rapid and whilst flooding cannot be tolerated, visually intrusive river engineering works are also highly undesirable. A solution to this dilemma is to accommodate the spates in landscaped floodplain areas where space permits, with waterlogging-tolerant vegetation maintained where submersion is predicted to be frequent (Clarke 1973; Hengeveld & de Vocht 1982, Ch. 5).

A major reason for maintenance dredging of rivers is to clear aquatic vegetation which is obstructing water movement. Other methods of water-weed control have been proposed to avoid both dredging damage to the channel communities and also the need to clear riverbanks for dredger access. Spillett (1981) used wooden groynes as flow deflectors in a small, 'flashy' river to direct water away from vulnerable banks and on to areas of weed and silt accumulation, thereby keeping the channel open against encroaching vegetation during periods of low discharge and maintaining a diversity of water depths and flows. This retained good conditions for fisheries by avoiding the need for dredging and also preserved a semi-natural appearance. A second approach is to use bankside and fringe vegetation to shade the channel and thereby reduce waterweed growth (Dawson & Haslam 1983), an interesting contrast with the engineering practice of clearing such vegetation, which sometimes admits so much more light to the water that weed growth is exacerbated. Where trees are used, the required half-shade is achieved by discontinuous planting, mostly on one bank only, so landscape enhancement may be substantial. Indeed visually-monotonous continuous planting is to be avoided because it creates excessive shade for fisheries and too much leaf accumulation in autumn (Dawson & Kern-Hansen 1979). Hinge & Hollis (1980) provide a useful discussion and bibliography on the whole subject of land drainage, aesthetics, fisheries and conservation, where these various developments in vegetation management offer considerable possibilities for river landscaping in the future.

Disturbance of riverbanks during engineering works has probably assisted the spread of some alien species, notably Himalayan Balsam (*Impatiens glandulifera*) and, more recently, Japanese knotweed (*Polygonum cuspidatum*). Whilst attractive in limited clumps, these plants become visually oppressive when they dominate long lengths of bank as dense, monospecific stands. Control is very difficult, other than by drastic herbicidal and mechanical methods, but tree planting and less ecologically-disruptive bank management may re-introduce the curb of effective competition.

In some rivers, however, problems arise quite unconnected with channel engineering. An example is the Cheshire Dee. This is regulated by headwater impoundment and severe spates have been eliminated. The channel is not engineered to any appreciable extent yet it is almost devoid of waterplants and the banks alternate between lengths of open, unsightly bare mud and much longer lengths densely and monotonously wooded with overhanging willow, alder and hardwoods which cut off the water as a landscape feature and as an amenity. Pearce (pers. comm.) attributed the absence of reed fringes on open banks to trampling by cattle, anglers and boaters, together with wave damage from passing pleasure-boats. This was investigated experimentally by erecting parallel lines of open mesh fencing along the top of the bank and at 1 m depth offshore, to prevent access without appreciably reducing incoming wave energy. Within a few months a diverse marsh vegetation developed on the mud down to, but not into the water. *Typha latifolia* appeared and spread rapidly, becoming dominant during the second season with willow saplings developing amongst it. By the end of the third season the original diversity had been replaced by a dense *Typha–Salix* fringe (Fig. 18.2). An adjacent fenced length planted with various reeds and sedges formed a tall and attractively diverse fringe down to water level in the first summer, though plants introduced below water level were washed away by boat waves. Over the next 2 years there was the same trend to *Typha–Salix* as in the unplanted fenced length. Plantings in unfenced areas failed, the introduced plants being destroyed by trampling, cattle grazing and, below water level, by washing away. Unplanted, unfenced lengths remained bare throughout the 3 years of observations.

These experiments demonstrate the susceptibility of emergent vegetation to trampling and grazing, and also the speed with which such vegetation can establish naturally once the disturbance is removed. Presumably the seed bank in the mud and waterborne viable fragment supplied the inoculum of species, some of which were otherwise found only in tributaries of the main river. The ease of establishing visually-attractive plantings was also shown, but so too was the rapidity of succession to woody associations. Boat traffic was important here because it restricted the reeds to the marshy zone where they were quickly replaced by willow, and prevented them from colonizing the open water where they could probably have retained dominance for a longer period.

Regulation of the river may have removed the periodic episodes of severe erosion which would clear out parts of the woody association and restart the succession. Thus, the dense tree fringe has been able to develop and spread undisturbed for many years, concentrating the trampling and grazing pressure onto ever-smaller remaining areas of open bank. Evidently periodic vegetation maintenance is needed to preserve landscape quality on regulated as well as on channelized rivers.

Boats also damage floating-leaved vegetation, but by propellor and hull impact rather than by their wash. Of the few beds of the lily *Nuphar lutea* which occur in the lower Dee, most exist with submerged leaves only. However, when light, floating booms were placed round some of them to keep boats out without significantly reducing wave action, the beds developed floating-leaved canopies (Figure 18.3). Unprotected beds lost their photosynthetically-important floating leaves by shredding or cutting off by passing

FIG. 18.2. Vegetation management on the River Dee near Chester. (a) Open riverbank, trampled and grazed. (b) Two summers after fencing to exclude trampling and grazing. Diverse flora includes clumps of *Typha latifolia* and *Salix* saplings. (c) Two summers after fencing and planting with *Typha latifolia*. Dense *Typha* with *Salix* saplings.

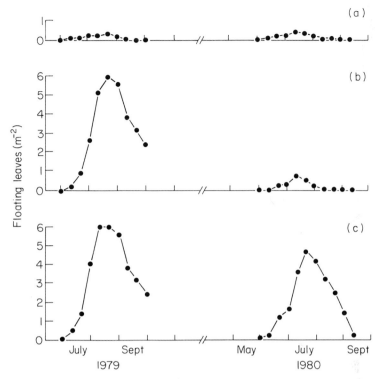

FIG. 18.3. Effect of boat traffic on floating leaf canopy density in a group of *Nuphar lutea* stands in the River Dee near Chester, 1979–80: (a) unprotected from boat access in both years; (b) protected by floating boom in 1979, unprotected in 1980; (c) protected by floating booms in both years.

boats, a loss which has probably contributed to the paucity of *N. lutea* in the lower Dee and to its failure to make a visual contribution to the river scene.

Navigable canals and land drainage channels

Many of the navigable canals constructed in the eighteenth and early nineteenth centuries no longer carry freight traffic but have been retained for amenity uses such as recreational boating, angling and towpath walking, as well as for water supply and land drainage. Amenity use requires strong emphasis on landscape considerations, and here canals offer special opportunities and challenges. The opportunities arise through the survival of historic architectural and engineering structures, mellowed by time to create a uniquely attractive waterway environment. The challenge is to adapt this

historic environment to the needs of amenity users (White 1977, 1984;
Gerritsen 1979).

Vegetation management criteria are more rigorous than on many rivers.
Though bankside vegetation enhances the canal environment, it has to be
selected and managed to avoid obstruction to waterway users and root
damage to embankments, cuttings and masonry, and to harmonize with these
engineering features (British Waterways Board 1984). The 'ideal' water
channel defined in British Waterways Board (1981) accommodates within its
very restricted width a central line of open water and compact fringes of
vegetation which soften the otherwise hard edges of the waterway (Fig. 18.4).
There is, however, a high risk of vegetation spreading across the channel and
blocking it, especially as water flow is usually much too slow to exert the
scouring effects which restrain vegetation colonization in rivers. Boat traffic
counteracts this colonization, and the range of traffic densities in approximate
equilibrium with desirable quantities of vegetation can be defined (Murphy &
Eaton 1983). Below these traffic densities colonization is often rapid (Twigg
1959), and although the visual effect may at first be judged picturesque, once
the channel is filled with reeds and the contrast between open water and
vegetation is lost, the canal is likely to be considered an eyesore, especially in
an urban area. Frequent and expensive weed control may then be require to
maintain an open channel (Murphy, Eaton & Hyde 1982).

At high boat traffic densities, aquatic vegetation is lost and the combination of frequent boat movements with the absence of a protective fringe of

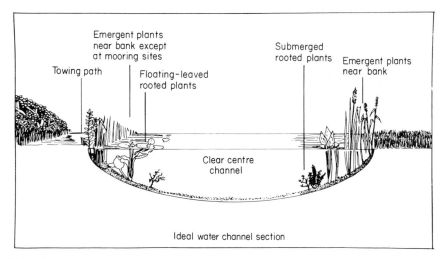

FIG. 18.4. Proposed 'ideal' canal vegetation (British Waterways Board 1981).

vegetation often leads to bank erosion. This is usually stopped by visually-intrusive metal or concrete piling of the banks (Hydraulics Research Station 1980). Here the alternative of reinforced planting with selected species may offer landscape, ecological and economic advantages if appropriate techniques can be developed (Bonham 1980; Bache & Macaskill 1981).

Artificial land drainage channels are usually straight and have to be kept clear of vegetation to give effective protection against flooding (Cave 1981). Their visual impact is correspondingly stark, though Hackett (1976) suggests that flat landscapes such as East Anglia more readily accept them than do undulating areas, and Hengeveld & de Vocht (1982, Chs 3, 4, 8) emphasize the landscape value of drainage canals and their carefully-controlled associated vegetation in urban developments in the Netherlands.

Lakes

In large lakes, most of the important environmental features such as exposure and depth operate on such a massive scale that they are beyond influence by normal techniques. Management is necessarily confined to the conservation of marginal vegetation, in whatever natural disposition it occurs, against man-made stresses which tend to reduce it. Trampling, grazing and boating disturbances occur on lake margins as on rivers and regulation of activities may involve difficult compromises between lake users and landscape requirements (Sukopp 1971; Sukopp *et al.* 1975; Rees & Tivy 1978; Tivy 1980).

Smaller lakes and ponds are more readily influenced by management. Shelter can be modified by shoreline tree planting (or felling), and depth profiles and outline shape can be adjusted in small waterbodies. Using a knowledge of depth and exposure tolerances of individual emergent species, it is then possible to design predictable vegetation patterns into the landscape and ensure their long-term stability by using sharp increases in depth to circumscribe the area colonized.

In previous centuries, lakes were often constructed solely for landscape purposes. Formal watergardens of the seventeenth and early eighteenth centuries use water, like grass, for uniform texture and outline shape, with aquatic vegetation often confined to carefully-planned clumps of floating-leaved species. Later, less formal lakes include emergent plants in their design. Many imported and cultivated species are used in these highly-managed waters, though a relatively short period of neglect permits more natural associations to establish. The role of water and its vegetation in landscape gardening are discussed by Sculthorpe (1967, pp. 506–514), Hadfield (1969) and Fleming & Gore (1979), while Perry (1961) and Hart (1976) provide useful design criteria for composing and stabilizing ornamental waterplant communities.

Nowadays it is rare for waterbodies to be constructed or restored solely for landscape reasons, with the possible exception of small pools (e.g. Brabham 1981), although even here angling needs often influence water and vegetation design (e.g. Lisney Associates 1981). Most larger lakes are managed as multipurpose amenities, and landscape design is inevitably modified by other uses of the water. New lakes have been created in recent decades by landscaping flooded mineral workings. Angling, boating and wildfowling are common uses. The shapes, depth profiles and raw mineral substrata of such lakes are often far from ideal for establishment of aquatic vegetation, though patterns of mineral extraction may be adjusted so that final lake shape is better suited to subsequent uses (e.g. Harrison 1974; Reich 1980; Kelcey 1984).

Another cause of lake construction is the need for stormwater balancing in new urban developments. Here requirements for vegetation management are rigorous, as the lakes are often in areas of great landscape sensitivity and amenity demand, yet have to accommodate the ecological stresses of varying level and poor water quality. Level variation is inevitable to their hydraulic function, but can be brought within limits tolerable to emergent plants by designing a large normal (i.e. non-storm) surface area wherever space permits. Hengeveld & de Vocht (1982, pp. 229–249) and Kelcey (1978) discuss how lake design and ecological engineering can contribute to the accommodation of these features.

Eutrophication is a common cause of management problems in lowland lakes (Moss, Chapter 24). Dense plankton blooms and unsightly floating weed mats are difficult to prevent where nutrient-rich water sources create high nitrogen and phosphorus loadings. Designers sometimes believe that biological problems will be lessened by 'keeping the lake flushed out with new water to prevent stagnation'. This is only true when complete water replacement occurs every few days, even in summer, and plankton is washed out before it can grow significantly. More commonly, with longer summer replacement times, management is better aimed at extending these times by reducing inflows. Balancing lakes can be kept offstream between spates, so that for most of the time the supplying river's nutrients (and also pollutants and silt) are not adding to the loading of the standing water. Likewise inflows to other types of lakes should be minimized, ideally to the rate which maintains water level against evaporation and leakage, and prevents major evaporative concentration of dissolved salts in the lake water in summer. Useful reductions in dissolved nitrogen can occur by denitrification during static periods. Phosphorus may be lost by incorporation into sediments, at least under well oxygenated conditions. Classen (1980) describes how the trophic status of a lake can be predicted and the extent to which management can reduce eutrophication effects, though it is difficult to apply some of the principles to lakes as shallow as those often favoured in landscape schemes. Nevertheless, weired inlets and bypass channels or culverts can be good capital investments if their operation curbs eutrophication and reduces recurrent costs of vegetation control and dredging. Additionally, silt and oil traps are especially needed on inflows receiving urban and other drainages liable to carry large quantities of these materials.

Surprisingly great on-site improvements of lake water quality can sometimes be achieved, with some landscape enhancement, if the inflow can be led through a reedbed. Reeds are quite effective removers of silt, pollutants and nutrients, especially during the crucial summer growing season (Toth 1972; Krotkevitch 1974; Jong & Kok 1978).

Aquatic weed control is a common problem in shallow lakes, as well as in the slowly-flowing rivers and artificial channels considered earlier. Cutting, dredging and chemical methods can be used (Fryer & Makepeace 1978; Barrett 1981; Kelcey 1981; Price 1981; Roberts 1982; Way 1983). Localized control, often needed in preference to total kill in landscape management, is now possible with herbicides (Barrett & Logan 1982; Barrett & Murphy 1982; Evans 1982) as well as with mechanical methods (Eaton & Freeman 1982; Eaton *et al.* 1981). However, such methods are expensive and may have to be repeated quite frequently, since they do no more than temporarily empty basically favourable niches which are then quickly refilled by the original species or invaded by often less desirable primary colonizers. Grass carp (*Ctenopharyngodon idella*), an herbivorous fish, may prove effective for longer-term control of submerged and free-floating weed (Buckley 1981; Riemens 1982), though in Britain temperatures are not always high enough for useful weed consumption to occur, especially in waterbodies fed by cool groundwaters (Mugridge *et al.* 1982).

It is better wherever possible to restrict the potential for plant growth at the design stage, by use of depth limitations. This may entail extra excavation or dredging, but the expense should be weighed against the reduction in subsequent maintenance costs and the generally more satisfactory appearance achieved, as well as any gains for fisheries and other uses.

Waterfowl in small numbers enhance any waterside scene and may help to control marginal submerged weed. However, large numbers of birds damage waterside vegetation and their excreta cause eutrophication problems. In the worst cases, such as ornamental lakes carrying very large bird populations sustained by artificial feeding, aquatic vegetation is wholly eliminated and the water becomes green and evil-smelling. Here improvements can only be achieved when waterfowl numbers are reduced.

Reservoirs

Reservoirs and artificially-regulated lakes present a special problem because of their unnaturally large fluctuations in water level. An ugly, often eroding 'drawdown zone' of bare stone and soil is uncovered as water is drawn off and is devoid of aquatic plants either because of desiccation and heating in summer in direct supply reservoirs where the level is lowest in that season, or because of desiccation and freezing in regulatory reservoirs which are drawn down in winter (Quennerstedt 1958; Geen 1974). Terrestrial species are largely excluded by the effects of submersion and waterlogging during the periods of high water level.

This drawdown zone creates a harsh visual impact, especially in upland direct supply reservoirs where it is conspicuous from high vantage points and its fullest development coincides with the main summer tourist season. These points often feature prominently amongst the environmental objections to new reservoir schemes in areas of high landscape value.

Swards of tolerant herbaceous species have been planted to counteract erosion in the drawdown zone (Little & Jones 1979) but these do not greatly alleviate the landscape problem since at most they replace a grey or brown zone by a green one. More useful visually is the introduction of large-scale vegetation in the form of flood-tolerant trees just above and below top water level. These break up the continuity of the rim and can be supplemented by clumps of trees of various textures (e.g. deciduous and evergreen) wholly above top water level to further break up the uniform appearance of the reservoir margin and lift the observer's eye away from the drawdown zone (Gill & Bradshaw 1971a, b, 1977).

CONCLUSIONS

Successful application of landscape design to a water system must therefore take account of the distinctive ecology of aquatic vegetation. In new developments this can be done at the design stage through depth, exposure and water quality specifications. In existing waters the feasibility of modifying these factors may be limited but at least these limits should be defined and taken into account. Although design aims may be stated quite easily in such terms as integrated attractive appearance, textural diversity, stability and minimum maintenance, water differs so much from land as an environment for plant growth that the means of achieving these aims may require special consideration separately from those employed for the surrounding land. This can be a major challenge to the skills of the landscape designer. The challenge is increased by the fact that nowadays most waterbodies have multiple uses so that landscape considerations may have to be set alongside the needs of such other functions as water storage, stormwater balancing, land drainage, effluent disposal, navigation, angling, nature conservation, and public access and safety, each of which imposes ecologically-important effects on aquatic vegetation. A few of these conflict almost intractably with landscape aims, but most can be accommodated or even used supportively. The rewards of success are great when ecologically-satisfactory 'waterscapes' are created which powerfully enhance the landscape.

ACKNOWLEDGMENTS

The author would like to thank Mr J. K. Hulme (Director, Ness Botanic Gardens, University of Liverpool), Mr P. White and Miss J. Grice (Architect/ Planners, British Waterways Board) and Dr P. Spillett (Senior Fisheries Officer, Thames Water Authority) for information supplied for this paper.

REFERENCES

Bache, D.H. & Macaskill, I.A. (1981). Vegetation in coastal and stream bank protection. *Landscape Planning*, **8**, 363–386.

Barrett, P.R.F. (1981). Aquatic herbicides in Great Britain, recent changes and possible future development. *Proc. Aquatic Weeds and their Control.* Assoc. Appl. Biol., Oxford, pp. 95–103.

Barrett, P.R.F. & Logan, P. (1982). The localised control of submerged aquatic weeds in lakes with diquat-alginate. *Proc. EWRS 6th Symposium on Aquatic Weeds*, Novi Sad, Jugoslavia, pp. 193–199.

Barrett, P.R.F. & Murphy, K.J. (1982). The use of diquat-alginate for weed control in flowing waters. *Proc. EWRS 6th Symposium on Aquatic Weeds*, Novi Sad, Jugoslavia, pp. 200–208.

Binder, W. & Gröbmaier, W. (1978). Courses of streams and rivers—their form and care. *Garten und Landschaft*, **1**, 25–30.

Bolas, P.M. & Lund, J.W.G. (1974). Some factors affecting the growth of *Cladophora glomerata* in the Kentish Stour. *Water Treatment and Examination*, **23**, 25–51.

Bonham, A.J. (1980). *Bank Protection using Emergent Plants against Boat Wash in Rivers and Canals.* Int. Report No. IT206, Hydraulics Research Station, Wallingford, Oxon.

Brabham, M. (1981). Upton Grey village pond. *Landscape Design*, **133**, 21.

British Waterways Board (1981). *Vegetation Control Manual.* British Waterways Board, Director of Engineering, London.

British Waterways Board (1984). *Waterway Environment Handbook.* British Waterways Board, Architect/Planner, Hillmorton, Rugby.

Brookes, A., Gregory, K.J. & Dawson, F.H. (1983). An assessment of river channelisation in England and Wales. *Science of the Total Environment*, **27**, 97–111.

Buckley, B.R. (1981). Practical problems with the use of grass carp for weed control. *Proc. Aquatic Weeds and their Control.* Assoc. Appl. Biol., Oxford, pp. 125–129.

Cave, T.G. (1981). Current weed control problems in land drainage channels. *Proc. Aquatic Weeds and their Control.* Assoc. Appl. Biol., Oxford, pp. 5–14.

Clarke, A. (1973). Flood plains—Bristol Polytechnic. *Journal of the Institute of Landscape Architects*, **104**, 14–17.

Classen, J. (1980). *The OECD Co-operative Programme for Inland Waters (Eutrophication Control). Regional Project: Shallow Lakes and Reservoirs: Final Report.* Water Research Centre, Medmenham, U.K.

Davison, W., Woof, C. & Rigg, E. (1982). The dynamics of iron and manganese in a seasonally anoxic lake; direct measurements of fluxes using sediment traps. *Limnology and Oceanography*, **27**, 987–1003.

Dawson, F.H. & Haslam, S.M. (1983). The management of river vegetation with particular reference to shading effects of marginal vegetation. *Landscape Planning*, **10**, 147–169.

Dawson, F.H. & Kern-Hansen, U. (1979). The effect of natural and artificial shade on the macrophytes of lowland streams and the use of shade as a management technique. *Internationale Revue der Gesamten Hydrobiologie und Hydrographie*, **64**, 437–455.

Dearden, P. (1982). Comparative risk assessment associated with the growth of Eurasian Water Milfoil (*Myriophyllum spicatum* L.) in the Okanagan Valley, British Columbia, Canada. *Proc. EWRS 6th Symposium on Aquatic Weeds*, Novi Sad, Jugoslavia, pp.113–122.

Dove, R. & Wallis, M. (1981). The 1980 Aquatic Plant Quarantine Project. *Studies on Aquatic Macrophytes Part XXXIV*. Ministry of Environment, Victoria, British Columbia, Canada.

Eaton, J.W. & Freeman J. (1982). Ten years' experience of weed control in the Leeds & Liverpool Canal. *Proc. EWRS 6th Symposium on Aquatic Weeds*, Novi Sad, Yugoslavia, pp. 96–104.

Eaton, J.W., Murphy, K.J. & Hyde, T.M. (1981). Comparative trials of herbicidal and mechanical control of aquatic weeds in canals. *Proc. Aquatic Weeds and their Control*. Assoc. Appl. Biol., Oxford, pp. 105–116.

Ellenberg, H. (1963). *Die Vegetation Mitteleuropas und der Alpen*. Ulmer Verlag, Stuttgart.

Evans, D.M. (1982). *Phragmites* control with glyphosate through selective equipment. *Proc. EWRS 6th Symposium on Aquatic Weeds*, Novi Sad, Jugoslavia, pp. 209–211.

Fleming, L. & Gore, A. (1979). *The English Garden*. Michael Joseph, London.

Fryer, J.D. & Makepeace, R.J. (Eds.) (1978). *Weed Control Handbook: Recommendations*. Blackwell Scientific Publications, Oxford.

Geen, G.H. (1974). Effects of hydro-electric development in western Canada on aquatic ecosystems. *Journal of the Fisheries Research Board of Canada*, **31**, 913–27.

Gerritsen, A. (1979). Van vaarweg tot recreatievoorziening het Apeldoorns kanaal, meerdere functies, meervoudig gebruik. *Recreatie*, **17**, 29–32.

Gill, C.J. & Bradshaw, A.D. (1971a). Some aspects of the colonization of upland reservoir margins. *Journal of the Institute of Water Engineers*, **25**, 165–173.

Gill, C.J. & Bradshaw, A.D. (1971b). The landscaping of reservoir margins. *Journal of the Institute of Landscape Architects*, **95**, 31–34.

Gill C.J. & Bradshaw, A.D. (1977). The landscaping of reservoir margins. *Landscape Design with Plants* (Ed. B. Clouston), pp. 198–208. Inst. Landscape Architects, London.

Hackett, B. (1976). Water in landscape architecture. *Water (Journal of the National Water Council)*, **6**, 11–12.

Hadfield, M. (1969). *A History of British Gardening*. Spring Books, London.

Harrison, J. (1974). *The Sevenoaks Gravel Pit Reserve*. Wildfowlers Association of Great Britain and Ireland. Chester.

Hart, A. (1976). Design with plants 5: water plants. *Landscape Design*, **113**, 10–15.

Haslam, S.M. (1978). *River Plants*. Cambridge University Press, Cambridge.

Haslam, S.M., Sinker, C.A. & Wolseley, P.A. (1975). British water plants. *Field Studies*, **4**, 243–351.

Hengeveld, H. & de Vocht, C. (Eds) (1982). *Role of Water in Urban Ecology*. Developments in Landscape Management and Urban Planning, 5. Elsevier, Amsterdam.

Hinge, D.C. & Hollis, G.E. (Eds) (1980). *Land Drainage, Rivers, Riparian Areas and Nature Conservation*. Discussion Papers in Conservation No. 37, University College, London and Thames Water Authority.

Hutchinson, G.E. (1975). *A Treatise in Limnology*, Vol. 3 *Limnological Botany*. Wiley, London.

Hydraulics Research Station (1980). *Report on Bank Protection in Rivers and Canals*. Hydraulics Research Station, Wallingford, Oxon.

Jellicoe, S. & Jellicoe, G. (1971). *Water: the use of water in landscape architecture*. Black, Huntingdon.

Jong, J. de & Kok, T. (1978). The purification of wastewater and effluents using marsh vegetations and soils. *Proc. EWRS 5th Symp. on Aquatic Weeds*, Amsterdam, The Netherlands. pp. 135–142.

Kelcey, J.G. (1981). Weed control in amenity lakes. *Proc. Aquatic Weeds and their Control.* Assoc. Appl. Biol., Oxford, pp. 15–31.

Kelcey, J.G. (1978). Techniques No. 31: Creative Ecology Part 2: Selected aquatic habitats. *Landscape Design*, **121**, 36–38.

Kelcey, J.G. (1984). The design and development of gravel pits for wildlife in Milton Keynes, England. *Landscape Planning*, **11**, 19–34.

Krotkevich, P.G. (1974). *Use of the water protection and purification properties of the common reed.* Ukrainian Scientific Research Institute of the Pulp and Paper Industry. Transl. from Vodnye Resursy no. 5, pp. 191–197, 1976. Plenum, N.Y., UDC 676.088: 628.35.

Lewis, G. & Williams, G. (1984). *The River and Wildlife Handbook.* Royal Society for the Protection of Birds and Royal Society for Nature Conservation. RSPB, Sandy, Bedfordshire.

Lisney Associates (1981). Restoration of Abbey Pond, Bishop's Waltham. *Landscape Design*, **133**, 22.

Little, M.G. & Jones, H.R. (1979). *The uses of herbaceous vegetation in the drawdown zone of reservoir margins.* Tech. Rep. TR105. Water Research Centre, Medmenham, U.K.

Marshall, E.J.P. & Westlake, D.F. (1978). Recent studies on the role of aquatic macrophytes in their ecosystem. *Proc. EWRS 5th Symp. on Aquatic Weeds*, Amsterdam, The Netherlands, pp. 43–51.

Mason, C.F., Macdonald, S.M. & Hussey, A. (1984). Structure, management and conservation value of the riparian woody plant community. *Biological Conservation*, **29**, 201–216.

Moss, B. (1977). Conservation problems in the Norfolk Broads and rivers of East Anglia, England—phytoplankton, boats, and the causes of turbidity. *Biological Conservation*, **12**, 95–114.

Mugridge, R.E.R., Buckley, B.R., Fowler, M.C. & Stallybrass, H.G. (1982). An evaluation of the use of grass carp (*Ctenopharyngodon idella* Valenciennes) for controlling aquatic plant growth in a canal in Southern England. *EWRS 2nd Internat. Symp. on Herbivorous Fish*, Novi Sad, Jugoslavia, pp. 8–16.

Murphy, K.J., Eaton, J.W. & Hyde, T.M. (1982). The management of aquatic plants in a navigable canal system for amenity and recreation. *Proceedings EWRS 6th Symposium on Aquatic Weeds*, Novi Sad, Jugoslavia, pp. 141–151.

Murphy, K.J. & Eaton J.W. (1983). Effects of pleasure-boat traffic on macrophyte growth in canals. *Journal of Applied Ecology*, **20**, 713–729.

Newbold, C. (1981). The decline of aquatic plants and associated wetland wildlife in Britain—causes and perspectives on management techniques. *Proc. Aquatic Weeds and their Control.* Assoc. Appl. Biol., Oxford, pp. 241–251.

Newbold, C., Purseglove, J. & Holmes, N. (1983). *Nature Conservation and River Engineering.* Nature Conservancy Council, London.

Perry F. (1961). *Water Gardening.* 3rd edn. Country Life Ltd., London.

Pfeffer H., (1978). Natural construction and care of the landscape exemplified by the Untere Vils. *Garten und Landschaft*, **1**, 31–36.

Phillips, G.L., Eminson, D.F. & Moss, B. (1978). A mechanism to account for macrophyte decline in progressively eutrophicated freshwaters. *Aquatic Botany*, **4**, 103–126.

Price, H. (1981). A review of current mechanical methods. *Proc. Aquatic Weeds and their Control.* Assoc. Appl. Biol., Oxford, pp. 77–86.

Purseglove, J. (1983). Integrating nature conservation with river engineering. *Landscape Design*, **146**, 39–41.

Quennerstedt, N. (1958). Effect of water level fluctuation on lake vegetation. *Verhandlungen der Internationalen Vereinigung für Theoretische und Angewante Limnologie*, **13**, 901–6.

Rees, J. & Tivy, J. (1978). Recreational impact on Scottish shore wetlands. *Journal of Biogeography*, 5, 93–108.

Reich, R. (1980). The Broad, University of East Anglia, Norwich. *Landscape Design*, 129, 14–16.

Reynolds, A.J. & Eaton, J.W. (1983). The role of vegetation structure in a canal fishery. *Proc. 3rd Brit. Freshwat. Fish. Conf.*, Liverpool, pp. 192–202.

Reynolds, C.S. (1971). The ecology of the planktonic blue-green algae in the North Shropshire meres. *Field Studies*, 3, 409–432.

Reynolds, C.S., Morison, H.R. & Butterwick, C. (1982). The sedimentary flux of phytoplankton in the south basin of Windermere. *Limnology and Oceanography*, 27, 1162–1175.

Reynolds, C.S. & Walsby, A.E. (1975). Water blooms. *Biological Reviews*, 50, 437–481.

Riemens, R.G. (1982). The result of grass carp stocking for weed control in the Netherlands. *EWRS 2nd Internat. Symp. on Herbivorous Fish*, Novi Sad, Jugoslavia, pp. 1–7.

Roberts, H.A. (Ed) (1982). *Weed Control Handbook: Principles*, 7th edn. Blackwell Scientific Publications, Oxford.

Scales, P. & Bryan, A. (1979). Transport of *Myriophyllum spicatum* fragments by boaters and assessment of the 1978 Boat Quarantine Program. *Studies on Aquatic Macrophytes Part XXVII*. Ministry of Environment, Victoria, British Columbia, Canada.

Sculthorpe, C.D. (1967). *The Biology of Aquatic Vascular Plants*. Arnold, London.

Seibert, P. (1968). Importance of natural vegetation for the protection of the banks of streams, rivers and canals. In *Freshwater, Nature and Environment*, Series 2, pp. 34–67. Council of Europe.

Spence, D.H.N. (1964). The macrophytic vegetation of freshwater lochs, swamps and associated fens. In *The Vegetation of Scotland* (Ed. by J.H. Burnett), pp. 306–425. Oliver & Boyd, Edinburgh.

Spence, D.H.N. (1982). The zonation of plants in freshwater lakes. *Advances in Ecological Research*, 12, 37–125.

Spillett, P. (1981). *Workshop on Conservation and Environment*. Lea Division, Thames Water Authority, Reading.

Sukopp, H. (1971). Effects of man, epecially recreational activities, on littoral macrophytes. *Hidrobiologia* (Bucarest), 12, 331–340.

Sukopp, H., Markstein, B. & Trepl, L. (1975). Rochrichte unter intensiven Grossstadtenfluss. *Beitrage zur Naturkunde Forschung im Südwestlich-Deutschland*, 34, 375–385.

Tivy, J. (1980). *The Effect of Recreation on Freshwater Lochs and Reservoirs in Scotland*. Countryside Commission for Scotland, Perth, UK.

Toth, L. (1972). Reeds control eutrophication of Balaton Lake. *Water Research*, 6, 1533–1539.

Twigg, H.M. (1959). Freshwater studies in the Shropshire Union Canal. *Field Studies*, 1, 116–142.

Vivash, R.M. (1982). Better drainage and better fisheries. *Proc. Inst. Fish. Mgmt 13th Annual Study Course*, Aberystwyth, Wales, pp. 92–97.

Way, J.M. (Ed) (1983). *Management of Vegetation*. British Crop Protection Council, Croydon, UK. Monograph no. 26.

Weeks, K.G. (1982). Conservation aspects of two river improvement schemes in the River Thames catchment. *Journal of the Institute of Water Engineers and Scientists*, 36, 447–458.

Wetzel, R.G. (1975). *Limnology*. Saunders, Philadelphia.

White, P. (1977). Utilitarian construction in the man-made landscape-waterways. In *The Man-Made Landscape*, Museums and Monuments, Vol. 16, pp. 53–71. UNESCO, Paris.

White, P. (1984). What is Conservation? In *Canals—A New Look* (Ed. by M. Baldwin & A. Burton) pp. 93–109. Phillimore, Chichester.

Wiegleb, G. (1981a). Probleme der syntaxonomischen gliederung der Potametea. *Berichte der Internationalen Symposien der Internationalen Vereinigung für Vegetationskunde*, pp. 207–247. Cramer, Vaduz.

Wiegleb, G. (1981b). Struktur, Verbreitung und Bewertung von Makrophytengesellschaften niedersächsischer Fliessgewässer. *Limnologica* (Berlin), 13, 427–448.

19. A NATURAL APPROACH IN CITIES

H. HENKE[1] AND H. SUKOPP[2]

[1]*Federal Research Centre for Nature Conservation and Landscape Ecology, Bonn, Fed. Rep. of Germany and* [2]*Institute for Ecology: Ecosystem Research and Vegetation Science, Technical University of Berlin*

SUMMARY

The original wild flora and fauna has been steadily reduced by the redevelopment of cities in the Federal Republic of Germany. Natural scientists and landscape architects are playing an important role in protecting nature and reintroducing it to cities. This is demonstrated using West Berlin as an example. The system of subject planning and strategic planning demanded by the Berlin Nature Conservation Act of 1979 is described. Landscape architects have helped to elevate nature conservation to a strategic planning level and to make it part of political thinking. The procedure used to develop the Nature Conservation Programme, which is one of four types of programmes included in conservation subject planning, is outlined. Systematic survey of wild habitats and their assessment, based on classification of development zones for different historical phases in the inner city, are suggested as tools for identifying and protecting wild species and habitats in urban areas. Planning policies which promote the long-term survival of wild natural communities are also dealt with. The Nature Conservation Programme, which is a part of the strategic plan being prepared for West Berlin, is a good example of how a natural approach can be achieved in urban planning.

INTRODUCTION

Conservation bodies and groups including, on the one hand, citizens' action groups and voluntary conservation organizations and, on the other hand, government authorities, are a moving force influencing the change in awareness toward nature and environment in the Federal Republic of Germany. In the government, nature conservation authorities, natural scientists, in particular botanists and zoologists, play a key role. An essential contribution to the improvement of environmental quality has also been provided by landscape architects. Since the 1930s they have continuously broadened the scope of their work and have developed a statutory landscape planning instrument, i.e. an independent subject plan (Henke 1979). Through the combination of natural science oriented nature conservation and landscape planning, the surveying, evaluating and protecting of wild plant and animal species and their habitats have been greatly improved. This combination is especially beneficial in cities since their redevelopment has left less and less room for natural communities to survive or develop. A natural science based nature conservation programme is being prepared for West Berlin.

THE PLANNING SYSTEM

With the enactment of the Berlin Nature Conservation Act in 1979, subject planning and strategic subject planning became mandatory. Subject planning consists of four types of programmes (Fig. 19.1):

  (a) nature conservation (areas, habitats, species);
  (b) ecological processes (water, soil, air);
  (c) landscape scenery (elements, landscape types, design units);
  (d) recreation (open spaces, neighbourhood recreation areas).

All four programmes feed into strategic subject planning, which consists of a landscape programme for the entire city, district landscape programmes and local landscape plans. At the same time, however, each of the four programme areas remains an individual subdivision and as such can continuously expand. The stage of development for each, with respect to its data base and supportive scientific disciplines as well as its administrative representation, is very different.

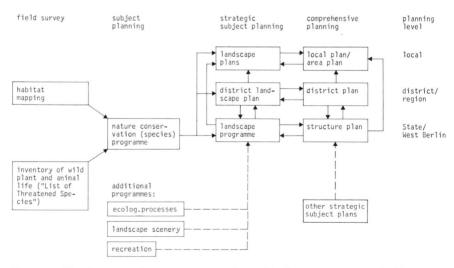

FIG. 19.1. Planning system for nature conservation and landscape management in West Berlin.

The nature conservation programme subdivision is the most advanced. It has a relatively well established data base, since it is backed by the Institute of Ecology at the Technical University of Berlin, which is where landscape planners receive their ecological training and where many of these students specialize in this branch of natural science.

The landscape scenery and recreation programmes have been developed through contracts with landscape planning offices. The ecological processes programme has yet to be developed. Landscape planners will probably attempt to fill in the information needed for this subject area with the help of the relevant natural science disciplines.

While subject-planning basically reflects the work of scientists, managers and landscape planners, strategic subject planning is primarily the domain of landscape planners. Landscape planners are responsible for harmonizing these four programmes with each other and with other subject plans, such as those for agriculture, forestry, housing and transportation (Henke & Krause 1980). Upon completion of this process the landscape programme is presented to the West Berlin Parliament where it is given statutory authority. The objectives and measures contained in the landscape programme thus have a strong supportive base for inclusion in the structure plan, in which economic, social and ecological interests are reconciled.

The planning system described is important for the effective implementation of nature and landscape conservation measures as well as for the provision of recreational facilities. In addition, strategic subject planning has certain advantages, i.e.:

(a) it promotes the development of objectives and a comprehensive concept for nature and landscape conservation;

(b) nature and landscape conservation is raised to a policy planning level where its objectives can be actively and more forcefully pursued;

(c) the responsibility of the Nature Conservation and Landscape Management Authority is extended to the city as a whole rather than being limited to individual protected areas;

(d) the Nature Conservation and Landscape Management Authority is consulted early in the planning process on all administrative levels.

The Nature Conservation Programme being devised for West Berlin is a good example of how to achieve a natural approach in planning for cities.

NATURE CONSERVATION PROGRAMME

The Berlin Nature Conservation Programme* is being developed under the

* The material presented on the nature conservation programme is based on the work of the Berlin State Advisor for Nature Conservation and the Institute of Ecology at the Technical University of Berlin.

direction of the Nature Conservation and Landscape Management Authority to prepare, implement and supervise objectives and measures for the protection, planning and management of wild flora and fauna and their habitats in the entire city. Such a programme is usually developed for a whole state like Bavaria or Lower Saxony, but is here applied to West Berlin as a City State.

Certain facts about the city are essential for a better understanding of the work of the Programme. West Berlin has an area of 480 km² (Greater London is, for example, 1580 km²) and has a population of 1·9 million. This city is an island within the territory of the German Democratic Republic and, therefore, has no hinterland. All city development problems must be solved within the city. The shortest transit route between Berlin and the Federal Republic of Germany is about 180 km. The topography and drift geology of Berlin are mainly the result of the last glaciation in Northern Germany which ended around 8000 BC. As the glaciers retreated river valleys and chains of lakes were formed, which are dominant features in the city landscape today. The glaciers eliminated most species of both the flora and fauna. In the short period since, only a relatively slight differentiation of species into races has occurred. Today West Berlin consists of about 25% forests and woodlands, including informal open space and parks, 7·6% farmland and a variety of urban zones in which distinct plant and animal communities have developed. Not only the knowledge of the actual state of nature and landscape, but also their history of development and their development potential are important bases for the Nature Conservation Programme. To gain this knowledge several studies of the natural and urban history of Berlin have been conducted (Sukopp et al. 1978; Ermer et al. 1980; Horbert et al. 1982; Sukopp 1982; Sukopp & Werner 1982, 1983).

Until the nineteenth century the larger part of what later became Greater Berlin was a collection of rural communities rather sparsely populated. During the second half of the nineteenth century a rapid increase in development began, accompanied by devastating effects on the living conditions of the growing working-class migrating from rural areas into the city. The sprawling blocks of flats huddled around small courtyards provided very poor living conditions. The villages surrounding Berlin were able to preserve their countryside character for quite a while but were later developed into villa type residential suburbs. In the workers' quarter, large public parks were established to compensate for the bad living conditions. Also, land was provided for allotments which arose from the former gardens of almshouses. The late Victorian (in Berlin this period is known as Wilhelminian) city belt with its backyard industry, storage depots and small-scale business, intermingled with the expanding outskirts and was very largely made up of three-storey buildings serving both for housing and business. In 1925 urban districts were divided into 11 building zones ranging from 50 inhabitants to about 550–700 inhabitants per hectare. Non-profit development companies began to build large housing estates with the help of government subsidies under social benefit schemes. During World War II the city was almost completely destroyed. The systematic redevelopment started about 1950 with large-scale housing programmes. From the estimated 80 million m³ of rubble, ten hills were built which serve as green open space and outdoor recreation areas.

Flora and fauna

The description of Berlin's urbanization process gives the impression that city and nature are opposed to each other and, because of this, the number of plant and animal species ought to be low. However, a city as a whole does have a high number of species compared to the surrounding countryside, as can be seen from investigations by Walters (1970) in Cambridgeshire and Haeupler (1974) in Southern Lower Saxony. Dividing the country in equal squares (10 × 10 km and 5 × 5 km, respectively) the squares which include larger cities, such as Cambridge and Hanover, are found to contain the highest number of species. This may be largely explained by the following arguments.

(a) The great diversity of the urban landscape, including different structures of settlements, varied uses of open spaces and many small-scale habitats, provides a great variety of ecological environments, so that plants from coastal as well as steppe regions can establish themselves.

(b) Cities are important immigration areas for alien plants introduced either with or without the help of man. The increase in the number of species by immigration is chiefly due to man's activity, either directly, through the use of ornamental plants, or indirectly, for example, through transported seeds. These species are called hemerochorics or anthropophytes.

The growing importance of cities for species' immigration has become very evident since the Industrial Revolution, starting about 1840, after worldwide traffic and trade increased tremendously. Most of the present-day neophytic flora spread best in cities and industrial areas, whereas many archaeophytes, which migrate as field weeds, spread best in rural areas. The species which are limited to settlements are classified as ruderal. The restriction of ruderal hemerochorics to human settlements is all the more pronounced, as well as increasingly confined to more specialized locations, the further they are away from their geographical place of origin. Kunick (1982), in an extensive study comprising both western and eastern Europe, shows that 15% of the species recorded are common to the nine cities which were compared, and that 25–30% of the flora in these cities belongs to a general urban flora. This percentage reflects the common city environment. The proportion would definitely increase if only the inner city is considered, where the native species of the surrounding natural areas appear less frequently. The tree population in Central European cities is uniform to a large extent, despite differences in the underlying strata and in climatic conditions.

The larger number of species from urban areas have to be brought into perspective and judged critically, when considering the effects of the urbanizing process on the composition and quality of flora. These effects can be measured by comparative and gradient analyses (Falinski 1971). Table 19.1 shows some of the relevant results.

The proportion of native species is progressively reduced under increasing urbanization and the proportion of hemerochoric species accounts for more than half the number of species in cities. The decline of the native flora is not only due to the destruction of the original vegetation but also partly to the competition with better adjusted neophytes.

Taking West Berlin as an example, it can be seen that rare plants have much less chance of survival when urban density increases. Changes in living conditions particularly affect species of oligotrophic habitats and wet biotopes; this is due to eutrophication of soil and water and drop in ground water level or sealing of soil through construction. In West Berlin 54% of the native flora is already threatened (Auhagen & Sukopp 1982).

The fact that cities still have a relatively high diversity of species is largely due to the urban fringes. Large areas of open spaces covered with vegetation are also very important for maintaining the variety of species in the inner city. The dissemination and naturalization of plants cannot only be explained in terms of environmental changes and in the transport of seeds. In addition, genetic changes must have occurred in the populations enabling species to spread over wide areas.

TABLE 19.1. Composition of the synanthropic flora of villages and cities in Poland (Falinski 1971)

	Native (%)	Alien (%)
Forest settlements	70–80	20–30
Villages	70	30
Small towns	60–65	35–40
Medium towns	50–60	40–50
Cities	30–50	50–70

Development of populations

Because it is traditional to eliminate native flora and fauna as well as urban species in the city, their protection and reintroduction requires convincing arguments for the presence of such wild plant and animal life in the inner city to the political decision makers and the public they represent. Such utilitarian arguments are inherent in the planning process. But also of importance are the ethical reasons for nature conservation; these have to be primarily pursued by voluntary organizations serving as an exemplary social movement.

The arguments for nature conservation can be divided into those of global validity and those of special application to the Federal Republic of Germany. While the reasons for nature conservation in the city fringes are the same as for the Federal Republic as a whole, only certain arguments apply to the inner city. The direct value of wild plant and animal life to man on a global and German level and particularly in the inner city is shown in Table 19.2. The aims and arguments are further described and defined in the Nature Conservation Programme. These should help the landscape planner present and evaluate the nature conservation case where it is in competition with the demands of other land uses.

Although reasoned arguments show the importance of wild plant and animal life for man, practical measures must be developed from them for effectively protecting and reintroducing nature in the city. All wild habitats of a city need to be systematically recorded and mapped. An inventory of selected habitats or areas is insufficient because the Nature Conservation and Landscape Management Authority can only claim competency for areas actually surveyed. Such recording and mapping of wild habitats on a scale 1:10 000 has been carried out for West Berlin. At the same time research projects and studies on specific sites are listed and reviewed, indicated on a map of 1:20 000 and indexed. Also for this purpose, information is collected from inquiries and the information from the site survey is used to document fully individual sites as well as the defined habitat types.

Rubble sites are one of the main habitat types. Large areas of derelict sites in Berlin consist of layers of rubble from the last War, one to several metres thick, which have since developed into soils with specific properties. The original material for this soil formation consisted mainly of unbroken bricks of poor quality, held together by porous mortar with a high carbonate content. The effects of frost, heat and root action have reduced the mortar to small pieces, while in the upper centimetres the poorly baked bricks have disintegrated. In addition, acids from soil organisms and acid rain have dissolved and leached the bulk of the carbonates, thus lowering the pH.

TABLE 19.2. Arguments for the conservation of wild plant and animal life (Auhagen & Sukopp 1983)

	World	Germany (Fed. Rep)	Inner City
Conservation of the functions of biological systems			
Production of food	+	+	
Stability of ecosystems	+	+	+
Biological pest control	+	+	(+)
Pollination of cultivated plants	+	+	
Biological filter and decontaminator	+	+	+
Humus production in soil for agriculture and forestry	+	+	(+)
Bioindicator potential	+	+	+
Obtaining of biochemical information			
Conservation of evolutionary adaptive potential	+	+	+
Breeding for new varieties or races and for resistance	+	+	+
Pharmacology	+		
Protection of research objects and sites			
Discovery of new food species	+	+	
Bionics	+		
Biotechnical energy production	+		
Bioengineering research	+		
Fundamental research in biology and ecology	+	+	+
Recreation and preservation of local natural and cultural heritage			
Phenological diversity	(+)	+	+
Diversity in spatial composition, landscape scenery	(+)	+	+
Diversity for the senses	(+)	+	+
Diversity of colour, form and pattern of movement	(+)	+	+

Besides this, semi-decomposed leaf litter has accumulated predominantly in the indentations on rubble mounds.

After the War an intensive and spontaneous development of vegetation began on the rubble, proceeding in more or less rapid succession from short-lived and perennial stages of herbaceous vegetation to shrub and forest-like stands (Fig. 19.2) (succession diagram Table 19.3; Sukopp 1973). This natural succession was generally disrupted by clearing and reconstruction work. The areas cleared of rubble or destined for redevelopment, which have often been recolonized by woody plants, are gradually disappearing in the course of on-going construction.

One of the characteristic pioneer colonizers is the summer annual

FIG. 19.2. A rubble heap originating from bombed buildings showing the growth of poplars and willows after 20 years of succession. Mugwort is colonizing the verge of the street (Berlin-Kreuzberg, Alte Jakob-Straße, Sept. 1964).

TABLE 19.3. Succession stages of ruderal vegetation on inner city derelict land in West Berlin

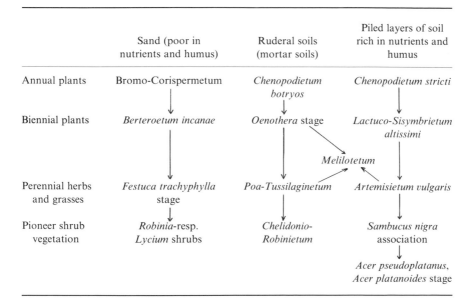

	Sand (poor in nutrients and humus)	Ruderal soils (mortar soils)	Piled layers of soil rich in nutrients and humus
Annual plants	Bromo-Corispermetum	*Chenopodietum botryos*	*Chenopodietum stricti*
Biennial plants	*Berteroetum incanae*	*Oenothera* stage	*Lactuco-Sisymbrietum altissimi*
		Melilotetum	
Perennial herbs and grasses	*Festuca trachyphylla* stage	*Poa-Tussilaginetum*	*Artemisietum vulgaris*
Pioneer shrub vegetation	*Robinia*-resp. *Lycium* shrubs	*Chelidonio-Robinietum*	*Sambucus nigra* association
			Acer pseudoplatanus, Acer platanoides stage

Chenopodium botrys (Sukopp 1971) (Fig. 19.3). This southern Eurasian–
Mediterranean plant has expanded its range due to human influence into
broad areas of central and western Europe, North America and Australia.
Introduced in 1889, it is now a characteristic and specific ruderal plant for the
heat island of inner Berlin. Natural habitats of the plant are sandy and stony
soils near river banks, rubble footpaths and at the foot of rocks, i.e. special
habitats with little competition. Accordingly, roadsides, cultivated areas and
fallow fields are colonized as secondary habitats. Under natural conditions
the area covered by such open, low-competition habitats in Central Europe is
quite small. The open calcium-rich sandy to gravelly habitats which have been
created by man have made possible the appearance of *Chenopodium botrys*.
None the less, large and lasting colonies of this species are found north of the
Alps only in Berlin, Mannheim and Lille. The colonies in Stuttgart,
Saarbrücken and Leipzig are unstable or have disappeared.

   Among the woody plants on inner city vacant land in Berlin, *Robinia*

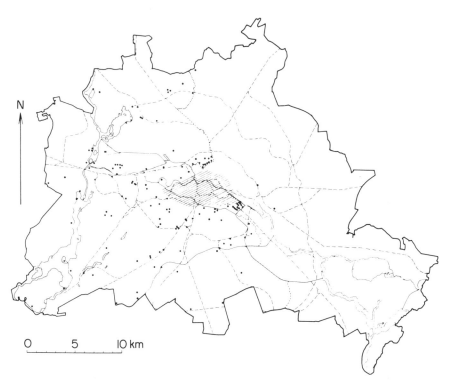

FIG. 19.3. Distribution of *Chenopodium botrys* in Berlin from 1947 to 1971. Area of
continuous distribution is hatched (Sukopp 1971).

pseudacacia stands cover the greatest area. On rubble–mortar substrate, a calcium-containing loose syrosem develops into a refuse pararendzina under *Robinia* in the course of vegetation and soil development (Fig. 19.4).

A vegetation map of the former Potsdam Railway Station in Berlin (Sukopp & Kunick 1976) shows dominant stands of aspen in habitats with a more favourable water regime, and sand birches in dry habitats. *Pinus sylvestris*, the main species of the Berlin forests, can be found sporadically in the inner city in very shallow topsoil habitats. In all other habitats the forest pine is out-competed by other more rapidly growing woody plants, becoming stunted under their leaf canopy. From the various ages of the pines it can be concluded that each year diaspores of *Pinus sylvestris* reach the inner city from the forest areas which are more than 6 km away.

Among the numerous neophytic woody plants, not only the variety of species but, in particular, their lush development is surprising. Besides *Robinia pseudacacia*, the tree of heaven (*Ailanthus altissima*) is spreading on inner city vacant land. This vigorous drought-tolerant tree settles on extreme habitats such as railway land and adjacent to buildings, but is also frequently found in green open spaces. In derelict areas *Ailanthus* is able to establish large

FIG. 19.4. Derelict land with soil consisting of rubble. Because of the loose stratification of the brick and the special climate of the city the sites are typically arid and colonized by *Robinia pseudacacia*. See also Table 19.3—ruderal soil (Berlin-Kreuzberg, Schöneberger Hafen, Sept. 1965).

connected shrubbery (polycormons) through suckering. Figure 19.5 shows the main distribution in the warmest areas of Berlin. Like *Chenopodium botrys*, *Ailanthus* is a typical thermophilous 'city plant' (Kowarik & Böcker 1984).

Buddleja davidii, which often occurs in rubble habitats in the subatlantic region of Central Europe, has been growing wild in Berlin only for the last 25 years. Presumably, there were not enough specimens in Berlin at the end of the War to induce mass dissemination. *Buddleja* was often planted for economic reasons as a rapidly growing, undemanding and inexpensive woody plant in the green areas of post-war housing developments.

FIG. 19.5. Distribution of tree of heaven (*Ailanthus altissima*) in West Berlin. Dots: *Ailanthus*, hatching: mean annual temperature—the darker the warmer (zone 1 8·1°, zone 5 10·5°, zone 6 11·3°) (Kowarik & Böcker 1984). Zones 1–4 are not differentiated; ▨ zone 5; ▨ zone 6.

The observations made so far enable us to summarize the spontaneous woody plant colonization of inner city vacant land in an ecological diagram (Fig. 19.6; Sukopp *et al.* 1974). It is evident that in the carbonate-containing and dry habitats that predominate in the inner city, neophytic species (*Robinia*, *Ailanthus*, *Clematis* and *Buddleja*) play an important role. On the other hand, in moist to damp habitats native woody plants such as *Acer plantanoides*, *Ulmus scabra* or *Sambucus nigra* dominate which, under natural conditions, are components of nutrient-rich mixed deciduous forests. After 1945 the extensive derelict lands in the destroyed city also allowed the development of a widely varying habitat for vertebrates.

Management

The criteria on the following page are used to evaluate habitats for their importance in preserving biological diversity.

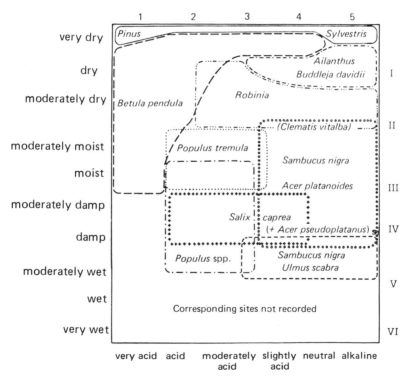

FIG. 19.6. Woody plants of the Berlin inner city. Range of tolerance of soil moisture and acidity for the most important trees and shrubs.

1   The assessment of existing species and biocenoses based on the degree of threat as shown in the *Red Data Lists* of endangered flora and fauna in West Berlin.

2   The suitability of habitats for a representative habitat network based on a biogeographical and urban zone classification of the city.

3   The suitability of habitats for a network of refuge and stepping-stone sites, whereby the size of sites and the distance between habitats plays an important role in the long-term survival of the species which they contain.

A biogeographical and urban zonal classification is fundamental for the evaluation of habitats in the inner and outer city. Similar habitat conditions are delineated as habitat types, and comparisons of individual habitats are only possible within a particular habitat type. The biogeographical units that are delineated for West Berlin's outer city are not applicable to the inner city. Since habitat conditions for species in the inner city are determined by the type, degree and age of developed areas, e.g. residential areas of particular densities, development zones have to coincide with habitat types. In addition, the species in these zones are influenced by previous land uses, e.g. agriculture or forestry. Residential areas, for example, are divided into the following development zones:

(a) flats dating back to the founder period—1880 to 1914;

(b) blocks of flats with an inner court and linear flats;

(c) high rise flats;

(d) flats in allotment and orchard settings;

(e) flats in woodlands or parklands.

Each of these zones has its own distinctive vegetation types and structures, site conditions and intensity of use and management by which the habitat type can be documented and assessed. Additionally, each zone has characteristic pressures responsible for species reduction as well as measures for protection, planning and management. Since possible threats to these habitat types are known in general, it is possible to devise planning strategies to counteract them. Detailed inventories, classifications and assessments are needed to determine the areas which will help secure the biological diversity of the city. Statutory protection and strategic planning are the means by which this can be achieved.

Decisions affecting the survival of wild flora and fauna are already decided at the highest level of structure planning. Therefore, it is essential that strategic subject planning has influence at this level. For structure planning in West Berlin the following policies have been established (Auhagen & Sukopp 1982).

Prevention of avoidable impacts on natural systems and landscapes. This means that further land demands for development have to be minimized; for instance, land sealing through buildings, roads, parking areas, etc. has to be reduced or reversed. Also, if open spaces in the city are needed for development, other open spaces must be provided elsewhere.

Establishment of priority areas for nature conservation. Since agriculture has been very much reduced in the outer city, the remaining larger areas with their hedges, ponds and particular flora and fauna need to be preserved. In the inner city, areas have to be preserved where the colonization of flora and fauna dates as far back as the founder stage. Of particular importance are old industrial sites, railway tracks and areas with pioneer vegetation.

Consideration of natural development in the city. By law, nature conservation applies to the city as a whole, consequently land where natural development can occur must be provided in the inner city.

Provision of historical continuity. Areas need to be given priority where the same land-use has occurred for long periods and in which flora and fauna are long established.

Preservation of different habitat conditions. All extreme site conditions—dry, wet, oligotrophic—should be maintained when development is planned and should be given priority, because the elimination of these site conditions is one of the main causes of species reduction.

Differentiation of use intensities. In all measures for use, protection, planting and management, the differences between areas of intensive use have to be considered. By concentrating use in one area, other areas are relieved of pressures.

Preservation of large undivided open spaces. The ecological functions of large undivided open spaces cannot be substituted by several small areas, due to greater disturbance and edge effects. The number of bird species on the relatively undisturbed Peacock Island (Pfaueninsel), for example, is greater than in the Central Park (Tiergarten) which is three times larger but divided by several broad avenues.

Establishment of an open space network. In order to reduce the isolation of insular habitats, open spaces need linear connections as well as stepping-stone sites. Of greater importance are disused railway tracks which provide a continuum from the outer to the inner city.

Preservation of the diversity of typical elements in the city landscape. Diversity of species and natural communities can only be preserved through a large variety of land uses.

Functional integration of buildings into the ecosystem. The disturbance and disruption of ecological interactions caused by buildings should be minimized. In densely developed areas, natural growth on roofs as well as facade vegetation increases the habitats available for wild flora and fauna.

These general planning policies will be further refined and linked to a spatial concept in the Nature Conservation Programme, which was finalized at the end of 1983. As an integrated part of the Landscape Programme and the Structure Plan, the objectives and measures will be given statutory authority. Further detailed objectives and measures will then be laid down in the local landscape plan.

Currently about twelve local landscape plans are being developed for West Berlin. One of the most important, and probably most interesting, is the Landscape Plan Anhalter Bahnhof. In the plan area lies the largest open space—without designated land uses—within West Berlin's inner city. Up to the end of World War II this area contained one of the largest rail cargo terminals and passenger terminals of the then capital city. After the War the right of use was handed over to the East Berlin Railway Authority, which partially used the tracks until the erection of the Berlin Wall. For approximately the last 20 years natural revegetation has taken place (Fig. 19.7). In the past few years, the East Berlin Railway Authority has indicated its willingness to give up its right of use under certain conditions. Consequently, there has been great speculation over the future use of this 68-ha open space.

At the end of the 1970s this area's vegetation was studied, and its

FIG. 19.7. Abandoned railway land. Gravel beds are colonized by trees, mainly birch; grasses and tall herbs have colonized sandy soils (Berlin-Schöneberg, Südgüterbahnhof 1982, photo by S. Stern).

importance recognized by the Nature Conservation and Landscape Management Authority. At first, overriding economic interests made the acceptance of large parts of this area for nature conservation purposes seem unrealistic. However, since a landscape plan with statutory authority is being developed for this area, there is a good chance that most of the area will remain an open space containing the habitats valuable for nature conservation. Without such a strategic planning instrument the chances for protecting these wild habitats would be slim.

CONCLUSION

The material presented in this paper indicates the need for a strategic plan, based on scientific arguments, for protecting and reintroducing wild plant and animal life into the city. Close cooperation between natural scientists and landscape planners is required for this task. Conservation and reintegration of nature into the city—that part of the environment that man has transformed to the greatest extent—is an important endeavour, not only for the improvement of environmental quality, but also for that vital change in man's view of himself in relation to nature.

REFERENCES

Auhagen, A. & Sukopp, H. (1982). Auswertung der Liste der wildwachsenden Farn and Blütenpflanzen von Berlin (West) für den Arten- und Biotopschutz. In *Rote Listen der gefährdeten Pflanzen und Tiere in Berlin (West)* (Ed. by H. Sukopp & H. Elvers), pp. 5–18. Landschaftsentwicklung und Unweltschutz 11.

Auhagen, A. & Sukopp, H. (1983). Ziel, Begründungen und Methoden des Naturschutzes im Rahmen der Stadtentwicklungspolitik von Berlin. *Natur und Landschaft*, **58**, 9–15.

Ermer, K., Kellermann, B. & Schneider, Ch. (1980). *Wissenschaftlich-methodische Grundlagen für ein Landschaftsprogramm Berlin.* Landschaftsentwicklung und Unweltforschung 2.

Falinski, J.B. (Ed.) (1971). Synanthroposition of plant cover. II. Synanthropic flora and vegetation of towns connected with their natural conditions, history and functions (Polish; English summary). *Materialy Zakladu Fitosocjologii Stosowonej, UW Warszawa-Bialowieza*, **27**, 1–317.

Haeupler, H. (1974). *Statistische Auswertung von Punktrasterkarten der Gefäßpflanzenflora Süd-Niedersachsens.* Scripta Geobotanica 8, Göttingen.

Henke, H. (1979). Role of the landscape architect in biological and ecological conservation. In *Landscape Towards 2000—Conservation or Desolation* (Landscape Institute) (Ed. by D. Smith), pp. 65–66, William Caple, United Kingdom.

Henke, H. & Krause, C.L. (1980). *Study on Environmental Effects for Conservation Planning* Special Report 5, (Ed. by National Committee), pp. 95–110. UNESCO Programme MAB, Fed. Rep. of Germany.

Horbert, M., Blume, H.-P., Elvers, H. & Sukopp, H. (1982). Ecological contributions to urban planning. In *Urban Ecology* (Ed. by R. Bornkamm, J.A. Lee & M.R.D. Seaward), pp. 255–275. Blackwell Scientific Publications, Oxford.

Kowarik, I. & Böcker, R. (1984). Zur Verbreitung, Vergesellschaftung und Einbürgerung des Götterbaumes (Ailanthus altissima (Mill.) Swingle) in Mitteleuropa. *Tuexenia*, **4**, 9–29.

Kunick, W. (1982). Comparison of the flora of some cities of the Central European lowlands. In *Urban Ecology* (Ed. by R. Bornkamm, J.A. Lee & M.R.D. Seaward), pp. 13–22. Blackwell Scientific Publications, Oxford.

Sukopp, H. (1971). Beiträge zur Ökologie von Chenopodium botrys L.; I. Verbreitung und Vergesellschaftung. *Verhandlungen des Botanischen Vereins der Provinz Brandenburg*, **108**, 3–25.

Sukopp, H. (1973). Großstadt als Gegenstand ökologischer Forschung. *Schrift. Ver. Verbreitung naturwiss. Kenntn. Wien*, **113**, 90–140.

Sukopp, H. (1982). Erfahrungen bei der Biotopkartierung in Berlin im Hinblick auf ein Schutzgebietssystem. *Schriftenreihe Deutscher Rat für Landespflege*, **41**, 69–73.

Sukopp, H., Blume, H.-P., Chinnow, D., Kunick, W., Runge, M. & Zacharias, F. (1974). Ökologische Charakteristik von Großstädten, besonders anthropogene Veränderungen von Klima, Boden und Vegetation. *TUB-Zeitschrift, Techn. Univ. Berlin*, **6 (4)**, 649–688.

Sukopp, H., Blume, H.-P. & Kunick, W. (1978). The soil, flora and vegetation of Berlin's waste lands. In *Nature in Cities* (Ed. by I.C. Laurie), pp. 115–132. John Wiley, London.

Sukopp, H. & Kunick, W. (1976). Höhere Pflanzen als Bioindikatoren in Verdichtungsräumen. *Landschaft + Stadt*, **8**, 129–139.

Sukopp, H. & Werner, P. (1982). *Nature in Cities*. Nature and environment series 28. Council of Europe, Strasbourg.

Sukopp, H. & Werner, P. (1983). Urban environments and vegetation. In *Man's Impact on Vegetation* (Ed. by W. Holzner, M.J.A. Werger & I. Ikusima), pp. 247–260. W. Junk, The Hague.

Walters, S.M. (1970). The next twenty-five years. In *The Flora of a Changing Britain* (Ed. by F. Perring), pp. 136–141. Classey, Middlesex.

20. ROADSIDES: A RESOURCE AND A CHALLENGE

J. R. THOMPSON

Estates and Valuation Department, Essex County Council, Chelmsford

SUMMARY

In the 25 years since the first motorway in the UK was opened, the role in the landscape of the verges and associated areas of land has changed and they are becoming increasingly recognized as reserves for wildlife. The processes of site preparation and planting leading to the established roadside verge are described and some examples given of the way sites have developed.

A brief review of the environment peculiar to roadsides is presented in the context of how this might limit their future wildlife potential. Steep concentration gradients of environmental factors, e.g. salt and lead, ensure that most major roads have considerable areas within their perimeters which are relatively unaffected. In the areas close to the carriageways, environmental effects are themselves proving of interest, with the development of saline plant communities and enhanced populations of certain species of insects.

The paper then extends the idea that if the role of verges is to include and emphasize nature conservation, then established landscaping techniques need modification, with far more attention to their present and future maintenance.

INTRODUCTION

Roads and some kind of edge to the road are as old as man's need to travel, but the increase in the numbers and sizes of roads in response to expanding demand has been a feature of the last 50 years. Modern roads, especially motorways, are designed with gentle curves and gradients, these being achieved by means of frequent cuttings and embankments, often of considerable size. Multi-lane roads, particularly motorways, have especially wide verges which in the UK vary from 5 m to more than 50 m, and today the total road network incorporates over 200 000 ha of central reserves and roadside verges.

These areas are contoured, sown with a mixture of grasses and frequently planted with trees in order to integrate the new road into the landscape through which it passes. The first section of motorway in the UK, the Lancaster by-pass was opened in December 1958: 25 years later seems an

The views expressed in this paper are those of the author and do not necessarily represent those of Essex County Council.

appropriate point to examine how well the landscaping objective has been achieved, whether other approaches might be more effective and whether other objectives such as nature conservation would be appropriate.

SITE PREPARATION AND PLANTING

In most major road schemes the topsoil is stripped off the site and stored for the duration of the roadworks. Newly created cuttings or embankments will become very compacted due to the repeated passages of heavy plant and the need to ensure the stability of embankments. After the slopes have been carefully graded the topsoil is then spread at depths varying from 10 to 20 cm. The quality of this topsoil may well have deteriorated during moving and storage. The whole surface is then usually sown with the British Standard seed mixture for roadside verges, the composition of which is shown in Table 20.1. This contains a high proportion of perennial ryegrass because of its ability to grow easily and vigorously and so stabilize the banks (Fig. 20.1).

Planting of trees and shrubs may take place at the same time, or a year or two later. Each tree or shrub is individually planted in a pre-prepared site, and any losses in the first 2 years are replaced. Weed control is also necessary in these early stages. Small, 2-year old forestry transplants are considered to be suitable in size, but stock about 2 m high may be used where a rapid screen is required or in small areas. Container-grown material may be used to good effect especially in difficult sites.

During the 3-year period between 1969 and 1972, when the modern road network was expanding rapidly, 3 228 000 trees and shrubs were used by the Department of the Environment. In general, native species have been planted, the choice being determined by site factors and the species already present in the landscape through which the road passes. Heavy use has been made of sycamores (*Acer pseudoplatanus*) and Scots pine (*Pinus sylvestris*) to act as a nurse crop for more desirable species. On older motorways the thinning of these two species has already been started but the process has far to go.

TABLE 20.1. British Standard seed mixture for roadside verges

	Species	Percentage
Perennial ryegrass (S23)	*Lolium perenne*	45
Red fescue (S59)	*Festuca rubra*	15
Smooth-stalked meadow grass	*Poa pratensis*	15
Crested dog's tail	*Cynosurus cristatus*	15
White clover (S100)	*Trifolium repens*	10

FIG. 20.1. M25 near Brentwood showing large area seeded with the standard mix: potential wasted? September 1983.

Dunball (1978) observed that schemes planted in the early 1960s have taken about 15 years to produce much visual impact and that improved planting techniques have reduced this period to 10–12 years. He considered that improvements in site conditions resulting from changes in engineering standards offered the best prospect for further reductions in this period.

THE POSITION AFTER 25 YEARS

There is little doubt that the tree planting during this period has already achieved much that was intended; successful plantings may be seen on the M1, M45, M50, and on the M6 in the older sections. Successful plantings also exist in a wide range of places on roads other than motorways. However, much work in thinning and general maintenance is required before mature trees develop. The situation in areas originally intended to be kept as grassland is far from satisfactory, especially on newer motorways. The main reason for this is lack of maintenance. In July 1975 the Department of the Environment instructed that grass-cutting on land forming part of trunk roads and motorways was to cease as a regular practice, and that grass was to be cut only in certain restricted places and circumstances. Since then, other than a narrow band close to the carriageways, the majority of verges have not been cut and their appearance has changed dramatically. As a result the newer roads have a very untidy appearance, with some of the more vigorous weeds appearing in abundance, e.g. spear thistle (*Cirsium vulgare*), dock (*Rumex obtusifolius*), nettle (*Urtica dioica*) and hogweed (*Heracleum sphondylium*). In contrast, on the older verges a wide variety of herbaceous species now occur including primroses (*Primula vulgaris*) and cowslips (*Primula veris*) in places on the M1 and wild daffodils on the M50 (Fig. 20.2). These sites would have been regularly cut before 1975. Also, the potential for invasion of more interesting species from adjacent farmland was probably greater than at present because of the continued use of herbicides and the depletion of the seed banks in these farmland soils.

For the majority of landscapes through which major roads pass, the verges, now rarely green and close cut, appear as discrete zones in themselves and quite different in appearance from the countryside which surrounds them. Without a return to increased cutting, which on economic grounds seems unlikely, the roadside verge will remain an entity rather than a link in the landscape.

In some areas there has been a conscious attempt to highlight the verge as a feature of the landscape. Examples of this approach are currently to be seen in the use of ivy and cotoneaster as groundcover on the Fletton Parkway in the new town of Peterborough, in some plantings in Germany and France, and in many examples in the USA associated with the concept of 'roads beautification'. On the whole such plantings tend to be nearer to suburban areas.

FIG. 20.2. M50 near Ross-on-Wye, spring 1974. Growth of wild daffodils and gorse on the verges.

For the majority of verges, their future role deserves serious consideration. Way (1977) has reviewed their importance as nature conservation areas. There is indeed much evidence showing the conservation value of verges on smaller roads with low traffic densities. Local naturalists and Naturalists Trusts often attempt to liaise with the various highway authorities to ensure that the value of a particular verge is recognized and that no operations are carried out to damage the site which could be avoided. In 1982, the Essex Naturalists Trust proposed to Essex County Council that fifty-six special verges be designated; these are marked with posts making it quite clear to maintenance crews that this is the case. The sites in question contain quite a number of local plant rarities and are mostly on small rural roadsides. Bradshaw & Chadwick (1980) list twenty-seven very rare plants in Britain which occur on roadside verges. Before advocating the use of verges on major, heavily-used roads as areas of potential nature conservation, it is useful to question whether their often rather severe environment would preclude such use.

THE ENVIRONMENT OF MOTORWAYS AND MAJOR HIGHWAYS

Much information about the environment of motorways in the UK has become available from a study undertaken at Imperial College under Professor A. J. Rutter, contracted to the Transport and Road Research Laboratory and at the instigation of the Secretary of State for the Environment's Advisory Committee on the Landscape Treatment of Trunk Roads (Colwill, Thompson & Ridout 1976; Colwill, Thompson & Rutter 1982). A Symposium organized jointly by the British Ecological Society and the Transport and Road Research Laboratory (Colwill, Thompson & Rutter 1978) contains many useful summaries.

The initial interest of the study was the feasibility of planting central reserves of motorways with shrubs to act as anti-dazzle screens as has been practised on the Continent, especially Germany, for many years. It was thought likely that such plantings would prove difficult to establish in view of the reported damage from the use of de-icing salt in Germany and elsewhere (Sauer 1967; Hanes *et al.* 1970). This led to a thorough analysis of the environment associated with major roads. The environmental factors concerned may be divided into those independent of traffic and those which are traffic dependent. The traffic-independent factors include the microclimate of the site, ambient pollution levels, construction effects such as poor soils and impeded drainage, and the use of salt as a means of snow and ice control. The traffic-dependent factors are primarily related to vehicle exhausts emissions. These include carbon dioxide, carbon monoxide, a range of hydrocarbons, the oxides of nitrogen and particles including compounds of lead and other metals. There are further products from the abrasion of tyres, the road surface, clutch and brake linings, and from oils.

Analysis of the soils and vegetation from various transects taken across motorways show that central reservations experience the highest concentrations of various pollutants. Concentrations decrease rapidly on the outside verges. A typical distribution pattern is shown in Fig. 20.3, which shows the concentrations of exchangeable sodium in soils sampled along a transect across the M62. In major roads or motorways with (in the UK) a hard shoulder of about 3·25 m, the concentration in soils at 2 m into the verge is less than half that at the mid-line of the central reservation. At a distance of 5 m from the hard shoulder the concentration of salt in the soil is negligible.

On the verges the transfer of salt from the carriageway is mainly by spray, but when snow ploughing is necessary the salt is deposited with snow at the sides. Saline spray may enter plants by two different pathways. Some is intercepted on the above-ground parts and, if not washed away by rain, may directly enter the plant. Salt-laden spray which is not intercepted by plants may enter the soil from where it can be taken up by the roots. These two ways of entering plants may have different significances at different distances:

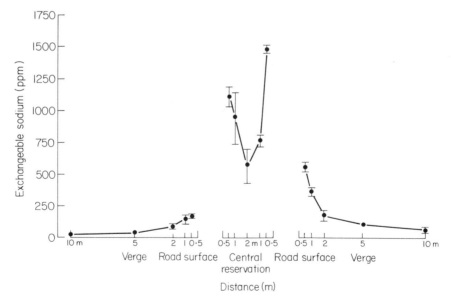

FIG. 20.3. Distribution of sodium in soils on a typical transect across the M62 motorway, sampled in April 1974. (From Colwill, Thompson & Ridout, 1976; bars represent 95% confidence limits.)

although most of the spray is deposited near the carriageway, small droplets may be blown to greater distances and spray damage to vegetation has been detected up to 40 m from the highway in other countries. Colwill, Thompson & Rutter (1982) report cases of severe browning of the foliage of Scots pine and spruce up to 15 m from the carriageway after the winter of 1978–79, and widespread damage to conifers and gorse (*Ulex europaeus*) after the very severe winter of 1981–82, but no significant spray damage to deciduous shrubs on the verges. The German reports of spray damage are perhaps a reflection of the larger proportion of conifers in their roadside landscapes.

It has sometimes been suggested that the use of de-icing salt will cause a progressive increase in the salinity of soils next to roads. But the results of this study indicated that although the salt application rates differed greatly from year to year and from area to area, with an average salt application rate a steady state is reached in the soil after 3–4 years for sodium content and after 1–2 years for chloride.

The extensive survey work which this study involved, allowed a model to be developed from a consideration of salt application rates, winter rainfall and trials on various shrubs. From this a map has been prepared (Fig. 20.4) identifying geographical areas where a salt hazard to vegetation in the central reserve may exist. Though not directly applicable to the verges in quantifiable terms, zones 3 and 4 would be the areas most likely to be affected, especially where there is no hard shoulder. On motorways in zone 4, verges have been successfully planted with a range of deciduous but hardy shrubs.

Thus, salt as a factor clearly limiting growth on roadside verges is largely limited to a band close to the edge of the road and especially where there is no hard shoulder. This would account for the interesting reports of the spread of maritime species along major roads from the A1 and M1 and south into Kent. Species reported include orache (*Atriplex hastata*) and the grasses *Hordeum marinum*, *Puccinellia distans* and *Puccinelia maritima*.

Of the traffic-dependent factors, the contamination of roadside soils by lead has been widely reported, but there is little evidence to suggest that the vegetation on the verges is being adversely affected. However, increased tolerance to lead in roadside populations of *Plantago lanceolata* has been described by Wu & Antonovics (1976). Lead can pass to herbivores and their predators, as reported in studies on roadside verges (Williamson & Evans 1972, 1973) and becomes concentrated along invertebrate food chains (Price, Rathke & Gentry 1974). However, lead as a factor affecting roadside habitats is likely to decrease in significance as lead concentrations in petrol are progressively reduced, and particularly if the Royal Commission on Environmental Pollution (1983) recommendations for lead-free petrol are implemented.

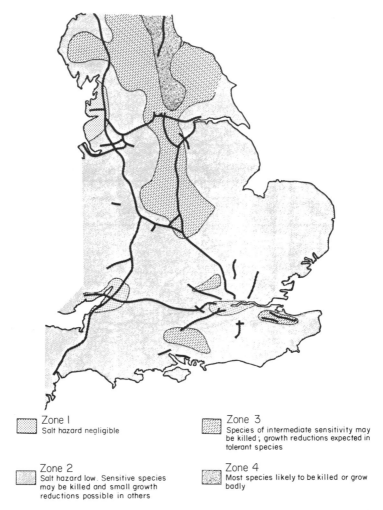

Zone I
Salt hazard negligible

Zone 3
Species of intermediate sensitivity may be killed; growth reductions expected in tolerant species

Zone 2
Salt hazard low. Sensitive species may be killed and small growth reductions possible in others

Zone 4
Most species likely to be killed or grow badly

FIG. 20.4. Predicted zones of salt hazard to shrubs on central reserves of motorways in England. (From Colwill, Thompson & Rutter 1982.)

A second factor which is dependent on traffic density is the concentration of oxides of nitrogen (NO_x). Concentrations of 80 pphm may occur during the daytime beside busy roads, compared with a mean maximum concentration of 5 pphm in clean air (Mansfield 1978). Plants take up both NO and NO_2 through the leaves, suggesting that near roads the nitrogen content of plants may be increased. In view of the importance of nitrogen in the diet of insect herbivores for growth and reproduction (McNeill & Southwood 1978), it is possible that any increase in the nitrogen level would stimulate growth of insect populations. Port & Thompson (1980) reported outbreaks of the buff tip moth (*Phalera bucephala*) on a hedge on the A423(M) and the gold tail moth (*Euproctis similis*) on hawthorn hedges on the M63. Flückiger, Oertli & Baltensweiler (1978) reported that plantings of hawthorn on a motorway in Switzerland are frequently attacked by the aphid (*Aphis pomi*) especially in the central reserve, and that the number and size of infestations by insects declines with distance from the motorway. They correlated large populations with high concentrations of free amino acids in roadside plants. Levels of total nitrogen were found to show the same pattern as that of other pollutants, when herbage taken along a transect across the M1 was analysed (Port & Thompson 1980) (Fig. 20.5).

POTENTIAL OF HIGHWAY AND MOTORWAY VERGES FOR NATURE CONSERVATION

From the information so far presented there would appear to be little reason why the verges of major roads should not become more useful as visual amenity and nature conservation areas, and examples already presented have

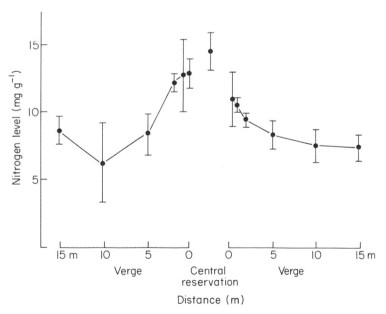

FIG. 20.5. Levels of nitrogen (mg g⁻¹) found in plants on a typical transect across the M1 motorway, sampled in September 1975 (bars represent 95% confidence limits) (Port & Thompson 1980).

hinted at their potential. The environment should in general cause few problems and the gradients in concentration of the various factors described should add to the diversity of the habitat. With nature conservation and amenity as an objective, there may often be opportunities for the landscape architect to create new habitats in the design phase. The creation of the small lake at the Tebay intersection on the M6 is a good example of what can be achieved (Fig. 20.6). Separation of the carriageways with wider central reserves, again as on the M6, is a useful way to reduce severance and a similar approach has been used in Hessen in Germany where a motorway passes through a block of established woodland (Fig. 20.7).

The conservation value of verges on established roads will in future depend to a large extent on their maintenance. The abrupt change in 1975 from very frequent cutting and spraying to virtually no cutting is profoundly significant. Normal successional change is already leading to the appearance of large areas of brambles on some motorway verges and grassland is being replaced by scrub. Gorse alongside the M3 in Surrey has become so well established and dominant that it has had to be cut to allow other shrubs and trees to grow. This is not a new problem. British Rail had to pay large sums for the clearance of scrub which resulted from the cessation of the mowing of railway banks in the mid-sixties (Way 1978). The cheapest and most effective means of bramble and scrub control are to prevent the plants ever establishing in the first place.

FIG. 20.6. M6 junction 38 at Tebay: creation of useful wetland habitat with early colonization by *Typha*.

FIG. 20.7. Planting in Hessen showing how the wide reservation passing through the woodland may help to reduce severance, and how plantings in the foreground soften the visual effect of dissecting the woodland.

Way (1969 and 1977) showed, in a series of experiments on the management of roadside verges, that it was possible with a single cut in late May, or at the most a second in June, to maintain the height of the sward at 30 cm. This is because the flowering stems, when cut, do not regrow and the height of the vegetation is then dependent on the development of the basal leaves. If a height of 15 cm is required, this can only be achieved by very frequent cutting of the leaves of the plants, perhaps every 3 weeks or so, at the beginning of the season. Low growing herbs were favoured where the vegetation was kept shorter throughout the season and taller growing herbs where it was kept longer. Dunball (1969) reported that the earlier policy for maintenance of motorway verges had been to keep grass cutting to a minimum to encourage the development of the native flora, but rapid invasion by weedy species has necessitated a change of plan. Revised instructions were to cut the grass as frequently as possible during the first 2–3 years (together with the use of herbicides) to produce a tight clean sward: 25 years later we are witnessing the vegetational response to a cessation of effective maintenance.

But it is not only the grasslands where maintenance is lacking. Although a start has been made with some thinnings alongside the M1, there are now a number of areas along the older motorways where weak trees await thinning, and are starting to lean out over the hard shoulders. Whilst the conservation value of these areas is not so directly affected, it would be a pity to see wind damage occurring and possible removal of woodland for reasons of highway safety. The task of maintaining this large number of trees would seem a daunting one: who should do the work and who should pay for it? There is probably no easy solution, but it may be possible to enter into maintenance agreements with adjoining landowners who might derive some benefit from the wood which will become available.

The idea of maintenance agreements might usefully be explored in various motorway verges. Local Naturalists Trusts might wish to maintain sites of some recognized conservation value. In other areas, where perhaps a motorway passes through land used for grazing by sheep or cattle, the verges could be re-incorporated into the farms from which they originally came. Steep verges would, of course, have no arable value but with the danger of lead contamination now likely to decrease, some useful grazing might be gained and the conservation value of the land enhanced.

Whether the purpose of the roadside verge is purely landscaping or wildlife conservation, a far greater degree of maintenance is required and yet seems unlikely to be forthcoming. Sadly, it is the processes by which the vegetation was established in the first place which are largely responsible for the maintenance need today. These processes are the use of topsoil on a routine basis, the use of a mixture of grasses with a high proportion of ryegrass and the use of traditional methods of plantation woodland establishment rather than an ecological approach.

NEW APPROACHES TO THE ESTABLISHMENT OF VEGETATION ON ROADSIDE VERGES

The routine use of topsoil on verges is considered to be of value in the establishment of vegetation. But since most modern major roads pass through agricultural land, this will usually be the source of any topsoil used. The soils are likely to be deficient in seeds of species of conservation value and may well contain a seed bank of agricultural weeds. Soils rich in nutrients will support very vigorous vegetation from which more interesting species are excluded by competition. There will also be a greater need for maintenance. Subsoils and underlying rocks are much less nutrient-rich, especially in nitrogen, and will, if left, become colonized by particular collections of plant species. Some of the most interesting and attractive areas of motorway environs are where the underlying rocks have been left exposed, such as in very steep cuttings, the best example being perhaps where the M40 passes through the chalk of the Chiltern Hills in Oxfordshire (Fig. 20.8).

Verges are routinely established without topsoil in many parts of the USA,

FIG. 20.8. M40 near Stokenchurch: a contrasting approach with steep banks through the chalk, creating an interesting habitat for the future (September 1980).

Germany, France and other countries. But careful attention must be paid to the preparation of the seed bed and to the fertility. In many cases it may be desirable to use a nurse crop such as a legume. Research on these techniques has been in progress for a considerable period in the USA, with a useful review by Blaser *et al.* (1961). Examples may be seen in France of erosion problems with this approach, and work on the early nutrient requirements has been carried out by Hénensal (1971, 1978). Mulches can often be used to good effect. The choice of seed mixes will also have a critical effect on the maintenance requirement; Bradshaw & Roberts (1978) have suggested a less aggressive alternative. This is shown in Table 20.2.

The roadside verge can certainly be regarded as an area where we should attempt to produce a grassland which is attractive and which also has conservation value. Wells, Bell & Frost (1981) have produced a list of seeds mixtures for various soils, but they too recommend the use of a nurse crop for roadside verges. They suggest the annual, Westerwolds ryegrass, but point out that it could swamp the other species on more fertile soils, where it should be omitted from the mixture. The seeds industry now has many of these desirable species available in commercial quantities. Any higher establishment costs will surely be repaid by the results and the reduced maintenance.

Tree and shrub seeding also offers considerable potential for creating vegetation which requires low maintenance, with consequent cost savings.

TABLE 20.2. A suggested less aggressive mixture for roadside verges (Bradshaw & Roberts 1978)

	Species	Percentage
Creeping red fescue, e.g. Dawson or Merlin	*Festuca rubra* var. *rubra*	30
Chewings fescue	*F. rubra* var. *commutata*	25
Creeping bent	*Agrostis stolonifera*	10
Common bent	*A. tenuis*	10
Smooth-stalked meadow grass	*Poa pratensis*	10
White clover, e.g. Kent (S184)	*Trifolium repens*	10
Bird's foot trefoil	*Lotus corniculatus*	5

This has been used with some success in parts of Germany in the absence of topsoil. In one particular example sea buckthorn, *Hippophäe rhamnoides*, was used successfully as a nurse crop. These techniques as they apply to the UK have recently been reviewed by La Dell (1983) and Luke & MacPherson (1983). Basic tree and shrub seeding is cheaper than planting, and even the more sophisticated techniques using mulches are only as expensive as the simplest planting techniques using forestry transplants. The supply of seed is rather specialized but some commercial sources are now becoming available. One of the attractions as outlined by La Dell is the possibility of reducing maintenance by utilizing a process in which self-thinning plays an important role. The original seedling density needs to be carefully controlled as does the composition of the mixture. The technique is perhaps in its infancy in the UK, but early results look promising, and the success in Germany is remarkable.

CONCLUSIONS

After 25 years of motorway planting, many valuable woodland and grassland habitats have been created. Maintenance is now the main hindrance to the development of more attractive and species-rich sites and the outlook for increased maintenance is bleak. The time is now ripe to put new techniques into practice which will require less maintenance and produce more attractive results. We need more acceptance of the idea that new roadside verges can be valuable reserves for wildlife and that they should be managed as such.

REFERENCES

Blaser, R.E., Thomas, G.W., Brooks, C.R., Shoop, G.J. & Martin, J.B. (1961). Turf establishment and maintenance along highway cuts. *Highway Research Board Publication*, **928**, 5–19.

Bradshaw, A.D. & Chadwick, M.J. (1980). *The Restoration of Land.* Blackwell Scientific Publications, Oxford.

Bradshaw, A.D. & Roberts, R.D. (1978). The ecological aspects of establishment on roadside verges. In *The Impact of Road Traffic on Plants* (Ed. by D.M. Colwill, J.R. Thompson & A.J. Rutter) Supplementary Report 513. Transport and Road Research Laboratory, Crowthorne.

Colwill, D.M., Thompson, J.R. & Ridout, P.S. (1976). *Studies of Conditions for Vegetation in the Central Reserves of Motorways: a Preliminary Report.* Supplementary Report 217UC. Transport and Road Research Laboratory, Crowthorne.

Colwill, D.M., Thompson, J.R. & Rutter, A.J. (Eds) (1978). *The Impact of Road Traffic on Plants.* Supplementary Report 513. Transport and Road Research Laboratory, Crowthorne.

Colwill, D.M., Thompson, J.R. & Rutter, A.J. (1982). *An Assessment of the Conditions for Shrubs alongside Motorways.* Laboratory Report 1061. Transport and Road Research Laboratory, Crowthorne.

Dunball, A.P. (1969). Discussion comment, p. 74. In *Road Verges: Their Function and Management* (Ed. by J.M. Way). Monks Wood Experimental Station, Abbots Ripton.

Dunball, A.P. (1978). The establishment of woody plants on motorways. In *The Impact of Road Traffic on Plants* (Ed by D.M. Colwill, J.R. Thompson & A.J. Rutter). Supplementary Report 513, Transport and Research Laboratory, Crowthorne.

Flückiger, W., Oertli, J.J. & Baltensweiler, W. (1978). Observations of an aphid infestation on hawthorn in the vicinity of a motorway. *Naturwissenschaften,* **65,** 654–655.

Hanes, R.E., Zelazny, L.W. & Blaser, R.E. (1970). Effects of Deicing Salts on Water Quality and Biota. Literature review and recommended research. NCHRP Report 91. Highway Research Board, Washington.

Hénensal, P. (1971). La végétation sur les routes et autoroutes. Role—Importance—Problèmes techniques. *Revue Générale des Routes et des Aerodromes,* **468,** 78–101.

Hénensal, P. & Arnal, G. (1978). Grass fertilization problems and experimentation on highway embankments in France. In *The Impact of Road Traffic on Plants* (Ed by D.M. Colwill, J.R. Thompson, & A.J. Rutter). Supplementary Report 513, Transport and Road Research Laboratory, Crowthorne.

La Dell, T. (1983). Techniques No. 44. An Introduction to Tree and Shrub Planting. *Landscape Design,* **144,** 27–31.

Luke, A.G.R. & MacPherson, T.K. (1983). Direct tree seeding: a potential aid to land reclamation in central Scotland. *Arboricultural Journal,* **7,** 287–299.

Mansfield, T.A. (1978). The effects of nitrogen on vegetation. In *The Impact of Road Traffic on Plants* (Ed by D.H. Colwill, J.R. Thompson & A.J. Rutter). Supplementary Report 513, Transport and Road Research Laboratory, Crowthorne.

McNeill, S. & Southwood, T.R.E. (1978). The role of nitrogen in the development of insect/plant relationships. In *Biochemical Aspects of Plant and Animal Coevolution* (Ed. by J.B. Harborne). Academic Press, London.

Port, G.R. & Thompson, J.R. (1980). Outbreaks of insect herbivores along motorways in the United Kingdom. *Journal of Applied Ecology,* **17,** 649–656.

Price, P.W., Rathke, B.J. & Gentry, D.A. (1974). Lead in terrestrial arthropods. Evidence for biological concentration. *Environmental Entomology,* **3,** 370–372.

Royal Commission on Environmental Pollution (1983). *Lead in the Environment.* HMSO, London.

Sauer, G. (1967). Uber Schäden an der Bepflanzung der Bundesfernstrassen durch Auftausalze. *Nachrichtenbl. Dtsch. Pflanzenschutzd. (Braunschw.),* **19,** 81–87.

Way, J.M. (1969). Road verges—research on management for amenity and wildlife. *In Road Verges: Their Function and Management* (Ed. by J.M. Way). Monks Wood Experimental Station, Abbots Ripton.

Way, J.M. (1977). Roadside verges and conservation in Britain: a review. *Biological Conservation,* **12,** 65–74.

Way, J.M. (1978). Roadside verges and their management. In *The Impact of Road Traffic on Plants* (Ed by D.M. Colwill, J.R. Thompson & A.J. Rutter). Supplementary Report 513. Transport and Road Research Laboratory, Crowthorne.

Wells, T., Bell, S. & Frost, F. (1981). *Creating Attractive Grasslands using Native Plant Species.* Nature Conservancy Council, Shrewsbury.

Williamson, P. & Evans, P.R. (1972). Lead: levels in roadside invertebrates and small mammals. *Bulletin of Environmental Contamination and Toxicology*, **8**, 280–288.

Williamson, P. & Evans, P.R. (1973). A preliminary study of the effects of high levels of inorganic lead on soil fauna. *Pedobiologia*, **13**, 16–21.

Wu, L. & Antonovics, I. (1976). Experimental ecological genetics in Plantago. 11. Lead tolerance in *Plantago lanceolata* and *Cynodon dactylon* from a roadside. *Ecology*, **57**, 205–208.

21. NATIONAL PARKS

W. HABER

Lehrstuhl für Landschaftsökologie, Technische Universität München (Munich), Weihenstephan, D-8050 Freising 12, Fed. Rep. of Germany

SUMMARY

The concept of the national park began in the USA as a way of preserving ageless monumental scenery together with its wildlife. Some economic production from the areas was permitted but the basic concept was of undisturbed preservation. The idea spread to other countries, but the difficulty in Europe especially was to find suitable areas. It seemed easier, in fact, to export the idea to Africa where it was thought that resource uses in balanced existence for thousands of years needed suddenly to be terminated. All this was accompanied by an enormous development of tourism which was seen as one of the objectives of all parks.

As a result, the original ecological balances within all National Park areas have been grossly disturbed and worrying changes have occurred. A re-examination of management practice has been essential. The emphasis has gone back to preservation, with the realization that national parks are an essential part of a global conservation system. This does not mean that the areas have to be locked away. Reinstatement of traditional economic uses and other management practices, as well as control over tourist numbers and circulation, is essential. The underlying need is to appreciate that natural ecosystems have a dynamic balance, and that it is to this that planning and management must be related.

THE BEGINNINGS IN USA

It is well known that the national park is a United States' achievement in conservation, which started with Yellowstone in 1872, and was followed by nine other parks within the next four decades, all of them situated in the western part of the continent. In his recent book, Runte (1979) discusses the particular cultural and social context of this rather early achievement in a young and expanding nation which was not very much conservation-minded. It was the influence of writers like Emerson and Thoreau who opened the way for early conservationists and landscape architects like John Muir or F.L. Olmsted. The national parks were also regarded as fulfilling the young country's deeply felt desire to have a national heritage. As Runte writes: '. . . by the 1860s many thoughtful Americans had embraced the wonderlands of the West as replacements for man-made marks of achievement. The agelessness of monumental scenery instead of the past accomplishments of Western Civilization was to become the visible symbol of continuity and stability in the new nation. . . . Here at last—in the blending of the eastern mind and the western experience—was the enduring spark for the American inspiration of national parks.'

But at the same time, establishing national parks was by no means a matter of course in those times when the west was being conquered. The pioneer ethic, pushing so many settlers and adventurers westward, committed them to occupy land and to make money from it. For these people, it was absurd to leave vast tracts of public land intentionally without any use; so the first National Park Bills were really exceptional. It had to be proven that these areas were economically worthless land. In the following decades until 1910, a principle of 'utilitarian conservation' got the upper hand. G. Pinchot regarded a sustainable use of timber, waterpower and other resources of national park lands as fully compatible with conservation objectives. So even a reservoir was built within Yosemite National Park.

The decisive change came when the railway lines made the young parks more easily accessible. Now they could be considered as attractive tourist travel destinations and potential holiday areas. An advertising campaign 'See America first' was launched. A congressman complained in 1915 that the amount of money Americans spent overseas on scenic travel had soared to an estimated $500 million yearly, 'a considerable portion of which goes to scenery that in no way compares with our own. The American people have never yet capitalized their scenery and climate, as they should, while Switzerland derives from $10.000 to $40.000 per square mile per year from scenery which is not equal to our national parks' (Runte 1979). Thus, a kind of cultural nationalism was channelled into both an aesthetic and economic defence of the national parks, which now received broadest support from the public. The principle of utilitarian conservation was given up in favour of comprehensive preservation, with, however, touristic use and accessibility.

In 1916, Congress voted the National Park Act, which states: '. . . the fundamental purpose . . . of the parks . . . is to conserve the scenery and the natural and historic objects and the wildlife therein, and to provide for the enjoyment of the same in such a manner and by such means as will leave them unimpaired for the enjoyment of future generations.' In the same year, to enforce the new act, the National Park Service was established, as a bureau of the US Department of the Interior. Yet the parks were in need of more public support; so visitations were encouraged, many visitor facilities were created, roads were built, and motor traffic with cars and boats was rather freely consented to. Perhaps unconsciously, or not fully realized, the emphasis was laid on enjoyment rather than on preservation of main natural features and scenery.

THE SPREAD OF THE IDEA

In the meantime, the American national park idea—which meant preserving magnificent natural scenery as national heritage and for people's enjoyment—had spread to other countries, but was received with certain reservations in Europe. There were no vast expanses of virgin land suitable for national parks of the American type. Only Switzerland established a national park during that period (in 1914), but its main purpose was long-term scientific research, tourism being discouraged, all the more as the park was set in a rather remote area, in no way part of the most scenic Swiss alpine regions.

In Germany, where the authorities ignored the national park idea, a private association was founded in 1908 with the aim of acquiring sufficiently large tracts of land with natural scenery, and of establishing parks within them; one was to be in the lowlands, one in the central uplands, and one in the Alps. Because of World War I, they succeeded only with the lowland park, a tract of heathland near Lüneburg—at that time regarded as pristine nature!—and called it 'Naturschutzpark Lüneburger Heide'.

The European countries, however, did consider the national park concept for their colonies abroad. But the main idea was not so much the preservation of scenic splendours for public enjoyment as the protection of the rich wildlife, in particular the immense animal populations in the semi-arid open forest and grassland areas of Africa. It was here that wildlife protection entered the national park scene. In 1890, the first game sanctuaries were already established in the then German East Africa which is now Tanzania. But the colonial administrations did not grasp the fundamental differences between American and the envisaged African national parks: (i) that wildlife depends much more on a balanced habitat or ecosystem complex than do scenic features; (ii) that the areas envisaged for national parks were, for a large part, inhabited or at least used by indigenous people who derived their livelihood from them; they would not, and could not, be just driven away, as was done with the Indians in many American parks.

When at last, from 1930 onwards, a considerable number of national parks became established in Africa, they were an outgrowth of western conservation needs, fears, and values rather than of the people whose livelihood was immediately affected. These people did not at all understand why hunting big game, which they had used as a resource for many centuries without exterminating it, was suddenly forbidden. In fact, the danger of extermination had been brought about by European colonizers, both by indiscriminate sport hunting or by clearing their big new farms of wildlife, regarded as vermin. Thus, African people could not but develop a reluctant or even hostile attitude towards their national parks.

THE DEVELOPMENT OF TOURISM

From the 1930s onward, and all the more after World War II, national parks everywhere experienced a dramatic new demand produced by travel, tourist, and recreation development, which has soared to unbelievable and unpredictable dimensions. Individual travelling facilities, provided by motor-cars, coaches, and aircraft, increasing leisure-time, and the growing desire for outdoor recreation of a more and more urbanized, affluent society, induced in all countries a rush to national parks, and therefore a real flood of visitors. The national park staffs were not well prepared to handle this—basically rather positive—new development. In the US national parks alone, in the period from 1940 to 1966, visitor numbers rose from 17 to 130 millions, while the National Park Service had only reckoned with 80 millions. For the year 2000, they envisage more than 350 million visitors and 1 billion visits per year. Thus, the national parks are obliged to receive far more visitors than their share of the national area and their recreation potential would justify.

An increase in area and number of parks became a necessity. Not only in North America, but also in many other countries, new national parks were created, but their status and objectives were different. In post-war West Germany, the only existing, privately-owned heathland park was trampled down by thousands of visitors. Its director, Dr Toepfer, realizing the need of the urban people for large outdoor recreation areas in scenic countryside, developed a programme for establishing twenty-five such areas throughout West Germany, which he called 'nature parks' (Haber 1972). The response of the authorities was overwhelming. Although the West German Nature Conservation Act did not provide for any parks before its amendment in 1976, today there are sixty-four nature parks covering 21·2% of the country's surface.

Essentially the same category of protected areas, with traditional farming and forestry land use, but supplemented by organized outdoor recreation and some conservation, was established in England and Wales by the National Parks and Access to the Countryside Act of 1949, but in this country there was no hesitation to call them national parks. They share with the West German nature parks their location in agriculturally marginal, mostly upland regions, and their character of 'working landscapes' where people reside and earn their living. So when park planners arrived and recreational facilities were created, the people living there were more eager to defend their local interests than to cooperate with the newcomers.

One of the major features of these areas was soon recognized to be the 'working landscape' itself. Because of the inherent difficulties imposed by the terrain on farming in these regions, the working landscape tended to be somewhat old fashioned and without the widespread mechanized simplification of the lowland.

In both countries, the problem of park management has therefore been somewhat unique: how to preserve the small scale of agricultural and other traditional land-use and how to maintain a leisurely pace of change. The objective must be to persuade the people living in the parks to accept economically suboptimal practices of land-use in exchange for additional revenues from park visitors.

In West Germany, however, two national parks have also been established in the original concept, the first having been created in 1969. A group of conservationists who were dissatisfied with the rather weak conservation component of the nature parks demanded a 'true' national park according to North American or African standards. Again there was the question of a suitable area. At last, a very remote, uninhabited, and extensive forest area of about 100 km² in the mountains at the Czech border was proposed as a national park. The German Council of Land Management asked the author to investigate the suitability of this area for a national park, which the Bavarian Government was willing to set aside. The author came to the conclusion that a national park of the American or African type could not be established, because many of the forests were artificial and rather intensively managed. Instead he recommended a kind of large nature park, but with a much stronger emphasis on conservation and wildlife protection, and more strictly controlled public access, which should provide not only recreation, but also environmental education through interpretation of nature (Haber 1969, 1971). The Bavarian Parliament accepted this recommendation and established the Bayerischer Wald Park under the name of National Park in 1969. Its main problems have been firstly a red deer population much exceeding carrying capacity, which had to be reduced against fierce opposition of the hunters in order to stop forest damage (Wotschikowsky 1977), and the large proportion of man-made spruce forests which can only gradually be changed into a more natural state (Haber 1974). Now the park is being hit severely by air pollution, in the form of acid rain.

A second West German national park has been created in the Berchtesgaden area of the Bavarian Alps, and two or three others may follow as a result of the recommendations of Henke (1976). Visitor pressure is, however, already their main problem and will be even more so in the future.

PRESENT MANAGEMENT PROBLEMS

If we return to the North American national parks, it is very clear that they are still suffering from past errors in management decisions and the lack of comprehensive planning because, 40–50 years ago, park management meant: protect from fire, chase out poachers, and build what roads, trails, and other facilities you could afford (McTaggart 1977). Wildlife management had not yet been invented. Given the difficulties of park travel, the official policy to promote tourism, and the lack of comprehension of negative landscape impacts, there is little wonder that in many of the older national parks tourist facilities were located near the sites most favoured by visitors. Added to these were administrative and supporting buildings. So the best scenic areas of the parks often became spoiled by visitor centres or facilities. As a result, automobile traffic and exhaust pollution may now be a worse nuisance than in London's Piccadilly. Round flights across Grand Canyon and Yellowstone increase the traffic noise considerably.

The consequences can be far-reaching. The boat parties on the Colorado River in the Grand Canyon leave about 20 tons of human faecal matter per year, and this amount is expected to double within 10 years. This has become a problem because the flash floods, which used to rush down the canyon as natural cleaning agents for all refuses and eroded matter, are caught behind Glen Canyon Dam upstream (Garrett 1978).

While such bad practices spoiling valuable park areas are being repaired in North American parks—even the removal of many buildings—they continue in the National Parks of Kenya. In Amboseli, for instance, new visitor lodges have been built at the only permanent water source in the whole area, thus threatening survival of the wildlife for which the park was created, because of site destruction and sewage pollution. Moreover, these lodges and visitor centres, though very carefully designed and styled, are so luxurious and expensive that the average African cannot afford to use them. As a result, they have become centres of prejudice, isolating tourists from local people for whom, in any case, national parks are alien reminders of colonial rule (Lusigi 1978). At the same time cars are the only means of transport allowed, due to the presence of dangerous animals, despite the pollution they cause.

Behind the many difficult problems of visitor pressure and consequent spoiling of parks, looms the basic question of the people's enjoyment of nature, as mandated by the US National Park Act of 1916. Certainly, the concept of enjoyment was understood to include such things as viewing, hiking, swimming, canoeing and horse-riding; all activities involving direct communication with nature and improved if that nature is unspoiled by man (Forster 1973). The general public of today expect quite different benefits from a national park: material conveniences and physical comforts, e.g. trailer campsites with laundry facilities, electrical connections and sewage outlets, hotels, shops, cocktail lounges, marinas, aircraft landing strips. Many visitors come to national parks with Disneyland-like expectations, and are disappointed by the realities, regulations, and restrictions necessary to preserve the parks' scenic qualities. And how many people visit national parks, not to experience nature's wonders, but because they are on holiday, have a car at their disposal, money to spend, and want the prestige associated with national park visits in North America?

This development of the national parks, although detrimental to the original aim, was promoted by the influence of politicians, park managers and planners, all without an ecological background. Amenity management has become so much more sophisticated than resource management that resource use threatens to overwhelm the resource preservation which is both the objective and the very basis of the national parks (Dolan *et al.* 1978).

THE MOVE BACK TO PRESERVATION

At the end of the 1960s, there was growing international awareness that national park environments were suffering irreparable damage, even becoming urbanized instead of offering a retreat from the pollution, unrest, and noise of urban environments. Many people, not only the experts, felt that parks should maintain the natural conditions which existed when or before they were established. Dasmann (1982), inquiring when nature conservation in Africa was most effective, gave the answer 'long before the words "nature conservation" were ever spoken. . . . In those days everybody lived in what we now call national parks and scarcely any plant or animal species could be called threatened . . .'

Thus, the emphasis on preservation is returning to the objectives of national parks and to a new awareness that they are, or can be, an effective conservation tool. IUCN's Commission of National Parks and Protected Areas (CNPPA) critically examined the proliferation of the world's national parks and tried to define certain ecological standards which an area was expected to meet before it could be designated as a 'national park'. It proved rather difficult to establish a generally valid definition of a national park. During the 1970s, IUCN shifted conservation philosophy away from national parks as the main desirable status for protected areas. Instead, a broad range of other types of conservation areas were introduced, better suited to a world that is characterized by great natural and political diversity. The categories voted by the IUCN General Assembly at Ashkabad in 1978 are given in Table 21.1. National parks occupy the second rank in this list. IUCN publishes

TABLE 21.1. Categories of conservation areas

Group A
Category I	Scientific reserve/Strict nature reserve
Category II	National park/Equivalent reserve
Category III	Natural monument/National landmark
Category IV	Nature conservation reserve/Managed nature reserve/Wildlife sanctuary
Category V	Cultural landscape/Heritage landscape

Group B
Category VI	Resource reserve
Category VII	Natural biotic area/Anthropological reserve
Category VIII	Multiple-use management area/Managed resource area

Group C
| Category IX | Biosphere reserve |
| Category X | World heritage site (natural) |

occasional lists of all national parks and equivalent reserves of the world. This list, however, does not always express the actual protection status and ecological situation of the parks, but rather how governments and park administrations judge them. The last IUCN national park list appeared in 1982 and contains 2383 national parks and equivalent protected areas covering more than 4 million km^2 in 124 countries. In the 1970s, they experienced a 46% rise in the number of sites and an 82% increase in the total area, a phenomenal growth. That means 2% of the earth's surface appears to be in protected areas; 10% would, however, be desirable.

Many ecologists and conservationists look at national parks with mixed feelings because of their tourist use. They would prefer a system of strict nature reserves, to protect ecosystems, communities and species habitats. This, however, requires the acquisition of large areas which, in turn, requires public support and sufficient funds. As a matter of fact, most existing nature reserves are either not strictly protected, or are much too small, or both, so that their purpose may be questionable. The new biosphere reserves are intended to serve as basic logistic resources for ecosystem research activities within the international MAB programmes, some of which may be incompatible with strict preservation aims, and may include areas which would not be suitable as nature reserves. On the other hand, biosphere reserves may often coincide partly with or incorporate national parks.

Only national parks, which are mostly rather large areas, can provide an effective ecosystem conservation. The fact that they are accessible to the public makes them politically more acceptable than strict nature reserves, and offers important socio-economic advantages. In many less developed countries—or less developed parts of developed countries—national parks are seen as encouragement to tourism or, more generally, as instruments for economic development. If national parks cannot be somehow exploited for the national economy, many of them simply would not exist. Hence, it may sometimes be necessary to set up elaborate plans for tourism in order to have large parks set aside (Forster 1973; Gilbert 1976). The same plans, however, if made by ecologically-minded planners, can serve to prevent from the outset: (i) over-exploitation and (ii) the installation of buildings, roads, cable-cars etc. which would spoil the landscape.

Thus, in spite of their exploitation for tourism, national parks are indispensable parts of our conservation system. Of course, commercial exploitation for tourism must be strictly controlled, and it will even be necessary to exclude certain aspects of tourism and intensive recreation. This results, of course, in a very difficult situation for ensuring effective preservation of natural features. On the other hand, if tourism pressures cannot be handled sensitively, the 'goose that laid the golden egg' will soon be lost to urbanization (Forster 1973). Conservationists, however, should weigh a bearable loss in ecological integrity, caused by visit or use, against a gain in size, diversity, and security of protected areas. As Gilbert (1976) points out, it must never be forgotten that conservation has not only natural and economic limits; it is also poised at the will of any generation, community, or government, and is likely to remain so. Moreover, it must be stressed that national parks are the longest, most continuous, successful and well-established achievement in the whole nature conservation field. In particular, the US national park system which is more than 100 years old is rather firmly rooted in the public; the US National Park Service, founded in 1916, is a stable and respected institution. Today, even in an atmosphere of environmental consciousness, the establishment of such an institution with such powers and rights would be hardly possible (Sax 1983). This is not to deny that the main reason for creating the US national parks was preservation of magnificent, pristine natural scenery. Although this implies protection of flora, vegetation, and fauna as well, wildlife protection—unlike the African national parks— was rarely an incentive for national park establishment. This type of protection is the task of the US Fish and Wildlife Service, another bureau of the US Department of the Interior, and is pursued in the National Wildlife Refuge System, created in 1903 (Doyle 1979).

FUTURE MANAGEMENT

From the ecologist's viewpoint, the growing national park problems must be tackled in two main ways. The first is assessment of sensitivity and carrying capacity of park ecosystems for visitor use, and to adapt this use to them. There are good examples, such as South Africa's Krüger National Park where access is controlled at seven entrance stations. Visitors are required to have reservations for overnight stays at one of the eleven developed areas, and even day visitors must make reservations. During peak periods, each station is assigned a quota, and once the quota is filled, the gates are closed. Strict observation of carrying capacity may prove the simplest aspect of national park use. This requires a different approach to assess the 'success' of a national park. Of course, in a growth- and number-oriented society, visitor numbers were considered the best indicator of success, more meaning better. The success of a national park, however, has to be assessed using the state of its flora, fauna, and ecosystems as indicators, as well as the effective supervision and education of visitors. If the park is an inhabited area, its success must also be derived from the well-being and satisfaction of the people living near or in the park. Visitor numbers, if used at all, should at best remain constant, if not recede!

The second way consists in regarding national parks—even large ones—not as separate spatial units but as larger entities, including the surroundings with which they interact. To give an example, the West German Berchtesgaden National Park has been established as part of a more extensive entity called 'Bavarian Alpine Park', consisting of three zones with different objectives: the national park *sensu stricto*, the recreation zone, and the zone of access and accommodation. This approach has been further elaborated by Western (1976) and especially Lusigi (1978) for Kenya's national parks which Lusigi considers parts or zones of larger 'conservation units'. They can also serve the interests of the pastoralist people living in these semi-arid regions, so their hostility towards the national parks may be overcome. In such a conservation unit, the people, their livestock, and the wild animals may coexist in a manner not too different from the pre-national park conditions. Wildlife can freely roam the whole area as before, but has an undisturbed habitat in the national park from which livestock is excluded. Outside the park, wildlife may be subjected to controlled management and hunting, the benefits of which go to the local people who have always partly depended on wildlife use, in particular during drought times which affect wildlife less than livestock; for example, 3% loss and 40% loss respectively in the 1973–74 drought in Amboseli (Western 1976). Furthermore, the local people receive Government indemnities for the loss of land incorporated into the park, and a fair share of the park revenues. They should also take over as many jobs as possible in the administration and tourist facilities which are to be located outside the parks in the so-called protected areas.

This approach to comprehensive national park maintenance should make the park and its preservation acceptable to local people, and integrate this originally alien institution into the everyday life of young Third World states. As a matter of fact, the approach has quickly found wide attention, and was recommended by the Third World Conference of National Parks held in Bali in October 1982 as a general strategy for the future. In a certain sense, this linking of protected areas and development concerns could be interpreted as a return to Pinchot's principle of 'utilitarian conservation'. It ensures that conservation becomes part of the fabric of human society especially in Third World countries. Thus, Nepal's Chitwan National Park supplies $1 million per year of thatch grass, Venezuela's Canaima National Park protects $20 million worth of hydroelectric energy, and a whole system of protected areas ensures the functioning of irrigation systems in Sri Lanka and Indonesia (IUCN Bulletin 1982). National parks need no longer be regarded as something 'set aside' and remote from everyday human life.

Of course, there are a host of particular problems worrying national park managers, which can only be mentioned briefly. There is the tourists' utilization of the African national parks, depending on the visitors' own, very selective viewing desires. In Amboseli, about 30% of the average game viewing time is spent watching lions alone, and about 80% watching lion, cheetah, elephant, buffalo, and rhinoceros. The resulting pressure on rare and timid animals such as the cheetah is causing concern. Moreover, 10% of the park is used 80% of the time, which results in congestion.

Park planning therefore must try to disperse visitor use of the area and diversify viewing interests. By planning circuit designs on the basis of animal numbers and distributions, on a seasonal and expected long-term basis, it may be possible to keep the level of visitor use in any park zone to well within the natural tolerance of animals, habitat, and terrain (Western 1976).

CONCLUSIONS

At the beginning of this century, in many countries a national park was likely to be part of a continuous wilderness, or at least of an extended scenic landscape of a natural character. Today there is a greater likelihood that a national park will be an isolated fragment of the once continuous natural or semi-natural ecosystems. For each area in a national park or nature reserve, there is a larger area outside of it which is degraded. Thus, national parks concern landscape design and planning in general. Many people yearn for large regions of beautiful and diverse scenery. Islands of parks surrounded by damaged or abused environment with disorderly industrial development, traffic noise and pollution, overgrazing, vast monocultures, or dereliction will not attract many visitors. It is the cared-for landscape both inside and outside the park boundaries—often inappropriately drawn—which is the common aim (Polunin & Eidsvik 1979; Lusigi 1978).

A sound use and management of both national parks and their surroundings requires the closest possible cooperation and mutual open-mindedness between ecologists, landscape designers and planners.

Conservation aims cannot normally be reached by non-management, and the opportunity for skilled management increases with the knowledge of natural processes. It is these processes, i.e. the dynamic state of nature, to which planning and management should be related, rather than to fixed ends. Accommodation of, or adaptation to the dynamic state of nature is by far the easiest, least costly, most effective and most appropriate response and way forward.

REFERENCES

Agee, J.K. (1980). Issues and impacts of Redwood National Park expansion. *Environmental Management*, **4**, 407–423.

Dasmann, R.F. (1982). Toward a dynamic balance of man and nature. The need for new life styles. *Parks*, **7 (3)**, 23–24.

Dolan, R., Hayden, B.P. & Soucie, G. (1978). Environmental dynamics and resource management in the U.S. national parks. *Environmental Management*, **2**, 249–258.

Doyle, R.E. (1979). Our national wildlife refuges: a chance to grow. *National Geographic Magazine*, **155**, 342–349.

Forster, R.R. (1973). *Planning for Man and Nature in National Parks.* IUCN Publications new series no. 26. IUCN, Gland, Switzerland.

Frome, M. (1981). The national parks. A plan for the future. *National Parks*, **Nov./Dec. 1981**, 10–13.

Garrett, W.E. (1978). Grand Canyon. Are we loving it do death? *National Geographic Magazine*, **154**, 2–51.

Gilbert, V.C. (1976). Biosphere reserves and national parks. *Parks*, **1 (2)**, 12–14.

Haber, W. (1969). Gutachten zum Plan eines Nationalparkes im Bayerischen Wald. *Schriftenreihe des Deutschen Rates für Landespflege*, **11**, 8–23.

Haber, W. (1971). 'Bayerischer Wald' National Park, Federal Republic of Germany. *Biological Conservation*, **3**, 313–314.

Haber, W. (1972). Nature parks and the new Bavarian national park in West Germany. In *Aspects of Landscape Ecology and Maintenance*. pp. 65–71. Wye College, Ashford, Kent.

Haber, W. (1973). Conservation and landscape maintenance in Germany: past, present, and future. *Biological Conservation*, **5**, 258–264.

Haber, W. (1974). Nationalparke—Wunsch und Wirklichkeit. *Garten und Landschaft*, **84 (3)**, 97–99.

Haber, W. (1974). Naturparke, Landschaft und Planung. *Naturschutz- und Naturparke*, **73**, 29–32.

Haber, W. (1976). *National Park Bayerischer Wald. Entwicklungsplan.* Landschaftsökologie Weihenstephan, Selbstverlag.

Haber, W. (1978). The future of the National Parks in Kenya. *Natur- und Nationalparke* **16 (62)**, 19–22.

Henke, H. (1976). Untersuchung der vorhandenen und potentiellen Nationalparke in der

Bundesrepublik Deutschland im Hinblick auf das internationale Nationalparkkonzept. *Schriftenr. Landschaftspflege und Naturschutz*, **13**, 9–154.

IUCN (1978). *Categories, Objectives, and Criteria for Protected Areas.* IUCN, Gland, Switzerland.

IUCN Bulletin (1982). The road from Bali parks congress: expanding the role of protected areas in society. *IUCN Bulletin, new series* **13**, 77–79.

Koepp, H. (1973). Über das Nationalparkwesen in den USA. *Landschaft + Stadt*, **5**, 5–17.

Lusigi, W.J. (1978). Planning human activities on protected human ecosystems. *Dissertationes Botanicae*, **48**. J. Cramer, Vaduz, Liechtenstein.

McTaggart Cowan, I. (1977). Islands in space and time. *Parks* **2** (2), 13–14.

Polunin, N. & H.K. Eidsvik (1979). Ecological principles for the establishment and management of national parks and equivalent reserves. *Environmental Conservation*, **6**, 21–26.

Runte, A. (1979). *National Parks: The American Experience.* Lincoln Univ. of Nebraska Press.

Sax, J.E. (1983). *National parks and equivalent reserves in North America—an introduction.* Lecture delivered at the International Working Conference 'New Directions for Conservation of Parks' at Niederhaverbeck (Naturschutzpark Lüneburger Heide), Fed. Rep. of Germany, on 6th June, 1983.

Western, D. (1976). A new approach to Amboseli. *Parks*, **1** (2), 1–4.

Wotschikowsky, U. (1977). Management of red deer in the Bayerischer Wald National Park. *Parks*, **2** (1), 5–7.

V
REPAIR AND RENEWAL

INTRODUCTION

So far it has been argued that if a landscape has been properly designed and managed it will be self-sustaining. In a perfect world this could be true. Unfortunately, we live in environments which are constantly changing, where proper management may be forgotten, where new savage stresses may arise, or where subtle changes accumulate to give problem situations. As a result we are increasingly faced with the need for repair and renewal, to replace what was there before the damage occurred.

In many situations the need for repair will be obvious from even a casual inspection of the landscape. But this will only reveal situations where damage has occurred. It would be far more valuable if we could detect damage before the symptoms became obvious. In this respect, the development of infra-red film techniques has been an outstanding step, and it is very significant to find that they are now being used both in industrial areas and on a routine basis for trees in cities. But any technique is only as good as the number of times it is applied. The implication of the first two contributions is that survey, of whatever sort, should be carried out on a regular basis with proper arrangements and staffing. We have hardly begun to accept such responsibility.

There is then the problem of the survey observations themselves. They merely show that something is wrong, without indicating what it is. The next step, therefore, is patient disentangling of causes. Sometimes these are self-evident, but often they are not. The penalty of assessing the causes of degradation or failure wrongly, is money wasted on the wrong treatment, as well as failure to achieve a cure. The work necessary to disentangle causes may be complex. This has already been shown by several preceding papers, but it is made very clear in the present contributions. Nowhere is the situation more difficult to disentangle than in the present problems of the Norfolk Broads, where both the symptoms and the possible causes are multifarious. As a result, it is only now that we are beginning to understand the chain of causation and can start to implement the repairs: the authorities responsible may seem dilatory, but it is because of their genuine concern not to carry out works which may be of no value. This is perhaps reasonable. But we are also entitled to expect the reverse, that once the causes are known, remedial measures are properly and fully implemented. It is therefore encouraging to see the steps that are now being taken in the Norfolk Broads and the positive effects which are resulting.

Sometimes the remedies for our environmental problems already lie in

357

front of us. At the outset, the degree to which nature is constantly on the move was stressed; plants and ecosystems have their own powers of growth and regeneration. Nature will develop and adapt to whatever environments are produced by man. This power of self-repair and self-renewal is nowhere more apparent than in the vegetation that colonizes derelict land in urban areas, as we have seen in other chapters. Certainly we should respect and capitalize on this, and incorporate it into our plans for the renewal for cities. It is an argument which is both practical and economical, yet in the past it has not always been heeded, as wilderness, classified as derelict land, has been unnecessarily 'improved'. Now, perhaps the pressures of public opinion and economy are beginning to make an impression, and urban renewal involves solutions incorporating wildlife.

All through this book the role of management and the contribution of people to landscape has been stressed. Without the contribution of people at both the design and the management stage, we would have incoherent, wild or brutalized landscapes; how wild or how brutalized depending on the degree of human impact. So repair and renewal involves the contribution of the community. There is no doubt that people will respect and cherish the landscapes with which they can identify. In a world where urban city dwellers feel less and less connected to what surrounds them, as it becomes more and more someone else's responsibility, a return to community involvement is crucial. Leaving everything to 'the Council' will not achieve landscapes which are cared for and appreciated.

But there is another aspect to the involvement of people which in some ways may seem alien to the subject matter of this book. This is the contribution of the original designer, which requires respect and understanding if repair is to be effective. Nowhere is this more apparent than in the repair of the great parks of the English landscape movement. Perhaps it is easy to appreciate and respect the contribution of the great landscape architects, but the lesson is surely that landscapes are not built in a day nor even in a year, so continuity of design and management is all-important. The crucial point is that this can and should apply to all landscape areas, from the grandest down to the most simple and mundane. Yet all too often it does not, and we see the imaginative designs of some unknown and unsung landscape architect of one generation ruined by the thoughtless repairs of the next generation.

Finally we come, as in all things, to matters of cost. In repair and renewal, remedial measures may sometimes be very simple and inexpensive—as in increasing wear resistance in amenity turf by fertilization—or they may be more expensive, as in the restoration of soil conditions for urban trees by aeration and drainage. Survey itself, as a prelude to repair, requires effort and staff time. All this has to be paid for, and is ultimately a charge on ourselves.

We can drag our heels out of meanness and retreat to the world of private affluence and public squalor—which at the time of writing so many politicians of the world seem to advocate—but we shall inevitably wake up to what we have done, and find it not to our liking. The repairs will then need to be more substantial and more costly. We should therefore put in the effort and the financial support now, as is well argued in so many contributions to this book, particularly that based on the experience of the city of Wiesbaden.

We only have the environments we are prepared to fight and to pay for. Perhaps this book, in showing not only what *needs* to be done, but also in many exciting ways what *can* be done by the combination of ecology and design, will stimulate not only concern, but also effective action.

22. LANDSCAPE UNDER STRESS

J. F. HANDLEY

The Groundwork Trust, 32/34 Claughton Street, St Helens, Merseyside
WA10 1SN

SUMMARY

The landscape manager seeks to maintain designed landscapes in a position of dynamic equilibrium. Scarcity of environmental resources or intensification of environmental factors working on the landscape may inhibit or disrupt the equilibrium. Such is the phenomenon of environmental stress. This paper explores the diagnosis, prognosis and treatment of landscapes under stress.

INTRODUCTION

The manager is normally required to maintain established landscapes around a pre-determined equilibrium position or to supervise the orderly attainment of such a position in newly created landscapes. Under favourable circumstances this management objective will be prescribed by a two-dimensional design and/or management plan.

An ecological approach to management perceives landscape as a qualitative attribute of a physical and biological interaction system: the ecosystem (Bradshaw & Handley 1982). Landscape quality is, in part, determined by the state of the system and is likely to be prejudiced when ecosystem function is disrupted. Stress ecology attempts to measure and evaluate the impact of natural and foreign perturbations on the structure and function of ecosystems and to provide a scientific basis for remedial treatment. The terminology and, more importantly, the underlying principles have something in common with the practice of human medicine. In both, a central question concerns which stressors the system can cope with by its own regulatory mechanisms, and which stressors lead to system breakdown (Rapport, Regier & Thorpe 1981).

THE NATURE OF LANDSCAPE STRESS

The ecosystem is in contact with its environment through input or output of energy, materials and information. The inputs constitute the system's requirements for resources, i.e. demand, whilst their availability in the environment constitutes a supply. Stress occurs when there is an excess or deficit of supply in relation to demand (Franz 1981). This principle may be illustrated by reference to a simple experiment which explores the response of sugar beet to nitrogen fertilizer under field conditions (Boyd 1961). In a low phosphate regime, yield increases in response to nitrogen applications up to an optimum level, beyond which excess nitrogen produces a marked reduction in yield (Fig. 22.1). The modification of this response curve in a high phosphate regime gives some indication of the complexities which are likely to be encountered in practice.

Growth and the need for management

Landscape management involves the maintenance of vegetation around a prescribed equilibrium position. So any factor which threatens to disrupt the equilibrium will be a matter for concern. The most obvious destabilizing factor is the ability of green plants to utilize solar energy for growth by photosynthesis, the very growth that is demonstrated in Fig. 22.1. With this always occurring in plant communities, management inputs will be required to maintain vegetation within defined limits. *Relaxation* of management therefore comes before any sort of stress as a potential destabilizing factor.

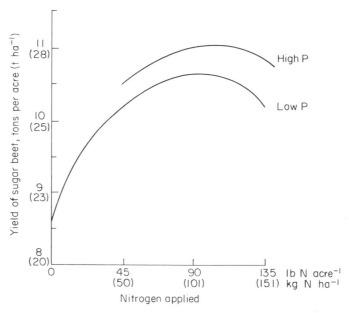

FIG. 22.1. Response of sugar beet to nitrogen fertilizer at two phosphate levels (after Boyd 1961).

Without management the structure of the ecosystem will inevitably change towards a new equilibrium position. If we don't cut the lawn we get a meadow; if we don't cut the meadow we get a wood!

One of the better documented examples of change in vegetation structure following relaxation of 'management inputs' is the impact of myxomatosis. The introduction of this disease to the UK in the early 1950s decimated the rabbit population. Thomas (1960) carefully recorded the floristic changes at a number of sites, but felt bound to observe that 'even more spectacular changes have been seen in the downland flowers, both common species such as *Helianthemum nummularium* (rock rose) and *Primula veris* (cowslip) and rarer plants such as pasque flower (*Pulsatilla vulgaris*); spots where a few (pasque) flowers used to show are now a sheet of purple'. Whilst this type of change proved rather ephemeral, the transition from grassy heath to scrub was more lasting (Table 22.1) and led to complex management problems (Grubb & Suter 1971).

From a more general point of view, relaxation of the management regime is a frequent cause of landscape change in the UK, because so much of our vegetation is held in check by management practice (plagioclimax communities). There have been widespread changes in the countryside due to the contraction of traditional practices, for example sheep-walk grassland, coppice woodland and mowing marsh, and we are currently experiencing a relaxation of management within urban amenity landscapes in the face of budget restrictions. Some of these changes have already been mentioned by Green (Chapter 12) and Emanuelsson (Chapter 15). Whilst the importance of landscape maintenance cannot be over-emphasized as a factor determining landscape quality, disruption of landscape equilibria by relaxation of management inputs should not, however, be equated with landscape stress.

TABLE 22.1. Vegetation change at Lullington Heath following cessation of rabbit grazing (after Thomas 1963)

		Year	
	1954	1957	1961
Vegetation height (cm)	14	21	56
Species abundance			
Agrostis spp.	58	39	17
Festuca rubra	25	36	5
Calluna vulgaris	13	25	47
Erica cinerea	66	37	13
Filipendula vulgaris	6	14	18
Ulex europaeus	2	7	58

Stress

In a systematic assessment of plant strategies and vegetation processes, Grime (1979) sought to distinguish between 'stress' and 'disturbance'. He proposes that stress be equated with phenomena which restrict photosynthetic production such as shortages of light, water, and mineral nutrients, or suboptimal temperatures. Disturbance, by contrast, results in partial or total destruction of plant biomass. It is related to the activities of herbivores, pathogens, man (trampling, mowing and ploughing), and phenomena such as wind damage, frosting, soil erosion and fire. The distinction is important not least because, under conditions of high stress but low disturbance, where the dominance of competitive species is resisted, we find species-rich plant communities of high conservation value.

Landscape stress, however, must be considered to occur when the trajectory or equilibrium of a managed landscape is disrupted by any environmental stress. The range of factors which may be involved as stressors on the landscape is indicated in Table 22.2. When tolerance levels are exceeded the landscape impact may be very considerable (Figs 22.2–22.6). The classification in Table 22.2 embraces both stress and disturbance phenomena as defined by Grime (1979). Whilst the disruption illustrated in Figs 22.2–22.6 is clearly unacceptable, it should be remembered that continuous or intermittent stress is an essential characteristic of some of our most cherished semi-natural landscapes such as heathland and downland. The particular

TABLE 22.2. Environmental factors which may be implicated in landscape stress

Factor	Examples
Atmospheric	Air pollution
	Salt spray
	Herbicide spray drift
	Exposure
Edaphic	Soil reaction (pH)
	Water relations:
	(i) drought
	(ii) waterlogging
	Nutrient deficiency
	Soil toxicity
Biotic	Fire
	Grazing
	Disease
	Trampling
	Wilful damage

FIG. 22.2. Localized damage to vegetation by regeneration of acidity in colliery spoil.

FIG. 22.3. Complete loss of vegetation on a 'reclaimed' colliery spoil heap at Welch Whittle, Lancashire.

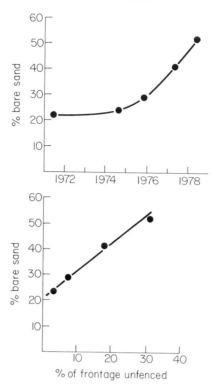

FIG. 22.8. Progressive erosion of a dune frontage at West Kirby, Merseyside is monitored as percentage bare sand.

required on plant and soil materials. This will permit correlations to be established between the pattern of damage and environmental variables.

During the assessment of small-scale pattern of success and failure of vegetation on a colliery spoil heap at Welch Whittle in Lancashire, several variables were recorded (Table 22.3a). The concentration of plant nutrients and spoil pH are positively intercorrelated, suggesting that nutrient availability is restricted at low pH. Moisture content is independent of these variables and is in fact negatively correlated with calcium. Vegetation was sampled along two transects and there was a significant first order correlation with moisture content in both cases but *not* with spoil pH (Table 22.3b). Whilst the failure of vegetation on colliery spoil, and indeed on other parts of this site, is frequently associated with regeneration of acidity, it is clear that in this case, moisture content associated with microtopography was a decisive factor. Moreover, it is likely that this factor was of critical importance during

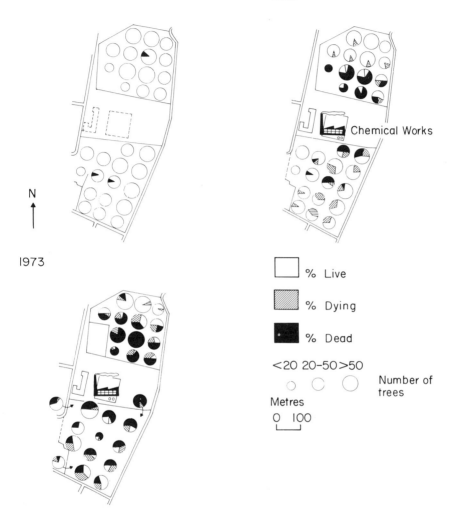

Inventory of Tree Damage at Kirkby Industrial Estate

1945   1962

N

1973

Chemical Works

☐ % Live

▨ % Dying

■ % Dead

<20 20–50 >50

○ ◯ ◯   Number of trees

Metres

0 100

FIG. 22.9. The progressive loss of trees around a phosphoric acid factory at Kirkby, Merseyside, is recorded from sequential aerial photographs (after Vick & Handley 1977). The woodland is illustrated in Fig. 22.6.

the establishment phase (Richardson & Greenwood 1967) rather than at the stage when observations were made.

Observation, combined with soil and plant analysis, can provide strong circumstantial evidence about the factor or factors responsible for stress.

TABLE 22.3. (a) Correlation between (a) environmental variables on colliery spoil, and (b) selected variables and plant biomass

(a) Variable	pH	H_2O	Ca^{++}	P
K^+	+0·93	+0·09	+0·47	+0·34
	***	N.S.	*	N.S.
P	+0·49	−0·06	+0·36	
	*	N.S.	N.S.	
Ca^{++}	+0·58	−0·66		
	*	**		
H_2O	+0·04			
	N.S.			

(b) Variable	Transect I	Transect II
H_2O	+0·73***	+0·48*
pH	+0·36 N.S.	+0·21 N.S.

Confirmation may well require diagnostic field trials (Bradshaw *et al.* 1975; Bloomfield, Handley & Bradshaw 1982). Figure 22.10 shows the results of a diagnostic field trial in which fertilizer treatments were applied in factorial combination to a moribund grass sward on china-clay waste. It is clear that nitrogen deficiency was the over-riding stress factor.

Where diagnosis is incorrect, a response pattern may occur which is independent of the experimental treatments. Just such a problem was encountered at Sefton Meadows, a domestic refuse landfill site in Merseyside, where a field trial was laid out with three seed mixes and two lime treatments (M. K. Brummage, pers. comm.). Seedling establishment was uniform but soon after this a striking pattern of success and failure appeared, which bore no relation to experimental treatments. Analytical work supported by transfer experiments in the field demonstrated that vegetation failure was associated with localized methane production (Table 22.4).

TREATMENT AND PREVENTATIVE MEASURES

The remedial treatment of landscapes under stress is dependent on proper diagnosis, supported where necessary by an experimental programme of field trials. But it should be remembered that landscapes have a degree of resilience and we should avoid over-reaction to what may be a transient stress response. Under stable equilibrium conditions the imposition of stress will bring into

FIG. 22.10. A diagnostic field trial on china-clay waste at St Stephen, Cornwall. Nitrogen deficiency is the over-riding stress factor; all sub-plots with regrowth received nitrogen fertilizer.

TABLE 22.4. Soil conditions and vegetation response on Sefton Meadows landfill site. (Data courtesy of M.K. Brummage, Merseyside County Council)

	Areas of satisfactory growth		Areas of poor growth	
	Mean	Range	Mean	Range
Vegetation cover (%)	83	(70–90)	26	(5–50)
Soil pH	4·6	(3·8–6·2)	5·3	(3·9–5·9)
Soil nutrient status (ppm):				
Phosphorus	9	(4–13)	10	(6–15)
Potassium	41	(23–91)	51	(25–80)
Magnesium	59	(24–184)	67	(15–252)
Soil temperature (°C) at				
15 cm depth	17	(16–19)	23	(17–35)
Soil atmosphere methane	No methane		All samples	
(determined by Drager tube)	detectable		> 5000 ppm	

play adjustment mechanisms which compensate for the stress (Holling 1981). At the other extreme where conditions are fundamentally unstable, disruption of the ecosystem and landscape failure may be almost inexorable. Landscape management is usually concerned with ecosystems which possess both characteristics, self-righting mechanisms operating close to the equilibrium, and large displacements initiating breakdown or rapid change to a new equilibrium position.

The form of treatment prescribed depends not only on the correct identification of the stress factor (or factors), but also prognosis, a careful appraisal of the state of the ecosystem. Whether or not the stress factor is alleviated, a variety of treatment options are available:

    (a) wait for self-adjustment;

    (b) accept new equilibrium position;

    (c) increase resistance to stress;

    (d) compensate for stress;

    (e) landscape surgery.

There is a further option, cosmetic repair of the landscape by replacement of what was there before. This is all too familiar in practice, but it cannot possibly be recommended! The simple re-establishment of vegetation in degraded environments similar to those illustrated in Figs 22.2–22.6, does nothing to alleviate or meet the stress. As a result, there will quickly be landscape failure again, with a concomittant waste of resources and loss in professional and political credibility. Proper treatment is essential and is best understood by some examples.

Urban soil acidity

One of the more insidious forms of environmental stress is associated with dry deposition and washout of sulphur compounds from the atmosphere and the consequent soil acidification. The process has been described by Christine Vick (1975), who assessed soil conditions along a gradient of increasing sulphur pollution from the remote countryside of North Wales to Liverpool, at the heart of the Merseyside conurbation (Fig. 22.11). In Victorian parks on sandy soils close to the city centre extremely acidic conditions (pH < 3·5) were encountered. The grass swards are poor and break up quickly when subjected to heavy use. The accumulation of a mat of organic matter at the soil surface is indicative of ecosystem disruption.

The treatment in this case is relatively straightforward. The acidity itself can be corrected very simply by applications of ground limestone (Vick & Handley 1975). These relieve the stress of acidity and increase resistance to wear and tear. Laboratory studies demonstrated the recovery of mineralization, one crucial step in the nitrogen cycle (see Bradshaw, Chapter 2), to aliquots of the lime requirement (Fig. 22.12).

Wear of sports turf

The pressure on sports and amenity turf is often such that it is necessary to

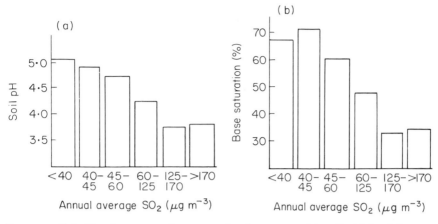

FIG. 22.11. Soil pH and percentage base saturation decrease as sulphur dioxide levels increase along a gradient from North Wales to Liverpool (after Vick 1975).

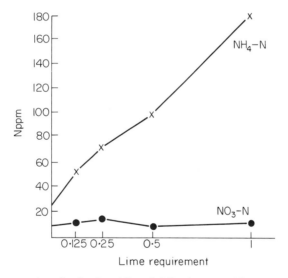

FIG. 22.12. Nitrogen mineralization in acidic soil following ground limestone application (after Vick & Handley 1975).

increase resistance of vegetation to wear and tear even in the absence of pollution. Research has shown that there are highly significant differences in wear-tolerance between species; ryegrass (*Lolium perenne*) is outstanding. There are also significant differences between different cultivars of the same species; recently produced cultivars may have 20% better wear tolerance than cultivars still in common use. Furthermore, wear resistance can be improved remarkably by ensuring that fertilizer, and especially nitrogen, is applied at the optimum level; these levels (250 kg N ha⁻¹ yr⁻¹) are much higher than those previously thought to be satisfactory (Shildrick & Peel 1984). The optimum nitrogen level for promoting wear tolerance in the ryegrass (*Lolium perenne*) cultivar 'Loretta' is shown in Fig. 22.13.

   However, whilst the wear tolerance of sports turf can be improved greatly by using resistant varieties and fertilizing correctly this still may not be sufficient to provide a satisfactory playing surface. Recent research has shown that the replacement of soil with a sandy growing medium can have greater effects than any of the previous treatments (Canaway 1984) (Fig. 22.13). Landscape surgery may then be the appropriate solution in cases of severe wear.

Sand dune erosion

It may also be possible to increase resistance to trampling in certain semi-natural environments by fertilizer application. The breakdown of a sand dune system under intense visitor pressure and the consequence for nearby pine woods is illustrated in Figs 22.14 and 22.15 respectively. This type of problem can only be solved by the development of an effective management plan. The first step may be to restrict access, reducing trampling to a level the ecosystem can stand. But then the ecosystem must be restored. This will require the

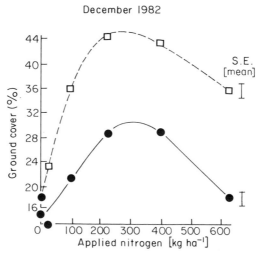

FIG. 22.13. Ground cover of turf in response to fertilizer nitrogen after 2 months' wear, December 1982: sand (O) and soil (●) (after Canaway 1984).

FIG. 22.14. Sand dune erosion at Formby, Merseyside is accelerated by severe visitor pressure.

FIG. 22.15. A pine woodland damaged by salt spray at Formby, Merseyside. The protective cuticle on the pine needles has been abraded by blown sand.

planting of marram grass (*Ammophila arenaria*). But its rate of establishment is crucial to the success of the landscape restoration. Experiments by Johnson and others have shown that growth can be promoted effectively—up to fourfold—by the use of nitrogen fertilizer, even in the absence of phosphate (Bradshaw & Chawick 1980) (Fig. 22.16). What is important is that such treatments have only a temporary influence on the ecology of the dune system and do not change the trajectory of dune succession.

CONCLUSIONS

The consequences of environmental stress can be expensive and difficult to remedy. Landscape surgery is likely to involve costly engineering and replanting measures which fall outside the normal range of financial provision for maintenance. It is therefore vital that preventative measures are used to ensure that, so far as possible, landscape surgery is avoided. For this to be achieved problems should be identified at an early stage, at low stress intensity, where a range of management options is still available. Fortunately, environmental stressors do not usually operate in a uniform manner. Acute symptoms of stress are likely to be encountered first of all in 'hot spots'. A properly conceived monitoring programme should identify warning signals at an early stage so that preventative treatments or management strategies can be applied.

For new schemes the site appraisal, which is a normal part of the landscape design process, should seek to identify potential stress factors before they occur. Then the design can be adjusted to compensate for what is likely to occur. Appropriate soils can be chosen, as we have already seen to be

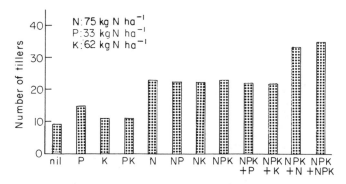

FIG. 22.16. Sand dune recovery can be promoted by fertilizer application: in this trial the growth of marram grass was tripled by addition of nitrogen; other nutrients had little effect (after Bradshaw & Chadwick 1980).

important in amenity turf. Plant material can be selected which is pre-adapted to the stress factors (e.g. Bradshaw, Humphreys & Johnson 1978; Gill & Bradshaw 1971). Indeed, the whole design should be planned to maximize resistance to the factors likely to occur or minimize their impact, thereby greatly reducing the likelihood of landscape failure. But there are certain stress factors which must be alleviated during or even before landscape construction can proceed. For example, in areas where vandalism and wilful damage can be anticipated, the chances of landscape establishment will be greatly increased by effective community involvement before, during and after the project (Figs 22.17 and 22.18).

The landscape designer should be alive to the possibility of utilizing stress factors in a creative manner. The variation in pH, water relations, fertility and other factors which is encountered in both natural and despoiled landscapes provides vital raw material for the designer. Given this at the design stage, and that plant material tolerant of the local conditions is selected, we may well produce landscapes which are visually interesting, biologically rich and cost-effective to produce and maintain.

FIG. 22.17. This woodland in Knowsley, Merseyside is over-mature and the death of trees is being accelerated by burning. (Courtesy of Merseyside County Council.)

FIG. 22.18. Community development, in which local children are invited to play their part in woodland management, is an essential prelude to selective felling and replanting. (Courtesy of the *Liverpool Daily Post and Echo*.)

But we cannot anticipate the full range of factors which may be at work on the landscape. Nor can we predict either their intensity, or the tolerance of biological materials to stress, with the kind of precision which is taken for granted in civil engineering. The aim of ecologically informed design is not only to predict or eliminate future surprises, but also to produce resilient landscapes which can absorb, survive and capitalize on unexpected events when they do occur.

REFERENCES

Bloomfield, H.E., Handley, J.F. & Bradshaw, A.D. (1982). Nutrient deficiencies and the aftercare of reclaimed derelict land. *Journal of Applied Ecology*, 19, 151–158.

Boyd, D.A. (1961). Current fertiliser practice in relation to manurial requirements. *Proceedings of the Fertiliser Society*, No. 65.

Bradshaw, A.D. & Chadwick, M.J. (1980). *The Restoration of Land.* Blackwell Scientific Publications, Oxford.

Bradshaw, A.D., Dancer, W.S., Handley, J.F. & Sheldon, J.C. (1975). The biology of land revegetation and the reclamation of the china clay wastes of Cornwall. In *The Ecology of Resource Degradation and Renewal* (Ed. by M.J. Chadwick & G.T. Goodman), pp. 363–384. Blackwell Scientific Publications, Oxford.

Bradshaw, A.D. & Handley, J.F. (1982). An ecological approach to landscape design—principles and problems. *Landscape Design*, **138**, 30–34.

Bradshaw, A.D., Humphreys, M.O. & Johnson, M.S. (1978). The value of heavy metal tolerance in the revegetation of metalliferous mine wastes. In *Environmental Management of Mineral Wastes* (Ed. by G.T. Goodman & M.J. Chadwick), pp. 311–334. Sijthoff & Nordhoff, Alphen aan den Rijn.

Canaway, P.M. (1984). The response of *Lolium perenne* (perennial ryegrass) turf grown on sand and soil to fertilizer nitrogen. 1. Ground cover response as affected by football-type wear. *Journal of Sports Turf Research Institute*, **60**, 8–18.

Chikishev, A.G. (Ed.) (1965). *Plant Indicators of Soils, Rocks, and Subsurface Waters.* Consultants Bureau, New York.

Cooper, J.I. (1978). *Virus and Virus-like Diseases of Trees.* Arboricultural Leaflet No. 4. HMSO, London.

Franz, H. (1981). A general formulation of stress phenomena in ecological systems. In *Stress Effects on Natural Systems* (Ed. by G.W. Barrett & R. Rosenberg), pp. 49–54. Wiley, New York.

Gill, C.J. & Bradshaw, A.D. (1971). The landscaping of reservoir margins. *Landscape Design*, **95**, 31–4.

Grime, J.P. (1979). *Plant Strategies and Vegetation Processes.* Wiley, New York.

Grubb, P.J. & Suter, M.B. (1971). The mechanism of soil acidification by *Calluna* and *Ulex* and the significance for conservation. In *The Scientific Management of Plant Communities* (Ed. by E. Duffey & A.S. Watt), pp. 115–133. Blackwell Scientific Publications, Oxford.

Handley, J.F. (1980). The application of remote sensing to environmental management. *International Journal of Remote Sensing*, **1**, 181–195.

Holling, C.S. (Ed.) (1981). *Adaptive Environmental Assessment and Management.* Wiley, New York.

Horler, D.N.H., Barber, J. & Barringer, A.R. (1981). New concepts for the detection of geochemical stress in plants. In *Remote Sensing in Geological and Terrain Studies* (Ed. by J.A. Allan & M. Bradshaw), pp. 113–123. Remote Sensing Society, London.

Horler, D.N.H., Dockray, M. & Barber, J. (1983). The red edge of plant reflectance. *International Journal of Remote Sensing*, **4**, 273–288.

Knipling, E.B. (1969). Leaf reflectance and image formation on colour infrared film. In *Remote Sensing in Ecology* (Ed. by P.L. Johnson), pp. 17–29. University of Georgia Press, Athens.

Leopold, A.C. (1961). Senescence in plant development. *Science*, **134**, 1727–1730.

Rapport, D.J., Regier, H.A. & Thorpe, C. (1981). Diagnosis, prognosis and treatment of ecosystems under stress. In *Stress Effects on Natural Ecosystems* (Ed. by G.W. Barrett & R. Rosenberg), pp. 269–280. Wiley, New York.

Richardson, J.A. & Greenwood, E.F. (1967). Soil moisture tension in relation to plant colonisation of pit heaps. *Proceedings of University of Newcastle Philosophical Society*, **1**, 129–136.

Shildrick, J.P. & Peel, C.H. (1984). Preliminary trials of perennial ryegrass cultivars (Trials A1, B2 and B3), 1980–3. *Journal of the Sports Turf Research Institute*, **60**, 73–95.

Thomas, A.S. (1960). Changes in vegetation since the advent of myxomatosis. *Journal of Ecology*, **48**, 287–306.

Thomas, A.S. (1963). Further changes in vegetation since the advent of myxomatosis. *Journal of Ecology*, **51**, 151–185.

Van Genderen, J. (1974). Remote sensing of pollution in Teeside. *Journal of Environmental Pollution*, **6,** 221–234.

Vick, C.M. (1975). *Air pollution on Merseyside*, Ph.D. Thesis, University of Liverpool.

Vick, C.M. & Handley, J.F. (1975). The correction of soil acidification in urban parklands. *Journal of the Institute of Parks and Recreation Administration*, **40,** 39–48.

Vick, C.M. & Handley, J.F. (1977). Survey of damage to trees surrounding a chemical factory emitting phosphoric and hydrochloric acid pollution. *Environmental Health*, **85,** 115–118.

Wallace, T. (1961). *The Diagnosis of Mineral Deficiencies in Plants* (a colour atlas and guide), 3rd edn. HMSO, London.

23. THE ECOLOGY AND PRESERVATION OF STREET TREES

H. DE LA CHEVALLERIE

6200 Wiesbaden, Nerotal 24, West Germany

SUMMARY

In Germany, the importance of green areas in cities and the need for their preservation has recently been realized, e.g. 4 million DM were spent on tree maintenance in Wiesbaden from 1974 to 1980. The environmental conditions to which trees are now subject in cities are appalling, the major factors being soil compaction, poor soil aeration, mechanical damage, gas leakage, salt, and air pollution. The effects of each of these factors must be properly understood before remedial measures can be taken. These measures are very specific and often involve careful construction work.

Tree maintenance is an on-going problem, and requires continuous monitoring. In Wiesbaden to provide critical early information for the permanent team of twenty-four maintenance staff, false colour infra-red surveys are taken at regular intervals. But all this care must be based on a proper understanding of the urban tree and its ecology.

INTRODUCTION

The aim of this paper is to provide an insight into the functions of one parks department with which you will be unfamiliar in Britain, i.e. the Parks Department of the City of Wiesbaden in Germany.

Parks departments in Germany are concerned with topical tasks. Within town planning and strategic planning they are responsible for ecological and biological matters, both in general and associated more specifically with landscape, settlements, parks, sport, playgrounds, schools, streets, allotments and cemeteries.

The dialogue 'ecology in cities' begins with the land-use plan. Since 1979 my department has been involved in constructing a landscape plan for Wiesbaden upon which every planning decision will be based. Much ecological information has been put together to work out this landscape plan. The analysis of infra-red thermographic photos and the tests of cold air circulation have, for instance, been the basis of a very important plan giving

383

information about those areas which have climatic–ecological effects on other parts of the city. Coloured infra-red aerial photos, the detailed mapping of landscape features, and studies of the flora, fauna, soil and water have been combined to produce a complete ecological analysis of the city. This has indicated, amongst other things, the areas of ecological worth and where to find places for recreation.

One detail of the landscape plan on which we have been working very intensively for 10 years is the provision of trees in the city and in the landscape. Although crucial to any city's landscape and environment there is one aspect of these trees which is extremely difficult to deal with, i.e. the planting and maintenance of trees in urban areas.

There have never been such unfavourable ecological conditions for trees in cities as exist today. Trees planted at the beginning of this century, growing in reasonable conditions, had the chance to fulfil their natural lifespan and now can be magnificent. But today, the sum of all unfavourable growing conditions has such a negative effect that life expectancy, at least for street trees, has been drastically reduced.

The preservation of green areas, especially trees in cities, is an obligation for all of us, for the welfare of the general public (de la Chevallerie 1976). But to understand the problem, we need to understand the importance of trees in urban areas and the reasons for their early mortality. We may then be in a better position not only to preserve them but also to encourage their better growth.

IMPORTANCE OF STREET TREES IN HOUSING AREAS

Most cities without street trees are like deserts without any life. For centuries the street tree has given the characteristic shape to the architecture of a city. Trees in streets, squares and gardens bring scale and space into the city. A view from a window into the crown of a tree will give the impression of being very close to nature. Even a single large-crowned tree next to a block of flats can be looked upon by many of its inhabitants as 'their house tree.'

Next to these aesthetic effects, it is the biological effects of street trees which increase the worth of living in a city. Cities with trees are living cities! 'Public trees' in streets, squares and parks have a special importance for people without their own gardens. They provide an environment of living things for people who otherwise may have none.

Street trees in cities are part of a very old tradition. It was the Italian Renaissance and later, more characteristically, the French Baroque, to whom we owe the idea of planting rows and avenues of trees. Even the eighteenth century English garden style kept the element of tree avenues. Then city planners took over the idea of avenue trees and built the famous boulevards which can be seen in Paris and other cities. In the nineteenth and at the beginning of the twentieth century, street trees belonged to the architecture of better residential areas in cities. The famous squares in London, all of which are planted with trees, still provide a model for modern architecture.

After the Second World War we forgot this good tradition of planting trees. In newly built housing areas many streets were built without any trees at all. In old urban quarters trees have been subjected to serious damage caused by asphalt, traffic and construction work. Complete avenues of trees have been cut down. The great plague of tree mortality began. At the same time, the cities were paying any price for the process of economic growth, and the quality of housing areas suffered great damage. In Germany, we did not start to change our understanding of our environment until the beginning of 1980. 'Green' lobbies came into being, and ever more frequently the word 'ecology' popped up in the vocabulary of city politicians. At last the relationship between plants and climate was recognized and discussed. It became realized that vegetation improves the unnatural climate of cities, changing our artificial environment into a more natural one. The important contributions of street trees to urban life can be classified as follows.

(a) *Their function for city planning.* Trees bring scale and space into the city. They are a good means of orientation and they increase drivers' security.

(b) *Their physical functions.* Trees can serve as dust filters, wind breaks, sun-shades. They are able to hide unwanted objects. They are traffic-safety factors.

(c) *Their physiological functions.* Trees produce oxygen and atmospheric moisture and reduce temperature. They assimilate carbon dioxide.

(d) *Their psychological functions.* Vegetation gives pleasure; green colours encourage a feeling of peace. Nature has a great effect on people's relaxation and can intensify their well being.

So it is of great importance to equip towns with trees, in gardens, parks, streets and squares. If we are now seriously interested in improving the environmental quality of our cities, city planning and city ecology must be regarded as equally important. The Parks Department of the City of Wiesbaden—a city with 260 000 inhabitants—was, in 1973, one of the first in Germany to develop a far-reaching programme for tree maintenance. Between 1974 and 1980, 4 million DM were spent on tree maintenance; 6000 young trees were planted; to improve the specific environment of many older trees, the soil was exchanged, systems for watering and ventilation were installed, and asphalt was removed. Moreover, a special tree maintenance team of twenty-four men was founded, to care for about 14 000 trees (de la Chevallerie 1974).

ECOLOGICAL CONDITIONS

The normal tree in parks and gardens has enough ground space in which to grow, usually an area as large as its crown. At the same time, the soil is porous, moist and rich in nutritive substances. However, in streets the situation for trees is different since they live in completely artificial conditions. Elements of their environment of vital importance, such as a high quality soil and adequate water, are rare. Tree beds are too small and mostly covered with asphalt or concrete; there is less water, less oxygen and fewer micro-organisms available, and the biological circulation in the soil is disturbed. This can be very unfavourable or even fatal for trees. In older streets conditions were better (Fig. 23.1).

In a normal, sufficiently aerated soil there should be about 21% oxygen, which is similar to the oxygen content of air. Oxygen is not only necessary to

FIG. 23.1. Extensive tree roots under the porous surface of granite sets of an older street in Weisbaden.

root metabolism; there are also aerobic micro-organisms, such as the soil bacteria and mycorrhiza, which are all dependent on oxygen in order to convert minerals and organic substances into material able to be assimilated by plants. For trees' roots and their effective absorption of nutritive substances, the mycorrhizal fungi are of particular importance. These live in symbioses with the trees' roots. Their mycelium covers the roots and, by covering them, enlarge the roots' surface and depth. As a result, the supply of nutritive substances can be increased substantially for the partner tree. In exchange the fungi obtain assimilated substances such as carbohydrates. But if there is lack of oxygen the fungi in the soil will react more sensitively than the roots.

The effects of oxygen deficiency in the soil are apparent if there is less than 14%. With 11% oxygen roots actually die (Ruge, 1979, 1980). At the same time the general biological activity in the soil will be upset. The normal conversion of nutritive substances is ended, and the soil changes to anaerobic processes. If, therefore, there is not enough oxygen in the soil the growth of the roots and consequently the growth of the trees are adversely affected. Organic materials may even be insufficiently oxidized and toxic acids and methane produced instead. It is not difficult to test for this (Fig. 23.2).

The main reasons for the lack of oxygen in soil are (i) because the soil is too compact with a low porosity, or (ii) because it does not have a natural ground

FIG. 23.2. Measuring oxygen and methane in a newly-planted paved area.

cover or leaves and therefore has an inadequate input of organic matter. When gardening, a serious mistake is to remove the natural covering of leaves. Leaves produce humus; they are food for soil organisms which disturb the soil particles and maintain an open structure; and they keep the ground from drying out. Maintenance of soil structure is one of the most important tree maintenance measures. This can be accomplished by mulching with bark-compost or covering with a ground vegetation.

At the same time, trees located on compacted soil may well suffer from a lack of water due to reduced soil porosity and hydraulic conductivity. This problem can be enhanced by overall deficiencies of soil water. The stomata of the leaves close; consequently they are no longer able to absorb enough carbon dioxide. As a result, photosynthesis is interfered with and the build-up of carbohydrate in the trunk and roots is slowed down. Because the consumption of carbohydrate during respiration goes on permanently, more assimilates are used than are being produced. As a result, such trees rapidly become senile. To compensate for this trees must be provided with additional water.

FACTORS CAUSING PREMATURE TREE DEATH

There are so many factors which can damage trees in towns that, once we list them all, it is very worrying. It is therefore not surprising that urban trees so often grow badly.

Compaction of soil

Not only parked cars, but also pedestrians ruin soil structure. Underground services and the enlargement of streets diminish the root-area of trees, resulting in negative growth, die-back, and even early death.

Damage by parasites

Trees which are not properly nourished can easily suffer from fungi and parasites of secondary rank. Likewise, plants weakened by other causes can be more easily damaged by insects and fungi which destroy the leaves. Species which have become most seriously in danger from these causes recently are plane trees (*Platanus*) and false acacia (*Robinia*).

Herbicides and their consequences

Herbicides not only destroy unwanted vegetation but, used incorrectly, may destroy life in soil as well, because it is so difficult to ensure their proper application. In Wiesbaden the use of herbicides is no longer allowed.

Mechanical damage

Tree trunks are often injured by parking cars. If the tissue of the wood is injured, transport of water and nutritive elements will be reduced and infections can often be found in the vascular system of the tree. Great damage is often also done to roots by construction work. This again can seriously weaken trees and causes secondary damage, such as fungus infection of the wounded roots.

Damage caused by gas

When moist coal gas is replaced by dry natural gas, leakage can occur into the soil because old pipe connections sealed with hemp become leaky. Although natural gas is not itself toxic it displaces the oxygen content of the soil and prevents normal root respiration. This destroys the root system and within a few weeks the tree dies.

Damage caused by using salt in winter

Sodium chloride damages the soil, ground water, animals and plants. Yet in 1981, for example, 2·6 million tons of salt were used for salting the streets in Germany (Klaffke 1980). The resulting symptoms caused by salt damage are very obvious. Leaves are smaller, their borders suffer from necrosis, they change their colour and fall off prematurely. Moreover, the chloride ions absorbed by the roots may provoke a toxic effect generally in the tree. If the internal concentration in the tree reaches a high enough level the tree will die. At the same time sodium ions, by replacing calcium ions in clays, cause the soil structure to collapse. Not only are the concentrations of nutrient ions therefore reduced, but their absorbtion made more difficult. Trees especially hard-hit by salt are maples (*Acer*), limes (*Tilia*) and chestnuts (*Aesculus*).

Damage provoked by polluted air

Sulphur dioxide concentrations have been very high in urban areas. Thankfully they are now falling. But instead we have increasing concentrations of ozone and nitrogen oxides caused by increasing numbers of motor vehicles. These can cause die-back of many species and the position appears to be becoming serious. Heavy metals, especially lead, also pollute the air. They fall directly on to the ground or are partly filtered and stored in the leaves. Later, in autumn, the leaves fall off, so eventually all such pollutants enter the ground and become available to plants.

Damage caused by reduced pH

If the pH in the soil is reduced, for example through acid rain, the normal circulation of ions within the ecological system of the soil is interrupted and aluminium, which is normally unavailable, becomes ionic and soluble and therefore toxic to plants. Acid precipitation is high in urban areas, and therefore soils may be affected. Fortunately however most urban soils, as the result of the presence of calcareous building material, are protected from extreme acidification.

MEASURES TAKEN AGAINST THE DEATH OF ESTABLISHED TREES

Against such a variety of possible damaging factors, it may seem well-nigh impossible to take sufficient protective measures. Indeed, mature trees which have been damaged within the area of their tree beds can be treated only to a certain extent. But many things are possible (Bernatzky 1981).

The treatments which are usually considered as the most important in aiding recovery and promoting more vigorous growth, are:

    (a) to remove the asphalt within the tree areas;
    (b) to loosen and open the compacted soil (Fig. 23.3);
    (c) to provide trees with irrigation;
    (d) to perform tree surgery;
    (e) to provide fertilization.

Measures which should be taken to maintain trees which are well established and not yet damaged can be divided into several categories:

    (f) contruction measures;
    (g) official tree preservation orders;
    (h) tree maintenance measures.

The best protection measures to be carried out in the constructional phase are:

    (i) to build tree beds as large as possible;
    (j) to maintain the soil structure and prevent compaction;
    (k) to protect the soil against pollution from toxic materials of all sorts;
    (l) to plant suitable ground cover in order to guarantee soil biological activity.

FIG. 23.3. Installation of aeration elements in a heavily paved area. These can also be used for irrigation.

    (m) to build mechanical protection against cars, e.g. low walls, protective hoops.

There is no protection against gas seeping into the ground. The only thing to do is to replace the old pipelines for new ones. Nor is there protection against salting in winter, the only possibility being raised curbs to reduce the quantities of salt-laden melt-water entering the tree bed. The most effective protection, however, is to stop using salt in winter. Last winter they did not salt in Berlin, and salting on Wiesbaden pavements in winter is no longer allowed.

A tree preservation order has been in operation since 1978 for the whole city of Wiesbaden. All trees, even privately owned ones, having a circumference of more than 60 cm and a diameter of more than 20 cm are taken into protection by the city. They only can be cut by permission of the Park Department and, as a general rule, a new tree must be planted whenever an existing tree has been removed.

However, it is often a problem to recognize in time that trees are suffering from damage and to take immediate counter-measures. One method is to discover damage by means of false colour infra-red photography as discussed by Handley (Chapter 22). This was carried out for the whole of Wiesbaden for the first time in 1978 and repeated in 1984. In these false colour photographs the chlorophyll of the leaves can be seen, according to the kind of tree they come from, in various shades of red. Divergences in colours within the normal red colour scale are an indication of sick leaves, allowing early recognition of damaged trees. Some results are shown in Fig. 23.4. But false colour photography does not show what is causing the damage; additional terrestrial photos are necessary for this.

For Wiesbaden it has been possible to arrange a very successful tree maintenance team consisting of twenty-four men. Their function is to water, fertilize and loosen the ground, to plant trees and perform tree surgery. Although the expenses for this team are rather high, their success has justified the cost. All tree treatment measures as well as the state of trees are catalogued in a special tree register. Furthermore, the trees which are being protected by the official preservation order are registered in zoning-plans.

GENERAL RULES FOR PLANTING AND MAINTAINING TREES

Within natural tree planting sites such as parks and gardens, the diameters of tree root areas are normally three times the diameter of their crowns (Meyer 1978). Consequently tree beds of large trees should be several hundreds m² in area. In reality however, matters are different. A tree bed of 4×4 m, which is now regarded as rather large, is therefore less than the minimum area for a large-crowned tree.

In streets and squares tree beds are, in fact, mostly limited to a size of 2×2 m. Although this quantity of ground is a good start for a young tree, growth will be halted unless the roots are able to spread out peripherally, when the reserves of the 'potted' plant in its small bed are used up. If this cannot happen the tree will become retarded or die. Street trees especially suffer from this. They are only able to grow normally and to reach their natural life span if they are located in front gardens or have the correct area and volume of satisfactory substrate.

Sufficient soil aeration is very important for the roots to be able to grow down into the lower layers. Deep rooting is very important for tree water supply. Tree planting sites which are not likely to be well aerated should be provided with drain-pipes, 40–50 cm below the surface, to allow air to penetrate into the lower layers. At the same time these pipes can also be used for watering (Fig. 23.3).

Ground preparation is of great importance. In light soils the content of clay colloids is not sufficient. Such soils have large pores, so that there is a reduced capacity to store water, the water seeps away too quickly, and the organic matter content is often too low. In heavy soils there is a deficiency of large pores and an excess of fine; the consequences are an excess of water and a

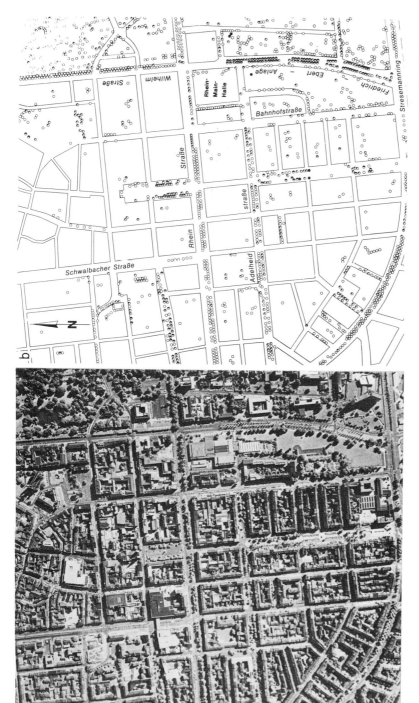

Fig. 23.4. False colour infra-red survey of part of Wiesbaden (a). From this, maps can be constructed indicating the growth condition of individual trees (b).

deficiency of air. Wherever building operations are compacting and destroying the natural ground, the ecological conditions of the soil have to be
carefully supervised.

Soil improvements can be carried out to increase both that part of the
texture which leads to strong, permanently draining pores and that part which
leads to storage of water and nutrients. To do this, an analysis of the soil,
carried out with care and precision, should be the obligatory first step. The soil
can be made more porous through the addition of polystyrol or other
granulates such as gravel. At the same time mineral colloids such as clay, or
chemical ingredients such as Hygromull can be used to hold the different
particles together in retentive aggregates.

Organic substances, such as compost, are obviously very important both
for the soil structure and long-term nutrient supply. But the material must not
to be dug too deeply into the ground, otherwise the soil will become
deoxygenated because of the oxygen demand of the organic matter. To
neutralize acidic soils with a very low pH, quicklime (CaO) or calcium
carbonate ($CaCO_3$) are useful. Within building sites the ground is often
extremely compacted by heavy machinery and wheeled vehicles. In these sites
at least 4 m³ of soil should be excavated in order to have tree pits of 2 × 2 × 1 m.

As far as possible fallen leaves should be left on tree beds because of their
mulching effect, or a ground cover of shrubs or wild herbs provided. By no
means, however, should poisonous substances be used to destroy the weeds.

Finally, street trees must be protected against parking cars. In sites which
are heavily used by pedestrians, e.g. bus stops, the ground surface should be
covered with steel grids or perforated concrete elements. But all this costs
money. The cost of just one unit is about 1000 DM. In addition there are the
costs of the tree, the construction work, the soil improvement, etc. These can
add up to another 1000 DM. Then the annual cost of maintenance of a single
tree is about 50 DM. All this is a large sum, but if we are to have trees at all it is
better to do the job properly, and produce a tree that survives and grows well.
If the costs seem high then it must be remembered that they are no more than
the costs of many other things, such as lighting standards, on which we spend
large sums of money in cities without being concerned.

STANDARD LIST OF TREES WHICH ARE KNOWN TO BE RESISTANT TO DROUGHT, POLLUTION AND PESTS

The air in city centres is warmer and dryer than in suburbs: it is to be compared
with that of south-east Europe, parts of China and North America. Therefore,
trees from these countries, such as *Robinia pseudacacia*, which is of North
American and Mexican provenance, grow better in our cities. Heat-tolerating

trees and shrubs can be recognized by the shiny surface of their leaves which reflect the sunshine more effectively, e.g. *Gingko biloba*, *Castanea sativa*, *Populus simonii*. Similarly, the red blooming *Aesculus oarnea* with its shiny leaves is more resistant than the white blooming *Aesculus hippocastanum*.

Trees with light hairy leaves are better protected against transpiration than those which are smooth and dark, e.g. *Tilia tomentosa* with light hairy leaves compared to *Tilia intermedia* with smooth leaves. Trees with pinnate leaves have better air circulation round their leaves, which prevents overheating and conserves more water than large leaves. Some examples suitable for cities are *Robinia pseudacacia*, *Sophora japonica*, *Ailanthus altissima*, *Gleditsia triacanthos*.

But the adaptation of tree species to their environment is complex with several different components, many of which are not properly understood. The pinnate habit, for instance, is also found in trees adapted to wind exposure, such as *Sorbus aucuparia*. So it is still necessary to rely on practical experience. As a basis for the use of tree species in urban streets the German 'Standing Conference of Parks Directors' has published a list of trees (Ständige Konferenz 1983). The shortened list is given in Table 23.1.

CONCLUSIONS

Not only in the most extreme planting sites in cities, but also in wider streets, pavements and courts, thousands of trees die because they can no longer find the minimum resources to maintain their growth between asphalt and concrete. In the last few decades, the factors which are of great importance for the environment of plants in cities—particularly soil, water and air—have become increasingly unfavourable. At the same time the chances of accidental death from motor vehicles, salting and pollution have increased enormously.

In many cases, the tree planting sites that are still available within street profiles or on compacted building sites can only be planted with trees if sufficient preparations have been made and if a minimum tree maintenance has been guaranteed. Because of the increasing emissions of pollutants of all sorts and the reduction of the components in soil, water and air crucial for plant life, it is, however, quite uncertain whether trees, even when well maintained and given a good start, will reach a similar volume to those of the nineteenth century. This is a challenge for all urban landscape architects, horticulturalists and ecologists. Fundamentally speaking, it is a fact that trees in cities, except those in parks and gardens, have to live in a very unnatural environment. Financial investment is therefore necessary in order to maintain them. So the question of vital interest is, how much are we willing to pay for trees and green areas? Up to now none of us has resented paying millions every

TABLE 23.1. Suitability of tree species for the urban environment (from Ständige Konferenz der Gartenbauamtsleiter beim Deutschen Städtetag 1983)

A. Tree species well fitted for urban streets
Corylus colurna
Gingko biloba and 'Fastigiata'
Gleditsia tricanthos 'Skyline'
Platanus hybrida acerifolia
Platanus hybrida 'Pyramidalis'
Quercus robur
Robinia pseudacacia 'Bessoniana'
Robinia pseudacacia 'Decaisneana'
Robinia pseudacacia 'Inermis'
Robinia pseudacacia 'Monophylla'
Robinia pseudacacia 'Monophylla fastigiata'
Sophora japonica 'Regent'
Sophora japonica
Tilia vulgaris 'Pallida'

B. Tree species fitted for urban streets
Acer platanoides
Acer platanoides 'Emerald Queen'
Acer pseudoplatanus
Acer saccharum
Ailanthus altissima
Alnus cordata
Carpinus betulus 'Fastigiata'
Fraxinus excelsior
Fraxinus excelsior 'Diversifolia Den Bosch'
Fraxinus excelsior 'Westhof's Glorie'
Gleditsia tricanthus
Gleditsia tricanthus intermis
Quercus petraea
Quercus palustris
Robinia pseudacacia 'Sandraudiga'
Robinia pseudacacia 'Stricta'
Sorbus latifolia
Tilia cordata 'Grennspire'
Tilia hybride 'Sheridan'
Tilia tomentosa

C. Tree species with limitations
Aesculus carnea (must not have too small tree beds)
Betula pendula (sensitive to heat; has very shallow roots)
Liriodendron tulipfera (requires very nutritive soil)
Tilia vulgaris (requires fresh soil)
Tilia americana (requires fresh soil)
Quercus coccinea (requires clean air; suffers easily from insect damage)

D. Trees not fitted for urban streets
Acer saccharinum (susceptible to wind damage)
Aesculus hippocastanum (sensitive to heat)
Fagus sylvatica (sensitive to heat and soil compaction)
Sorbus aucuparia (sensitive to heat)

---

year for streets, illumination, street cleaning, etc. It should be possible for us, surely, to invest a few thousand of pounds for trees and green areas in cities. In Wiesbaden at least we have tried to do this, but the basis has been a proper understanding of the urban tree and its ecology.

REFERENCES

Bernatzky, A. (1981). *Tree Ecology and Preservation.* Elsevier, Amsterdam.
De la Chevallerie, H. (1974). Studie zur Erhaltung der Wiesbadener Strassenbäume. *Das Gartenamt*, 2, 70–83.
De la Chevallerie, H. (1976). *Mehr Grün in die Stadt.* Bauverlag Wiesbaden u. Berlin.
Klaffke, K. (1980). Winterschaden an Strassenbäumen und Verkehrsgrünlächen. *Das Gartenamt*, 11, 699.
Meyer, F.H. (1978). *Bäume in der Stadt.* Eugen Ulmer, Stuttgart.
Ruge, V. (1979). Nachwirkungen der Winterstreuung 'Januar 1979' auf die Strassenbäume und mögliche Eigenmassnahmen. *Das Gartenamt*, 3, 131.
Ruge, V. (1980). Ursache des Baumsterbens und mögliche Eigenmassnahmen. *Das Gartenamt*, 1, 34.
Ständige Konferenz der Gartenbauamtsleiter beim Deutschen Städtetag (1983). Beurteilung von Baumarten für die Verwendung im städtischen Strassenraum. *Das Gartenamt*, 12, 665–672.

OTHER READING

Breyne, A., Eckstein, D. & Liese, W. (1980). *Holzbiologische Untersuchungen über Entwicklung und Zustand der Hamburger Strassenbäume.* Ordinariat für Holzbiologie der Universität Hamburg (Untersuchungsbericht als Umdruck).
Leh, H.O. (1973a). Untersuchungen uber die Auswirkungen der Anwendung von Natrium-Chlorid als Auftaumittel auf. die Strassenbäume in Berlin. *Nachrichtenblatt Deutscher Pflanzenschutzdienst*, 25.
Leh, H.O. (1973b). *Untersuchungen über die Standortbedingungen der Strassenbäume in Berlin und Möglichkeiten zu inrer Erhaltung unter besonderer Berücksichtigung der Schäden durch Streu- und Auftausalze.* Forschungsauftrag des Senators für Wirtschaft an das Institut für nicht parasitäre Pflanzenkrankheiten.
Meyer, F.H. & Höster, H.R. (1980). Streusalzschaden an Bäumen in Hannover. *Das Gartenamt*, 3, 165.
Meyer-Spasche, H. (1980). *Salzschaden an Hamburger Strassenbäumen, NaCl-Akkumulation in den Strassenrandböden und deren Auswirkung auf die Strassenbäume.* Ordinariat für Bodenkunde der Universität Hamburg (Abschlussericht als Umdruck).
Müller, H.J. (1980). Entwicklung umweltfreundlicher Streumittel. *Das Gartenamt*, 1, 22.
Säuer, G. (1976). *Untersuchung der Zusammenhänge zwischen Absterbeerscheinungen in Waldbeständen und der Zufuhr salzhaltiger Schmelzwasser von der Strasse.* Forschungsgesellschaft für das Strassenwesen, Information über Strassenbau- und Strassenverkehrsforschung.
Spirig, A., Zolg, M. & Bornkamp, R. (1978). *Wasserhaushalt von Strassenbäumen in Abhängigkeit von hohen Salzkonzentrationen im Boden.* Forschungsvorhaben des Bundesministeriums für Forschung und Technologie (BMPI—PNSW 10).
Ständige Konferenz der Gartenbauamtsleiter beim Deutschen Städtetag (1980). Resolution gegen die Vernichtung des Strassengrüns durch Streisalz. *Das Gartenamt*, 1, 25.
Wentzel, K.F. (1973). Salzstaub- und Salzspritzwasserschäden an Strassenbäumen. *Der Forest-Und Holzwirt*, 22.
Woick, H. (1977). Schutz von Strassenbäumen in Auffüllungsbereichen und Verkehrsflächen. *Das Gartenamt*, 2.
Woick, H. (1980). Ist ein unweltfreundlicher Winterdienst möglich? *Das Gartenamt*, 3, 176–180.

24. RESTORATION OF LAKES AND LOWLAND RIVERS

BRIAN MOSS

School of Environmental Sciences, University of East Anglia, Norwich
NR4 7TJ

SUMMARY

Human activities may greatly change lowland freshwater bodies through alterations in water quality, engineering operations, introduction of exotic species and recreation. The Norfolk Broadland includes a waterway comprising shallow lakes, rivers, undrained and partly drained wetlands which has been influenced in all four of these ways. Its ecosystems have become consequently less diverse. Attempts to restore some of the former diversity are discussed in the contexts of technical, practical and political constraints.

INTRODUCTION

Change is normal in freshwaters. Lakes fill with sediment and may succeed to wetlands or drier communities; rivers track across their flood-plains leaving cut-off meanders to provide a variety of lakes and swamps. This paper, however, is concerned with waterbodies changed by human activities where the diversity of communities, and often also the aesthetic appearance and value of the ecosystem for recreation, have been reduced (Dunst *et al.* 1974; Environmental Protection Agency 1979). Reversal of such changes (restoration) often involves economic and political factors as well as ecological ones.

There are three groups of ways in which freshwater bodies can be severely altered by human activities. The first is through development of the catchment area so that the chemical composition of the drainage water is modified. This may be through increases in dissolved nutrient (particularly nitrogen and phosphorus) concentrations derived from agriculture or, as sewage effluent, from settlement; through suspended particles resulting from erosion after cultivation; or through industrial pollution, either direct or indirectly as acidic rain. The second is modification of a waterbody by 'river engineering', canalization, embankment and piling to reduce the risk of flooding of what were originally natural flood plains, or through the introduction of exotic species, often higher plants, mammals or fish. And the third is by direct human impact such as recreational boating, and angling.

In the developed world, most waterbodies have been affected, albeit sometimes mildly, by at least one of these agents. Where populations are relatively high, several agents often act simultaneously and it is impossible to maintain waterways in a pristine state. Some waterways have been very considerably changed, however, and some restoration is often felt desirable. What is possible is generally a compromise between a habitat damaged to the extent that the Dutch would call it a 'cadaver ecosystem', and its original state, or at least that prior to the major changes of the last few decades.

The Norfolk Broadland in East Anglia, UK is a system of rivers, small lakes, undrained and partly drained wetlands (Fig. 24.1) on which many of the agents of change listed above have acted (Fig. 24.2). It is an area to whose problems the Countryside Commission have drawn national attention, so that considerable interest has been aroused in restoration of it. It thus forms a useful case study of the techniques and problems of restoration.

THE NORFOLK BROADLAND

Broadland is one of the best-known areas for water-recreation in Britain and, by its boat-eroded river banks, shows symptoms of recreation damage to such an extent that the casual observer might think that this was the main, if not the sole, problem.

The key problem in Broadland, however, is that of eutrophication (the input of plant and algal nutrients), which has caused the clear water and submerged plant communities to be replaced by turbid water and extremely dense suspended algal (phytoplankton) populations (Moss 1983). In turn this has resulted in loss of habitat for invertebrates, reduction in fish community diversity and perhaps reduction in survival of larger fish (Moss, Leah & Clough 1979). Sedimentation rates have increased so that the already shallow waterway is losing depth at a cumulative 1–2% per year. The sediment is mostly derived from the production of the algae, to a large extent through chemical precipitation of marl (calcium and magnesium carbonate) as a result of photosynthesis-induced pH change. Bank erosion has increased with the loss of the wave-absorbing barrier once provided by the plants. And there are perhaps more severe outbreaks of avian botulism (western duck sickness) than would normally occur because the causative bacterium (*Clostridium botulinum*) finds a ready environment for the hatching of its spores in the algal sediment.

Nor are eutrophication and recreation the only problems. Coypu (*Myocaster coypus* Molina), a South American rodent, escaped from fur-farms in the 1930s and bred to form a large population in Broadland. It may have been responsible (Boorman & Fuller 1981) for loss of substantial areas of

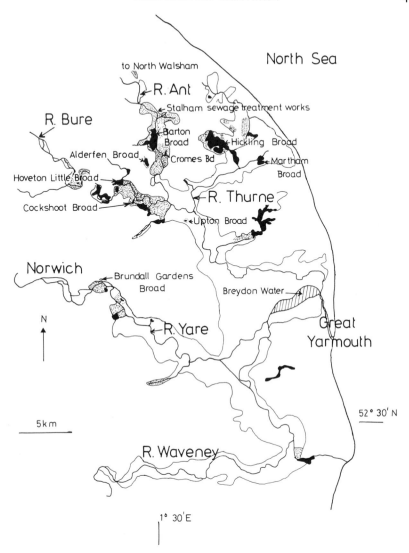

North Sea

to North Walsham

R. Ant

Stalham sewage treatment works

R. Bure

Barton Broad

Hickling Broad

Alderfen Broad

Cromes Bd

Marthcm Broad

Hoveton Little Broad

Cockshoot Broad

R. Thurne

Upton Broad

Norwich

Brundall Gardens Broad

Breydon Water

Great Yarmouth

N

R. Yare

52° 30′ N

5 km

R. Waveney

1° 30′ E

FIG. 24.1. Map of the Norfolk Broadland, with places mentioned in the text indicated. The thin lines bordering the river indicate the limits of valley wetlands, almost all of which are now partly drained to form the grazing marshes. Increasing areas are now being more deeply drained for arable cultivation. The stippled areas are those of undrained fens and carrs.

reedswamp which fringed the rivers and Broads but, with control of the coypu population, the reed has not recovered.

In extensive areas of the lower parts of the Broadland valleys, partial drainage in the past has given an open pastureland (the grazing marshes, Fig.

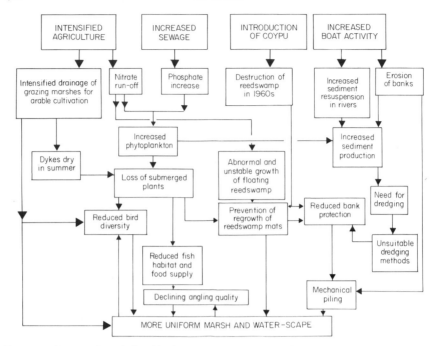

FIG. 24.2. Some major relationships between cause and ecological and landscape effect in the Norfolk Broadland.

24.1) intersected with dykes (ditches) containing a plant and invertebrate community similar to that now lost from the main waterway. The EEC Common Agricultural Policy financial arrangements now favour more intensive drainage and cultivation of wheat rather than grazing of cattle. This is resulting in a more uniform landscape (Broads Authority 1981), with dykes too dry in summer to support an aquatic community, and loss of winter feeding for migrating geese and of nesting sites for other birds. Figure 24.2 shows some of the main problems and their consequences.

Restoration of Broadland

The difficulties of restoring Broadland to a state resembling that occurring when G. Christopher Davies first publicized it in the 1880s (Davies 1883; Fig. 24.3) or even to that prized by sailors and naturalists prior to the Second World War, are probably too great to be overcome, for they concern social changes of national extent. A reversal of the extreme changes of the last three decades, however, is not impracticable. Three aspects will be considered—

FIG. 24.3. (a) A photograph taken probably on the R. Bure by P.H. Emerson and published in *Life and Landscape on the Norfolk Broads* (1886). Good stands of emergent vegetation, dominated by *Typha*, can be seen fronted by water lilies and other water plants. (b) Edge of the R. Bure near St Benet's Abbey in 1982. Severe bank erosion can be seen and the river edge no longer supports much aquatic vegetation. The remains of old piling in the water nearest the closer clump of rushes indicate the bank edge about 20 years ago (photograph Brian Moss).

eutrophication, loss of reedswamp, and boat damage—to illustrate the variety of problems encountered. The changes in the grazing marshes are severe and of direct interest to landscape managers, but their reversal is entirely subject to arbitrary political decisions and is of lesser technical and ecological significance.

Eutrophication

Solution to the eutrophication problem, with restoration of submerged plant communities instead of the present dominance by phytoplankton, must at least involve reduction of the nutrient concentrations in the water. Two tactics have been favoured: isolation of Broads from effluent-rich water, and a combination of treatment and diversion of sewage effluent. Isolation has been very successful. At Alderfen Broad (Fig. 24.1), an effluent-rich stream was diverted, in 1978, through a pre-existing dyke system in the surrounding wetland, to discharge below the Broad. As a result, the Broad now receives water only by seepage from the land. Concentrations of both nitrogen and phosphorus compounds in the Broad have been greatly reduced (Fig. 24.4) and, in the fourth year following isolation, an aquatic plant community had been restored, covering which covered almost all of the Broad and was dominated by *Ceratophyllum demersum* L., similar to that described for the Broad in the 1960s.

Phosphorus accumulates in sediments of eutrophicated lakes and under certain circumstances may be released back into the water, thus confounding measures to restrict the supply from outside sources. Such release had occurred in Alderfen Broad in years prior to isolation (Phillips 1977) and continued after isolation. The summer peaks of phosphate concentration resulting from it, however, progressively diminished and net release was not found in 1982 and 1983. The mechanism of release depends on a continual supply to the sediment surface of easily decomposable organic matter; this was ultimately provided from growth of algae in the water above. If this growth is reduced by restriction of the external nutrient supply, a progressive fall in the amount of phosphorus released from the sediment might be expected, as organic matter in the surface layers of the sediment is oxidized but not replaced.

However, such a fall did not occur over 10 years in a Swedish lake, L. Trummen (Björk 1972; Bengtsson *et al.* 1975; Lettevall & Svensson 1977), following reduction of phosphorus inputs from sewage treatment works, and clear water was not restored until several decimetres depth of sediment had been pumped from the lake. The Lake Trummen example, in general, alarms lake restorers because such pumping and disposal of the sediment is very

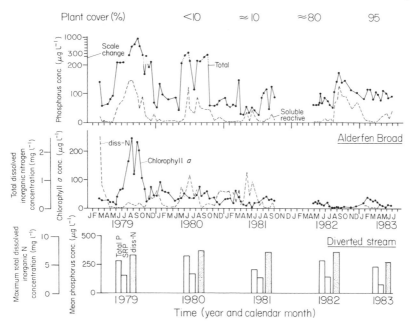

FIG. 24.4. Seasonal changes in total phosphorus concentration (P, μg l^{-1}, upper panel, solid line), concentration of soluble reactive phosphorus (P, μg l^{-1}, upper panel, hatched line), concentration of phytoplankton chlorophyll a (μg l^{-1}, middle panel, solid line) and total dissolved inorganic nitrogen ($NH_4-N + NO_3-N$, mg l^{-1}, middle panel, hatched line) in Alderfen Broad, Norfolk, after diversion of an effluent-rich stream from the Broad in January 1979. Lower panel shows the mean total phosphorus and soluble reactive phosphorus concentrations (as P, μg l^{-1}) and the maximum dissolved inorganic nitrogen concentrations (as N, mg l^{-1}) for the stream that was diverted, in the calendar years 1979–83.

expensive. However, Alderfen Broad may be a better indicator of the British situation because, unlike L. Trummen, it is not frozen for several months each year. In L. Trummen, all of the dissolved oxygen under the ice was rapidly used up in winter. The anaerobic conditions then favoured phosphorus release without a great accumulation of organic matter at the sediment surface and inhibited continuous winter oxidation of any organic matter that did accumulate. A complication in Alderfen Broad has occurred. After 1982, the organic matter produced by the plants themselves re-activated the release mechanism whilst nitrogen-fixing algae colonized. A switch back to phytoplankton dominance has occurred (1985)—perhaps as part of an alternating cycle of plants and algae (Moss *et al.* 1986).

   A second isolation experiment is being carried out at Cockshoot Broad (Fig. 24.1), which has been isolated from the R. Bure by dams. Cockshoot Broad had filled almost to the water surface with sediment and it was necessary to remove this simply to create a suitable depth of water. The Broad is now fed by a small stream of low phosphorus concentration, but moderately high nitrate concentration (Fig. 24.5). The water has become extremely clear, a range of submerged plants is present in part of the Broad, and the prognosis for successful restoration is very good.

A few other Broads (though, of course, none of the rivers) might also be successfully restored by isolation, but the technique cannot be widely used. This is because the waterway is tidal, with ancient rights of navigation guarded originally for freight traffic and now, by the Great Yarmouth Port and Haven Commissioners, for pleasure boats.

The only general options available, then, are phosphorus precipitation from sewage treatment plant effluent, or effluent diversion. In either case it should be necessary to remove only one key nutrient, phosphorus but not the other major nutrient, nitrogen. In theory this should reduce algal production in the same way that denial of either oil or petrol will effectively immobilize a

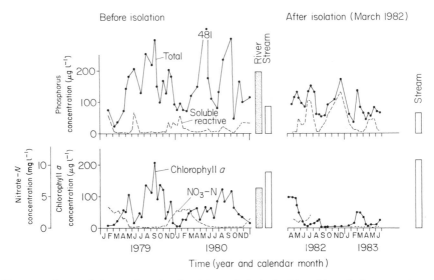

FIG. 24.5. Seasonal changes in water chemistry and phytoplankton chlorophyll a concentration in Cockshoot Broad, Norfolk, before and after its isolation by dams from the effluent-rich River Bure. Left-hand panels are for 1979–80, prior to isolation. Right-hand panels give data for the period after isolation in April 1982. Upper panels, solid line and symbols, show total phosphorus concentration (as P, μg l^{-1}) and hatched line, soluble reactive phosphorus (P, μg l^{-1}). Histograms show the mean total phosphorus concentrations for the nearby River Bure (stippled) and for a small stream which still flows into the Broad, during the periods indicated. Lower panels, solid line and symbols, show phytoplankton chlorophyll a concentrations before and after isolation, and hatched line the concentration of nitrate-nitrogen (as N, mg l^{-1}). Ammonium-N concentrations were negligible. Histograms show the maximum nitrate-N concentrations for the river (stippled) and stream in the periods covered.

car. At least two-thirds of the phosphorus entering most Broads comes from effluent.

Diversion of effluent to the sea is much too expensive for all but a few of the two hundred or so sewage treatment works potentially affecting Broadland, most of which lie far inland. Emphasis is thus being placed on precipitation by iron II salts of phosphate from the effluent. Institution of a widespread programme of phosphorus removal ('stripping') rests on two things: first, minimization of costs by installing such treatment at the least number of works necessary to achieve the desired results, and, secondly, attainment of successful results from the scheme which has been established on the R. Ant by Anglian Water with a view to restoration of Barton Broad.

To minimize costs a target concentration enabling submerged plants to persist must first be established, followed by a budget study of phosphorus sources in a given area and calculations of where the budget might best be reduced. Comparison among those Broads which still have and those which lack submerged plants suggests that, if the mean annual total phosphorus concentration in the water is less than about $100 \mu g \, l^{-1}$, plants can persist. Martham, Upton and Brundall Gardens Broads have well developed plant communities and mean total phosphorus concentrations of 50, 30 and 48 μg P l^{-1} respectively; Hickling Broad, with dense phytoplankton and about $114 \mu g$ l^{-1}, still has some submerged plants; but the Broads of the Bure, lacking plants, have mean annual concentrations of between 164 and 356 $\mu g \, l^{-1}$.

The $100 \mu g \, l^{-1}$ target assumes a single transition point for both directions of change, plants to plankton and the reverse; it is also a compromise between a much lower value which would be more reliable, and the need to keep the target as high as possible for financial reasons. Existing budgets show that the target is achievable through selective phosphorus removal, even though the background phosphorus concentrations from the normal catchment run-off are high (total P up to $70 \mu g \, l^{-1}$).

On the R. Ant, sewage from one works, North Walsham, has been diverted to the sea, and phosphorus is being precipitated from effluent at Stalham and at two smaller sewage treatment works discharging to the river (Fig. 24.1). The $100 \mu g \, l^{-1}$ target for water entering the Broad has been reached since 1982. There was little response in Barton Broad, for release of phosphorus from the sediments in summer still caused a mean annual concentration well above $100 \mu g \, l^{-1}$ (Anglian Water Authority 1983). Net phosphorus release had ceased by 1985 with a mean phosphorus concentration round $100 \mu g \, l^{-1}$ but the phytoplankton crop, though reduced to a half of that before phosphorus stripping, is still high. Aquatic plants have not recolonized. Restoration of aquatic plants has been achieved in Crome's Broad (Fig. 24.1), which receives water from the R. Ant for part of each year, and this is attributable to reduction in phosphate concentrations in the river.

Leakage of effluent from a nearby farm has temporarily set back the restoration.

Despite success of phosphorus stripping elsewhere (Forsberg 1979) and a well established justification for it, Anglian Water regards the phosphorus stripping as 'experimental' and funds the project on 3-year bases. There is a danger that a lack of rapid results in Barton Broad may jeopardize both continuation of the treatment there and establishment of other schemes (Moss *et al.* 1982). However, funding from the Countryside Commission, B roads Authority and Nature Conservancy Council will extend phosphorus removal to works affecting the upper Broads of the R. Bure, where circumstances are more favourable.

A fundamental difficulty may lie in the validity of the 100 μg l^{-1} target. Although submerged plants can grow in water with this phosphorus concentration, and even at much higher values (Leah *et al.* 1980), they only do so if the phosphorus is to a large extent prevented from being incorporated into the potentially-shading algae, which can occur as (i) suspended plankton, (ii) as epiphytes on the plant surfaces or (iii) as tangles of filamentous algae. An established submerged plant community harbours grazers, particularly Cladocera (Timms & Moss 1984), which prevent build-up of algae whilst receiving protection in the architecture, at least, provided by the plants from fish predators which would otherwise reduce their numbers and effectiveness. The changes which occur when the plant-dominated ecosystem is replaced by a phytoplankton-dominated one are not fully understood, but once the switch has occurred, the most efficient grazing animals—the larger Cladocera— become very scarce through fish predation. Without grazing impact, substantial phytoplankton populations can exist even at phosphorus concentrations of 100 μg l^{-1} and below. The 100 μg l^{-1} target may thus be too high unless grazing can be re-established. In practice it may be difficult to reduce Broadland phosphorus concentrations much below 100 μg l^{-1} through phosphorus stripping alone, and the husbandry of grazing zooplankton may have to be contemplated. This means initial provision of refuges against fish predation previously provided by the aquatic plants. Bundles of alder twigs cut from the surrounding wetlands and staked in the water may be one means of providing this but, for a moderate-sized Broad with even a low-density provision of 1 bundle per m^2, several hundred thousand bundles would be needed. A variety of possible artificial refuges will be tested in Hoveton Great Broad in 1986.

Reed regression

Successive Ordnance Survey and other maps show a progressive colonization of the open water of the Broads by reedswamp between the mid-nineteenth century and the 1940s. Maps drawn from aerial photographs taken since then have shown a reversal of this and a substantial increase in open water (Boorman & Fuller 1981). The main period of loss of reedswamp coincides with the greatest population densities of coypu. Coypu are herbivores and were seen to destroy reed beds (Ellis 1963) as well as root crops in the area. A control programme began in the 1960s and the coypu population was greatly reduced. Plants like the great water dock (*Rumex hydrolapathum* Huds) which were almost eliminated from Broadland by coypu grazing, have re-established their former status, but the reedswamp fringes have failed to grow back; reasons other than coypu grazing must therefore be sought.

Grazing by geese and swans can severely damage reedbeds if it is intensive, but exclusion of vertebrate grazers by wire-netting cages has failed to show that this could be a widespread cause of failure of reed growth (Crook, Boar & Moss 1983), although the loss of submerged plants, their preferred food, has diverted geese and swans to the reed edges. Erosion damage by boats is obvious along the rivers, with clear evidence of slicing of the banks during mooring, but reedswamp has disappeared also from Broads closed to boats for many years.

Reedswamp in Broadland is of two main forms—littoral and hover. The former is rooted in peat or clay on gently sloping banks, particularly along river channels. Hover is a floating mat, held together by inter-tangled rhizomes and roots. It is anchored in firm ground at the landward edge but floats out to its growing edge. Hover is particularly associated with the edges of the Broads, where the mediaeval peat cutters (who dug out the pits which form the basins of the Broads) left a vertical or stepped bank. Most of the reedswamp that has been lost has been hover; the littoral reed has been much less susceptible (Fig. 24.6). In Broadland there is a high correlation ($r = 0.71$, $P < 0.05$) between the extent of hover reedswamp loss in a particular Broad and the maximum nitrate concentration found in the water. Most nitrate is derived from leaching of agricultural land. Glasshouse experiments with hydroponic culture have shown that in response to increasing nitrate concentrations (up to 12 mg l^{-1} as N), shoot growth is increased to a greater extent than that of root and rhizome growth. Remaining hover mats in the field also have relatively high shoot to root and rhizome quotients. This top-heavy distribution of biomass may make the hover more vulnerable to capsizing if the mat is broken by natural or boat erosion or modest grazing damage, and provides a likely explanation for the failure of reedswamp to grow again even where coypu numbers have been reduced.

Restoration of the reedswamp fringes may thus be difficult. In the rivers erosion barriers will certainly be necessary and difficulties with these are discussed below. Even where boats are absent, some form of mechanical protection will be needed to reinstate hover, but there is little that can be done

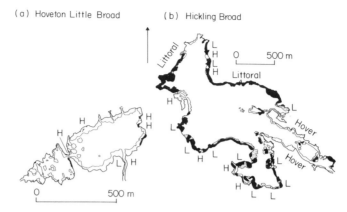

FIG. 24.6. Changes in area of fringing reedswamp between 1946 and 1982 in two of the Norfolk Broads: Hoveton Little Broad (left) and Hickling Broad (right). Areas were determined from tracings of aerial photographs. Open areas represent reedswamp which has disappeared, black areas that which was present in 1946 and is still present, and stippled areas, reedswamp that was present in 1982 but not 1946. The nature of the reedswamp (hover, H, a floating mat and littoral, L, rooted in solid substrata) was checked by ground survey at 50-m intervals along the outer limit of remaining reedswamp in 1982.

directly to restrict the nitrate concentration. The relationship between amount of run-off and the rates of application of nitrogenous fertilizer is a complex one, dependent on time of application, weather, and the natures of both soil and fertilizer. The prime attack on this problem might come from efforts to reinstate submerged aquatic plants. Communities of these are efficient at absorbing nitrate, either directly or through denitrification by suitable bacteria on the plant surfaces and in the associated sediment.

Damage caused by boats

Traditionally, local opinion has blamed all Broadland problems on the increasing activity of the weekly- or daily-hired motor boats though in numbers, if not in activity, the private fleet is the greater. Even in recent publications (Perring & Farrell 1983) boat damage is erroneously blamed for the disappearance of rare plants, though evidence of direct effects of eutrophication has been available for some time. The dense populations of phytoplankton are frequently mistaken for resuspended sediment, yet the latter is a negligible problem in the Broads themselves and, in terms of the relative amount of light absorbed, only a small problem compared with that of the phytoplankton in the lower reaches of the rivers. The undercutting of banks and their eventual collapse is much more serious, for this erosion removes any fringing vegetation with serious consequences for wildlife and for protection of flood embankments. Reinforcement of these by piling is very expensive, as well as ugly.

The erosive effect of wash of Broads cruisers increases greatly above 5 m.p.h. (8 km h^{-1}) (Payne & Hey 1982) and a reduction of the present 7 m.p.h. limit (11·2 km h^{-1}) on most rivers would reduce erosion rates. However, the Broadland navigation includes an estuary, Breydon Water (Fig. 24.1) where tidal currents are greater than 5 m.p.h. and governing of boat throttles to allow maximum speeds of only 5 m.p.h. might be dangerous. Adequate policing of the speed limit is impracticable and reliance on personal discipline of all boat users is thwarted by a Walter Mitty-like streak (Thurber 1965) which seems a frequent element in the taking of a boating holiday.

Another way of reducing erosion would be by restriction of the numbers of boats. A further increase can be avoided through refusal of planning permission for expansion of boatyards or for new ones; numbers could also be limited by the Great Yarmouth Port and Haven Commissioners by rationing of boat licences. However, the latter is a navigation authority with a statutory duty to promote, not restrict the navigation. The degree to which numbers or activity of boats should be restricted is not clear, because the precise relationship between boat activity and damage caused is difficult to obtain. Figure 24.7 shows the relationship between boats passing per hour on part of the R. Bure and the amount of sediment kept in suspension in excess of background (pre-dawn) concentration. The data suggest that a reduction by one-half in the damage caused (assuming that sediment resuspension is an index of other forms of erosive damage) would need a lowering of boat activity to one-third of the present amount. For a local industry providing a significant number of jobs and contributing about £17M to the local economy (Countryside Commission 1983), this is probably neither possible nor desirable. The current recession, it is claimed by the hirers' organization, has reduced activity by about one-third between 1978 and 1982 (Broads Hire Boat Federation 1983), but this may not be a sustained trend. Some positive effects might come from the redesign of boat hulls to give lower wash disturbance, but, because of the capital involved, this would be a long-term solution.

Treatment of the symptoms is possible. Underwater barriers, based on old tyres threaded on poles, to minimize undercutting, are being tested and a change in the method of dredging the channels may be beneficial. Dredging is now continually necessary and is probably not keeping pace overall with the rate of sedimentation; the present bucket-dredging close to the banks gives a vertical bank profile which is readily undercut. Suction dredging could give a more gently graded profile less susceptible to erosion and favouring colonization by littoral reedswamp. A policy of zoning the waterway is also

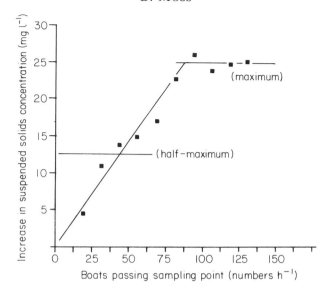

FIG. 24.7. Effects of number of boats passing per hour on the concentration of suspended solids above pre-dawn background concentration in the R. Bure near Malthouse Broad. Points are three-point moving averages of hourly-determined suspended solids concentrations made between dawn and late afternoon on 7 days in the summers of 1979 and 1981 ($n = 80$). The number of boats passing was continuously monitored and numbers expressed as passages per hour. These values were then grouped into successive frequency categories of 12·5 passages per hour for plotting. Powered boats constituted 97% of total passages. Trend lines are drawn by eye.

feasible, with perhaps certain areas restricted to sailing and rowing, which cause little erosion. None the less, it will not be possible to deny lengths of river to powered boats, in this essentially linear waterway, without excavation of costly by-pass channels. Greater flexibility may come from transfer of navigation responsibility to the Broads Authority by a Bill soon to be considered by Parliament.

GENERAL CONSIDERATIONS

Because lakes and rivers are open systems—they continually receive and export volumes of water and amounts of other substances, that are significantly large relative to the amounts stored in them at any one time—they are particularly prone to change. But, because of this open nature, a restoration—a change back—is probably easier to achieve here than in relatively closed, self-contained ecosystems like forests. A waterbody, however, cannot be restored to any chosen state so potential reversibility should not be used as an excuse for abuse. Three factors stand against unrestricted reversibility.

(a) The particular species composition of an aquatic ecosystem cannot be faithfully reproduced once it is lost, because we generally do not fully understand why a community is composed as it is. It will usually be possible to restore the overall structure of an aquatic ecosystem (e.g. plant dominance versus plankton dominance), but not to determine reliably in advance which particular organisms will form the structure.

(b) Although the reasons underlying some deleterious change to an aquatic ecosystem may be substantially understood, restoration of the original state may not be possible simply by removal of the cause of change alone. Change may proceed from one point to another by different pathways in different directions—in analogy with a magnetic hysteresis loop. Additional measures may be necessary as well as removal of the main cause. For example, in Broadland, surface sediment removal or husbandry of grazing zooplankton may be needed in addition to phosphate precipitation.

(c) Although restoration is a technical possibility, perception of its need and the degree of restoration felt desirable are political matters. Some lobbies in Broadland, e.g. the sailing clubs and boat-hire companies, indeed might view the loss of aquatic plants as an improvement, for it has given more open-water in which to navigate.

Different nations also have widely differing goals. In Canada, lakes with 20 μg P l^{-1} and water that is not extremely clear and able to support trout fisheries, may be considered severely eutrophicated and subject to considerable public concern (Dillon & Rigler 1975). In Britain, eutrophication is perceived mostly as a problem for the domestic water supply industry in terms of costs of filtration of algae during treatment (Collingwood 1977). Lakes like those of the Southern English Lake District which would be considered desirable subjects for restoration, in Canada and Scandinavia are not seen as problems, and a recent report (Lund 1980) singled out only Broadland and Lough Neagh as particular problem areas. Because widespread removal of phosphorus from sewage effluents (or replacement of the phosphorus compounds in domestic detergents, which contribute about half of that phosphorus) would not greatly reduce filtration costs in southern England, such measures receive little support from the water industry (Collingwood 1977). Yet growth of blanket weed (*Cladophora* sp.) in lowland rivers, another symptom of eutrophication by sewage effluent (Whitton 1970; Bolas & Lund 1974), costs large sums when it has to be manually cleared to prevent flooding.

Classification schemes for river quality in Britain depend largely on criteria of organic pollution (National Water Council 1981) and ignore other forms of significant environmental change so that claims of a general improvement must be examined very closely. The return of salmon (*Salmo salar* L.) to the R. Thames as a result of improvement in oxygenation of the water through better sewage treatment, must be viewed in the context that the fish are hatchery raised and cannot maintain a self-supporting breeding population. Siltation has removed most possible spawning sites in accessible reaches.

Restoration of lakes and rivers is sometimes practically limited, even if technically feasible, in a crowded country and economic restraints cannot be ignored. However, casual observation does suggest that, on balance, the waterways of Britain are becoming much less diverse, attractive and interesting. It would be unfortunate, therefore, if low water quality criteria, which are modest enough to be politically expedient, were to become widely acceptable through a general lack of availability of high quality habitats to act as standards.

REFERENCES

Anglian Water Authority (1983). *AWA research and investigations in Broadland during the year ended 31st March 1982*. Broads Authority, Broads Consultative Committee Minutes BA/BCC/B7. April 28 1983.

Bengtsson, L., Fleischer, S., Lindmark, G. & Ripl, W. (1975). Lake Trummen restoration project I. Water and sediment chemistry. *Verhandlungen internationale Vereinigung für theoretische und angewandte Limnologie*, 19, 1080–1087.

Björk, S. (1972). Swedish lake restoration program gets results. *Ambio*, 1, 153–165.

Bolas, P.M. & Lund, J.W.G. (1974). Some factors affecting the growth of *Cladophora glomerata* in the Kentish stour. *Water Treatment and Examination*, 23, 25–51.

Boorman, L.A. & Fuller, R.M. (1981). The changing status of reedswamp in the Norfolk Broads. *Journal of Applied Ecology*, 18, 241–269.

Broads Authority (1981). *Report of Landscape Working Group*. Broads Authority BASMP 6, Norwich.

Broads Hire Boat Federation (1983). *What future for Broadland? The views of the boat hire business*. Norwich.

Collingwood, R.W. (1977). *A survey of eutrophication in Britain and its effects on water supplies*. Water Research Centre Technical Report 40. Medmenham.

Countryside Commission (1983). *The Broads: A Review. A Consultation Document*. CPP 158, Cambridge.

Crook, C.C., Boar, R.R. & Moss, B. (1983). *The Decline of Reedswamp in the Norfolk Broadland: Causes, Consequences and Solutions*. Final Report to the Broads Authority, Norwich.

Davies, G.C. (1883). *Norfolk Broads and Rivers*. Blackwood, London.

Dillon, P.J. & Rigler, F.H. (1975). A simple method for predicting the capacity of a lake for development based on lake trophic status. *Journal of the Fisheries Research Board of Canada*, 32, 1519–1531.

Dunst, R.C., Born, S.M., Uttormark, P.D., Smith, S.A., Nichols, S.A., Peterson, J.V., Knauer, D.R., Serns, S.L., Winter, D.R. & Wirth, T.L. (1974). *Survey of Lake Rehabilitation Techniques and Experiences*. Dept. of Natural Resources, Wisconsin, Technical Bulletin 75.

Ellis, E.A. (1963). Some effects of selective feeding by the coypu (*Myocaster coypus*) on the vegetation of Broadland. *Transactions of the Norfolk and Norwich Naturalists Society*, **20**, 32–35.

Environmental Protection Agency (1979). *Lake Restoration.* EPA 440/5–79–001. Washington D.C.

Forsberg, C. (1979). Responses to advanced wastewater treatment and sewage diversion. *Archiv für Hydrobiologie, Ergebenisse Limnologie*, **13**, 278–292.

Leah, R.T., Moss, B. & Forrest, D.E. (1980). The role of predation in causing major changes in the limnology of a hyper-eutrophic lake. *Internationale Revue der gesamten Hydrobiologie*, **65**, 223–247.

Lettevall, U. & Svensson, S. (1977). *Sjön Trummen i Växjö: Förstörd—Restourerad—Panyttfodd.* Länsstyrelsen i Kronobergs Län och Växjo Kommun, Växjö.

Lund, J.W.G. (1980). *Eutrophication in the United Kingdom.* Soap and Detergent Industries Association, Hayes.

Moss, B. (1983). The Norfolk Broadland: experiments in the restoration of a complex wetland. *Biological Reviews*, **58**, 521–561.

Moss, B., Balls, H.R., Irvine, K. & Stansfield, J. (1986). Restoration of two lowland lakes by isolation from nutrient-rich water sources with and without removal of sediment. *Journal of Applied Ecology*, **23**, 391–414.

Moss, B., Booker, I.R., Manning, H.R. & Manson, K. (1982). *Study of the R. Bure, Norfolk Broads.* Final report to the Dept. of the Environment and Anglian Water Authority.

Moss, B., Leah, R.T. & Clough, B. (1979). Problems of the Norfolk Broads and their impact on freshwater fisheries. *Proceedings of the First British Freshwater Fisheries Conference*, pp. 67–85, University of Liverpool.

National Water Council (1981). *River Quality: The 1980 Survey and Future Outlook.* London.

Payne, S.J. & Hey, R.D. (1982). *River Management to Reduce Bank Erosion.* Broads Authority Research Series, 4, Norwich.

Perring, F.H. & Farrell, L. (1983). *Vascular Plants. British Red Data Books*, 2nd edn. Royal Society for Nature Conservation, Lincoln.

Phillips, G.L. (1977). The mineral nutrient levels in three Norfolk Broads differing in trophic status, and an annual mineral nutrient budget for one of them. *Journal of Ecology*, **65**, 447–474.

Thurber, J. (1965). The secret life of Walter Mitty. *The Thurber Carnival* pp. 69–74. Penguin, Harmondsworth.

Timms, R.M. & Moss, B. (1984). Prevention of growth of potentially dense phytoplankton populations by zooplankton grazing, in the presence of zooplanktivorous fish in a shallow wetland ecosystem. *Limnology and Oceanography*, **29**, 472–486.

Whitton, B.A. (1970). Biology of *Cladophora* in freshwater. *Water Research*, **4**, 457–476.

25. URBAN OPPORTUNITIES FOR A MORE NATURAL APPROACH

LYNDIS COLE

Land Use Consultants, 731 Fulham Road, London SW6 5UL

SUMMARY

Over the last 10 years increasing interest has been shown in urban nature conservation. In the past, lack of detailed survey work in urban areas gave the impression that little of natural interest existed there. More recent work has demonstrated a tremendous diversity of urban wildlife habitats, particularly as part of areas of encapsulated countryside and on urban derelict sites: in certain instances these sites are providing niches for regionally rare or localized species. A number of measures are now being taken both to conserve existing sites of interest and to create new semi-natural habitats as part of corner plot face-lift projects and major reclamation schemes. This paper provides examples of the natural wealth of urban habitats, and the means being adopted to retain these sites and create new habitats of ecological interest.

INTRODUCTION

Since the last war there has been a growing realization that that bastion of naturalness 'The British Countryside' is being laid seige by increasingly efficient and industrialized agriculture and forestry. Stirred by this realization, nature conservationists have become aware that it is not only rare species which are under threat but also the common—those species which to the British public are the very hallmarks of the countryside (Mabey 1980). Concerned to retain the common and improve our general understanding of these issues, conservationists are now seeking to promote nature conservation in urban areas and to encourage community involvement in local conservation schemes. At the same time, many within the landscape profession, inspired by the European examples of created natural landscapes, have turned to an ecological or naturalistic approach to urban landscape design. Combined, these two parallel and overlapping movements have encouraged a re-appraisal of the conservation, design and management of urban open spaces. Therefore, in summary, urban nature conservation is now concerned with:

(a) conserving areas of existing ecological interest;
(b) creating new wildlife habitats, either on a temporary or permanent basis, on down-graded and derelict sites within the inner city;
(c) encouraging community participation and public enjoyment of urban natural areas.

THE NATURAL WEALTH

In the past there was a commonly-held belief that little of natural interest remained in urban areas or, to be more precise, that there was nothing sufficiently rare to be worthy of the attention of conservationists. This belief was based on lack of observation rather than on proven fact. Today exhaustive surveys within urban areas—such as Bunny Teagle's survey of the West Midlands conurbation, *The Endless Village* (1978)—are beginning to reveal the wealth of semi-natural habitats which can be found within these areas.

Within all towns and cities there are many areas of relic countryside caught within the urban fabric. These survive as indelible reminders of the vegetation which once clothed the area before urbanization. These areas of encapsulated countryside survive for a range of reasons. Some remain on ground inhospitable to development, such as river washlands; some survive because of an exclusive land use such as cemeteries; some remain owing to some quirk in ownership; while others remain as designated areas of public open space. A few survive owing to the positive intervention by a group to hold an area in perpetuity from development, such as Perivale Wood, managed by the Selbourne Society since 1904 and acquired by them in 1921. Finally a few survive by chance, fortuitous escapes from the pressures of urban development.

To take a few examples from London: there are the open heathlands and commons including Richmond Park and Wimbledon Common. Wimbledon Common, despite tremendous recreational pressure, survives as open grass, heath and scrub and the remaining bog and pond areas support a wide diversity of wild flowers including *Hydrocotyle vulgaris* and *Scutellaria galericulata*.

The ancient woodlands of London include Kenwood, Hainault Forest with its hornbeam pollards, Dulwich Wood dominated by freely regenerating sessile oak, *Quercus petraea*, and Oxleas Wood in south-east London which still contains magnificent specimens of the wild service tree, *Sorbus torminalis* and patches of butcher's broom, *Ruscus aculeatus*, both indicator species of ancient woodland. Small patches of meadow flora can be found within London's cemeteries, from the spring display of cow parsley, *Anthriscus sylvestris*, in Highgate, to the discovery in 1980 of 419 spikes of the green winged orchid, *Orchis morio*, growing on a relict piece of damp meadow land in a south London cemetery: for obvious reasons the name of this cemetery is not being publicized. Likewise the canals of London harbour such species as flowering rush, *Butomus umbellatus*, and water plantain, *Alisma plantago-aquatica*.

When looking at the potentials of creative ecology, it is equally instructive to see how nature has adapted to and colonized urban derelict and vacant sites which are awaiting the next cycle of development. The Department of Environment's definition of derelict land, 'land so damaged by industrial or other development that it is incapable of beneficial use without treatment', conjures up a picture of biological poverty. This is by no means the case. Urban derelict sites range in type and size from large slag heaps, subsidence pools and open quarries, to the vacant corner plot. Looking first at the small urban derelict plot, these sites are usually associated with the more typical urban ruderal weeds such as Oxford ragwort, *Senecio squalidus*, and rosebay willowherb, *Chamaenerion angustifolium*, which in London frequently give way to dense sycamore scrub. But the range of species on these sites can be very much greater. A survey undertaken in central London between 1952 and 1955 (Jones 1958) revealed 342 plant species associated with these derelict sites. Of this total, many were ruderals, some were garden outcasts but, perhaps more interestingly, some were more typical rural species which had moved in to exploit a specific niche, e.g. calcicole flora including mignonette, *Reseda lutea*, viper's bugloss *Echium vulgare*, majoram, *Origanum vulgare*, and bromegrass, *Bromus erectus*, growing on the lime-rich mortar of building rubble.

At a slightly larger scale, the potentials of urban disturbed ground are exemplified by the Gunnersbury Triangle, West London. This 2·5 ha (6·5 acres) site lies north-east of Gunnersbury Station in Chiswick and is bounded by railway lines. Gravel was extracted from the Triangle for ballast in the 1900s and some topsoil dumping occurred about 35 years ago. However, since this time the area has remained relatively undisturbed. Today the site consists of a mosaic of woodland, scrub, grassland and damp patches with birch and willow forming the main tree cover and many oak saplings growing in the shrub layer. In total some 149 flowering plants have been recorded to date, including perennial species such as wild carrot, *Daucus carota*, wood avens, *Geum urbanum*, and meadow vetchling, *Lathyrus pratensis*, and a number of plants which are uncommon in this part of London, such as hemp agrimony, *Eupatorium cannabinum*, celery leaved crowfoot, *Ranunculus sceleratus*, wild pansy, *Viola tricolor* and ferns such as *Athyrium filix-femina* and *Dryopteris carthusiana*. Lack of disturbance has obviously allowed the development of this open woodland, but the predominance of birch and oak and the general species diversity, including a large number of perennial herbs more typical of rural areas, is no doubt associated with the site's location directly adjacent to two railway lines (Chiswick Wildlife Group 1983).

Derelict sand, gravel and chalk quarries contain a wealth of species (e.g. Kelcey 1975; Davies 1976) as do the smaller subsidence pools of mining areas. Taking one example, a rich marshland community has developed in a subsidence flash at Stubbers Green, Walsall District. Here the plant community includes such species as arrow grass, *Triglochin palustris*, reed mace, *Typha angustifolia*, cotton grass, *Eriophorum angustifolium*, marsh orchids, *Dactylorhiza* sp. and bog moss, *Sphagnum* sp. Many of these species are hardly known in the West Midlands County and all are extreme rarities within the truly urban context (Teagle 1978).

The waste products of mining and industry are a different matter, and are frequently inhospitable to plant growth. Within urban areas the most common wastes are the colliery shales of the Midlands and North of England, blast furnace slag found in the Welsh industrial valleys and also the Midlands, non-ferrous metal smelting wastes concentrated in the Lower Swansea Valley, pulverized fuel ash waste (PFA) from coal-burning power stations, and toxic wastes from the chemical industries primarily concentrated in the Midlands and North of England. Extensive literature (including Gemmell 1977; Bradshaw & Chadwick 1980) describes in detail the problems of plant establishment on industrial waste sites. Nearly all waste sites are deficient in nitrogen and, owing to the inorganic and free draining nature of the material, nutrient retention is often extremely poor. Many wastes suffer from extremes of pH. Colliery shales in the Durham and Northumberland coal fields may, in certain instances, have a pH as low as 1·5–2·0, while PFA may have a pH as high as 12·0. In addition, chemical wastes frequently suffer from heavy metal contamination. While colliery spoil is usually relatively free from heavy metals, it may, like PFA, suffer from high concentrations of soluble salts. In certain extreme conditions, as in the case of non-ferrous smelting wastes and certain chemical wastes, conditions are so inhospitable that few plants will establish readily without soil treatments. On the other hand, there are a number of slow-growing grasses such as wavy hair grass, *Deschampsia flexuosa*, and bent grass, *Agrostis tenuis*, which are adapted to low nutrient status and acidic conditions and which, over time, will naturally colonize acidic colliery shales.

However, perhaps one of the most exciting developments has been the discovery by Greenwood, Gemmell and others (Greenwood & Gemmell 1978) that within Greater Manchester the 70-year old tips of Leblanc process wastes and the lime beds from the Solvay process have formed calcareous habitats supporting very open species-rich communities, including many orchids. Calcicole plants of these habitats include wild flax, *Linum catharticum*, fleabane, *Erigeron acer* with the orchids including *Dactylorhiza fuchsii, D. incarnata* and *D. purpurella*. Similarly, PFA tipped within the last 25 years, such as that at Wigan Power Station, has been rapidly colonized to form willow scrub, marshland flora and grass communities. Like the Leblanc process wastes, these are extremely orchid-rich communities including such species as *D. incarnata, D. praetermissa* and *Epipactis palustris*. Not only are many of the species of these waste tips rare to West Lancashire, but the apparent hybridization between a number of the orchid species found on these sites has led Greenwood and Gemmell to suggest that the provision of these waste tips may be contributing to the breakdown of isolating mechanisms between species.

SITE CONSERVATION

This very brief review of some of the semi-natural habitats of urban areas demonstrates that there are many existing habitats worthy of conservation. Areas of encapsulated countryside and regenerating wasteland, such as the Gunnersbury Triangle, not only provide the urban dweller with an emotional link with the countryside—that tangible feeling of naturalness which is brought into sharp focus by the contrasting urban surroundings—but also provide immense opportunities for community involvement and formal and informal environmental education.

At the same time, it is obvious that certain areas of industrial dereliction are making a direct contribution to the natural wealth of urban areas. And, as suggested by Davies (1976), they may be performing one of a number of valuable functions by: (i) replacing habitats which are rapidly disappearing in rural areas, as in the case of Stubbers Green marsh; (ii) providing alternative habitats for regionally rare or localized species, as in the case of the West Lancashire alkaline tips; or (iii) they may, as in the case of domestic refuse tips supporting a range of culinary delights, be offering suitable habitats for new immigrant species (Cole 1983).

Certain of these sites now have statutory recognition as Sites of Special Scientific Interest (SSSIs). Within London alone there are twenty-six SSSIs, covering 1779 ha or 1% of the GLC area. But there are many other urban sites of natural interest which do not measure up to the general criteria by which SSSIs are judged, and to date the NCC has been against devaluing the national currency of SSSI notification by making particular exceptions for urban sites. However, these sites are of immense value to the urban population, and to try and provide a wider perspective on urban nature, a number of initiatives have been set in train. To take one example, the South Yorkshire County Council have published *A Review of Nature Conservation in South Yorkshire*. This was compiled with the aid of NCC, the Yorkshire Naturalist Trust, local natural history societies and the three County data banks, and lists a series of non-statutory sites, termed for convenience as Sites of Scientific Interest (SSIs), which the County would wish to protect through planning and development control procedures. A large number of the sites listed lie within urban areas.

At the same time many local sites have been adopted by the community and, through campaigns mounted by voluntary and locally formed pressure groups, an increasing number of urban sites have been saved from development. Such campaigns include that mounted by the Chiswick Wildlife Group to prevent the development of warehouses on the Gunnersbury Triangle, the 'Save the Marshes Campaign' aimed at preventing gravel extraction on the biologically-rich Walthamstow Marshes in the Lee Valley Regional Park, London, and representations by the Welsh Harp Conservation Group to prevent development on the fringes of the Brent Reservoir SSSIs. All these campaigns have been successful in stopping the proposed developments.

CREATION OF NEW SITES OF NATURAL HISTORY INTEREST

The other side of the coin in urban nature conservation is the creation of new sites of ecological interest. With nature's apparent ability to colonize derelict and dormant land, albeit over time, it might be argued that she should be left to design and implement our land reclamation programmes. Indeed Bradshaw (1979) and others have questioned whether 'the cleaning up has gone too far'. It is by no means unheard of for so-called 'environmental improvement schemes' to be the main instrument in the destruction of semi-natural habitats. But there are a number of reasons related to time and image which make 'designed' reclamation schemes a necessary requirement of twentieth-century life.

    (a) Certain waste materials, e.g. colliery spoil, take many years to develop a natural cover, and in the case of certain chemical wastes no plants will regenerate without extensive amelioration works.

    (b) Many urban derelict plots are only temporarily vacant. They may have an open life of only 2–5 years. Within this time-span there will be limited opportunity for natural development. Yet with intervention these sites can serve a useful temporary function.

    (c) Large areas of dereliction lower the status of an area in the eyes of both residents and potential employers, causing much-needed industry to shun urban areas for more prestigious green field locations.

(d) The frequent early colonizers of urban waste and derelict land, the ruderal weeds which are the delight of botanists and children, are to many urban residents the very symbols of past urban blight—a strong image which they wish to erase from their memories.

In addition, it is important to remember that gardening is one of the most popular outdoor recreation pursuits of the British public. We are a nation that loves to meddle with growing material. Indeed, it is man's past meddling and intervention which has diversified our natural habitats and has provided the rich heritage of local features which for many of us are the quintessence of countryside: the local coppice woodland and the village pond. So it follows that in urban areas, local residents should be given every opportunity to meddle with and manage small sites and, by so doing, learn about natural processes at first hand, provided they do not destroy areas of existing ecological importance in so doing.

AN ECOLOGICAL APPROACH

What guidelines have emerged for the restoration of urban derelict land to open space? There will always be a place for formal designs in urban areas, especially within housing and for formal recreation. However, the ecological or naturalistic approach to landscape design, which forms the theme of much of this book and which is now an increasingly accepted part of landscape design (see, for example Cole & Keen 1976; Ruff 1979; Tregay & Gustavsson 1983) offers an alternative approach. As set out in Nan Fairbrother's much acclaimed book *New Lives, New Landscapes* (1972), planners and designers should work *with* rather than *against* nature; they should use ecology as the cornerstone to design. In simple terms an ecological approach requires designers to act as accelerators in the natural successional process, by carrying out amelioration work to promote plant growth and by speeding up plant introductions. At the same time, the designer and land manager may be required to influence the development of introduced plant communities to meet specific user and aesthetic requirements. The outward expression of an ecological approach, therefore, will be the creation of a range of semi-natural plant communities adapted to the prevailing site conditions and proposed site use.

Disregarding the benefits to nature and nature conservation, the ecological approach has one major advantage over formal landscape designs, namely that of cost. Natural woodland blocks and meadow land can initially be established at 25% of the cost of formal planting schemes (Corder & Brooker 1981). It is also argued that similar financial arguments exist for the long-term management of these new ecological landscapes. For example, the slow-growing native grasses such as *Deschampsia flexuosa*, which are best adapted to the low nutrient status of most wasteland sites, require only limited cutting by comparison to a ryegrass sward. Making a specific site comparison, Crowe (1956) estimated that the costs of maintaining Wimbledon Common were less than 3% of those of Kensington Gardens. But these comparative costs can be misleading. Under current methods of urban open space management, with work studies and bonus schemes on the one hand and restrictive work rotas on the other, cheap maintenance often equates with unskilled, easily pro-grammed repetitive mechanized tasks, such as grass cutting, and not with infrequent but intensive tasks, such as hand scything or woodland thinning. In addition, aesthetic requirements, especially on small urban sites, may require constant intervention and management to maintain species diversity and the general attractiveness of the site.

Aesthetics

Much of the support for and revolt against an ecological or naturalistic approach rests on considerations of aesthetics. To the supporters, an ecological approach represents the glories of nature; to the antagonists, it represents an unruly weed patch. But the point which is often missed is that what is considered acceptable in the public eye frequently relates to the size of the site and its stage in the natural successional cycle. The attractiveness of an area of open heathland grasses, such as are found on Wimbledon Common or a colonized slag heap, relates to the total size of the site—that feeling of space. But create a heath grassland block on a small inner city gap site and it would look remarkably uninteresting and would, if complemented by patches of typical urban ruderal weeds, such as Oxford ragwort, *Senecio squalidus*, be declared by many an environmental eye-sore.

This suggests that on small-scale sites emphasis should be placed on high diversity and interest, either through the creation of a microcosm of semi-discrete minihabitats (such as have naturally developed on sites like Gunnersbury Triangle and Stubbers Green) or through the intentional 'window dressing' of one or two habitat types (for example, the creation of a highly species-rich grassland). Such sites will probably require sensitive and personalized management in the early years of establishment, to ensure that the distinction between habitat types, diversity and interest are maintained and enhanced, and to ensure that the spread of ruderal weeds is inhibited. However, it should be stressed that if a local site is adopted and managed by a local school or community group, considerations of aesthetic acceptability may become less important, as the *raison d'être* of the site and, therefore, its general acceptability will rest on the emotional commitment of the adopting body to the site in question.

On larger sites, continuing the Wimbledon Common example, the aesthetic appeal is likely to rest on the creation of broad zones and blocks of different vegetation types merging one with another. Following such an approach, local diversity need not and should not be ignored.

Two examples

An early example of the ecological approach practised at a large scale was the major land reclamation programme of Stoke-on-Trent. One of the focal points of this reclamation programme was Central Forest Park designed and implemented by Land Use Consultants between 1966 and 1974. Before reclamation, this site consisted of three large unvegetated colliery shale tips and surrounding colliery tailings and slurry lagoons. During the reclamation process the site was extensively regraded to provide stable slopes, although the two major shale tips were left substantially unaltered. The major aim of the reclamation design was to create an area of informal public open space reflecting the typical vegetation which would, over time, have naturally colonized, namely low nutrient demanding grasses such as *Festuca rubra*, *Agrostis stolonifera* and *Deschampsia flexuosa* and woodland blocks of birch and oak with associated shrub species.

During the reclamation process a range of soil ameliorants was tested, including sewage sludge and inorganic fertilizers. In each case the aim was to improve the nutrient status of the substrate sufficiently to encourage grass and tree establishment, but not to the extent that aggressive nutrient-demanding species would be favoured in the long term.

In developing the seed and plant mixes it would obviously have been best to identify all the different microhabitats of the site and draw up separate planting and seeding plans for each identified habitat. However, this was not possible owing to (i) the size of the site and the need to create an overall integrated plan suitable for a 5-year planting programme; (ii) the difficulty of predicting how substrate conditions would change during the earth-moving programme, i.e. the extent to which fresh unweathered shale would be brought to the surface; (iii) the financial limitations of the DOE reclamation grant, which only provided sufficient funds to cover a single overall planting treatment. Therefore, relatively universal mixes, which included a sufficient range of species adapted to the range of expected site conditions, were chosen for both grass and trees. Today, after some 10 years of establishment, the site consists of a mosaic of open grassland communities and woodland scrub with birch dominating on the original slag heaps (Fig. 25.1).

Turning to a more recent example, the 1 ha William Curtis Ecological Park, London, represented one attempt to create a range of microhabitats

FIG. 25.1. Central Forest Park, Stoke-on-Trent, created from old colliery workings.

including a pond, open meadow and scrub and woodland blocks, on a small urban gap site. For several years before the park closed in 1986 detailed monitoring was carried out by the Ecological Parks Trust, the managing body, to assess how 'created' semi-natural communities develop. In total some 300 species of vascular plants were recorded on the site. Some were introduced, some colonized from wind-borne seed, but many appear to have been brought in with the imported London subsoil which was spread to form a growing medium over the hard core surface of this once disused car park site. At the outset this site was planted with some 1000 native tree whips and seeded at 5 g m^{-2} with a grass mix consisting of:

    25%   *Festuca rubra*
    60%   *Festuca tenuifolia/ovina*
    9%   *Agrostis tenuis*
    3%   *Agrostis stolonifera*
    3%   *Anthoxanthum odoratum*

Four years after establishment, *Festuca rubra* had come to dominate the whole park, with all the other sown grass species at most representing only a small percentage of the sward. And as the sward developed so the number of naturally-colonizing ruderal species declined (Grime 1979) as shown in Table 25.1. Of particular interest, however, has been the natural spread of legumes (see Table 25.2). After 4 years of establishment twenty-seven legume species had been recorded of which only three are known to have been purposely

TABLE 25.1. Annual abundance of ruderals at the William Curtis Ecological Park

| | Number of grid squares* | | | | | |
	1977	1978	1979	1980	1981	1982
Atriplex hastata (orache)	45	34	8	7	2	5
Atriplex hortensis (orache)	1	1	0	0	0	0
Atriplex patula (orache)	17	2	0	0	0	0
Bromus mollis (soft brome)	1	7	2	3	0	0
Chenopodium album (fat hen)	40	13	6	4	6	1
Chenopodium ficifolium (goose foot)	35	7	2	1	0	0
Chenopodium polyspermum (all-seed)	12	0	0	0	0	0
Chenopodium rubrum (red goosefoot)	34	1	1	0	0	0
Galium aparine (cleavers)	1	7	11	7	1	5
Tripleurospermum maritimum (mayweed)	39	20	38	30	23	15
Medicago lupulina (black medick)	26	33	33	36	33	32
Poa annua (annual meadow grass)	44	45	21	14	10	5
Polygonum aviculare (knot grass)	45	38	18	2	0	0
Polygonum persicaria (persicaria)	2	1	0	2	1	1
Senecio vulgaris (groundsel)	44	45	26	10	5	5
Stellaria media (chickweed)	42	33	22	7	7	4
Trifolium dubium (yellow trefoil)	0	5	4	1	0	0
Veronica persica (large field speedwell)	3	8	5	4	2	1
Total	431	300	197	128	90	74

* The park has been divided into a grid of eight-five 10 × 10 m squares and the distribution of individual plant species has been recorded in terms of presence or absence in each grid square. Of these, forty-five squares are largely covered by grassland (Tyler 1982).

introduced (*Lotus corniculatus, Onobrychis vicifolia* and *Ulex europaeus*). By comparison, a wide range of non-leguminous plants typical of grassland communities (many of which have colonized naturally) have done little more than hang on, e.g. hardheads, *Centaurea nigra*; others, after an initial increase have begun to decline, e.g. beaked hawk's beard, *Crepis taraxacifolia* and wild carrot, *Daucus carota*. This suggests that the dense unmanaged grass sward, encouraged by the nitrogen-fixing legumes, is now inhibiting the growth of other less competitive perennial herbs and threatens to oust many of the more delicate introduced species such as Jacob's ladder, *Polemonium caeruleum*. If species diversity on this site is to be maintained, management is now required to remove the dense mat of grass litter, to generally open up the sward and to prevent the domination of the site by grass and leguminous species.

However, by far the most successful feature on this site, both in terms of natural development and public use and attraction, was the small 480 m² pond. As no natural fresh-water bodies exist in the area as a seed source, some

TABLE 25.2. Annual abundance of legumes at the William Curtis Ecological Park

	Number of grid squares				
	1977	1978	1979	1980	1981
Anthyllis vulneria (kidney vetch)	0	0	0	1	0
Galega officinalis (goat's rue)	2	4	4	3	4
Lathyrus latifolius (everlasting pea)	0	6	9	9	13
Lathyrus nissolia (grass vetchling)	0	0	1	1	0
Lathyrus pratensis (meadow vetchling)	0	0	1	1	1
Lotus corniculatus (bird's foot trefoil)	0	5	5	6	8
Lotus uliginosus (marsh bird's foot trefoil)	0	0	1	0	0
Lupinus polyphyllus (American lupin)	0	6	6	5	3
Medicago lupulina (black medick)	33	46	46	45	41
Medicago sativa (lucerne)	4	32	42	50	55
Medicago × varia (lucerne hybrid)	0	0	0	1	0
Melilotus alba (white melilot)	2	9	2	7	1
Melilotus officinalis (common melilot)	23	32	8	34	12
Onobrychis vicifolia (sainfoin)	0	0	2	3	1
Sarathamnus scoparius (broom)	0	1	1	1	1
Trifolium dubium (suckling clover)	0	6	4	2	0
Trifolium hybridum (alsike clover)	2	5	2	4	4
Trifolium lappaceum (clover species)	0	1	0	0	0
Trifolium pratense (red clover)	3	24	28	39	39
Trifolium repens (white clover)	18	56	48	58	58
Ulex europaeus (gorse)	5	6	6	7	8
Vicia angustifolia (narrow-leaved vetch)	0	0	2	1	0
Vicia cracca (tufted vetch)	0	0	0	4	0
Vicia hirsuta (hairy tare)	2	4	1	3	0
Vicia sativa (common vetch)	0	6	13	24	31
Vicia tetrasperma (smooth tare)	0	1	2	4	0
Vicia villosa (bush vetch)	0	3	9	4	11

thirty-four plant species were deliberately introduced. Of these some, such as marsh marigold, *Caltha palustris*, yellow flag, *Iris pseudacorus* and broad-leaved pondweed, *Potamogeton natans*, just managed to hang on while other more aggressive species such as *Juncus effusus*, hairy willow herb, *Epilobium hirsutum*, bur-reed, *Sparganium erectum*, greater reed mace, *Typha latifolia* and greater spearwort, *Ranunculus lingua*, spread rapidly. The spread of *Ranunculus lingua* was particularly impressive. Introduced in 1978 as a small clump 30 cm in diameter, by 1983 it had spread to cover an area of some 30 m² despite large numbers of plants being removed on several occasions to stock other urban ponds. Also surprising was the survival and spread of bog bean, *Menyanthes trifoliata*, in the face of heavy trampling. From two plants in 1979 it increased to over twenty plants in 1982 and spread from its original location on the north side of the pond to other patches on the south side. What is particularly interesting about this pond is that, unlike the grassland, it would appear that trampling held in check the spread of aggressive species, a natural balance being maintained through public pressure. For example, up until 1980 water-milfoil, *Myriophyllum*, was the dominant submergent and covered all but a small tongue of water. Yet by 1981, trampling in the northern half of the pond had become so intense that *Myriophyllum* had been almost completely eliminated in this area.

Seen in summer, this site demonstrated the attractions of creating species-rich natural areas at a small scale (Fig. 25.2). However, as demonstrated by the grassland development, a watchful eye is always required to ensure that the species diversity of such sites is retained.

Management and adoption

Following the argument that there is a need for diversity and interest at the small scale and, given the relative instability of small-scale habitats, it follows that these areas require personalized care from inception. This suggests that all ecologically-based designs at this scale are best created and managed by voluntary groups or MSC teams sponsored by voluntary groups or local authorities. There are an increasing number of precedents for this, such as the work of the Rural Preservation Association Green Site Projects in Liverpool,

FIG. 25.2. William Curtis Ecological Park, London, created from an urban clearance site.

and the work of the British Trust for Conservation Volunteers in Avon and the North of England.

If these sites are to be used for informal and formal environmental education, it is also important that a permanent warden is on site, able to act as a catalyst to local involvement, and able to provide a personal interpretation of the site's interest. Unfortunately, voluntary groups are finding it increasingly difficult to find funds to employ professional staff able to act as wardens and interpreters. This suggests that there should be a marriage between public funding and community management, as has happened at Lavender Pond Ecological Area, Surrey Docks, London. This 1 ha site was designed by Land Use Consultants and forms part of the open space network of the Surrey Docks Redevelopment. It consists of a large pond, reed beds, a winter flood-meadow and a small woodland belt. Although the basic reclamation and landscape works were undertaken by landscape contractors, it was agreed from the outset that once the basic landscape contract was completed, the site would be handed over to the Ecological Parks Trust to run and manage under a 7 year licence with the London Borough of Southwark. As part of this licence it was agreed that the funds which would normally have been used by the Parks Department to run the site should, in this instance, pass to the Trust to employ a full-time warden. It was also agreed that the work of the warden would be supervised by a Management Group including representatives from the Borough, the education authority and the local community. At the end of 7 years it is hoped that a local community group will take over the full-time management of the site.

On larger sites it is probable that the management of these created landscapes will fall to the Local Authority Parks Department or equivalent. If these schemes are to succeed in the long term, however, it is essential that there is sufficient flexibility to allow countryside management techniques to be adopted. These schemes will only reach maturity if sensitive and appropriate management is applied.

CONCLUSION

Urban nature conservation may be dismissed as a popular movement. But this is precisely what it is and it is for this reason that it should *not* be dismissed. In the past conservationists have often been bad at selling their product. Now urban nature conservation has brought conservation to the modern market place. It shows us the opportunities which are already present within our old degraded urban areas to use them, at low cost, both for the maintenance of wild life and for public enjoyment, in areas which have, otherwise, often little to offer. With some manipulation the opportunities are endless. It is an idea which is gaining popular support and our fast-dwindling natural heritage will only be retained if we do all we can to encourage it.

ACKNOWLEDGMENT

I would like to thank members of LUC staff and all those at the Ecological Parks Trust for their helpful comments and guidance.

REFERENCES

Bradshaw, A.D. (1979). Derelict land—is the tidying up going too far? *The Planner*, **3**, 85–88.

Bradshaw, A.D. & Chadwick, M.J. (1980). *The Restoration of Land: the Ecology and Reclamation of Derelict and Degraded Land*. Blackwell Scientific Publications, Oxford.

The Chiswick Wildlife Group (1983). *The Gunnersbury Triangle as a Local Nature Reserve*. Unpublished.

Cole, L. (1983). Urban nature conservation. *Conservation in Perspective* (Ed. by A. Warren & F.B. Goldsmith), pp. 267–287. Wiley, Chichester.

Cole, L. & Keen, C. (1976). Dutch techniques for the establishment of natural plant communities in urban areas. *Landscape Design*, **116**, 31–34.

Corder, M. & Brooker, R. (1981). *Natural Economy: an ecological approach to planting and management techniques in urban areas*. Kirklees Metropolitan Council, Huddersfield.

Crowe, S. (1956). *Tomorrow's Landscape*. Architectural Press, London.

Davies, B.N.K. (1976). Wildlife, urbanisation and industry. *Biological Conservation*, **10**, 249–291.

Fairbrother, N. (1972). *New Lives, New Landscapes*. Penguin, Harmondsworth.

Gemmell, R.P. (1977). *Colonization of Industrial Wasteland*. Arnold, London.

Greenwood, E.F. & Gemmell, R.P. (1978). Derelict land as a habitat for rare plants: S. Lancs. (v.c. 59) and W. Lancs (v.c. 60). *Watsonia*, **12**, 33–40.

Grime, J.P. (1979). *Plant Strategies and Vegetation Processes*. Wiley, Chichester.

Jones, A.W. (1958). The flora of the City of London's bombed sites. *London Naturalist*, **37**, 189–210.

Kelcey, J.G. (1975). Industrial development and wildlife conservation. *Environmental Conservation*, **2**, 99–108.

Mabey, R. (1980). *The Common Ground*. Hutchinson, London.

Ruff, A.R. (1979). *Holland and the Ecological Landscapes*. Deanwater Press, Stockport.

Teagle, W.C. (1978). *The Endless Village*. Nature Conservancy Council, West Midlands Region.

Tregay, R. & Gustavsson, R. (1983). *Oakwood's new landscape: designing for nature in the residential environment*, Swedish University of Agricultural Sciences, Uppsala.

Tyler, J. (1982). *William Curtis Ecological Park: development of the grassland areas 1977–81*. Ecological Parks Trust, London.

26. ON THE REPAIR OF THE GREAT PARKS OF THE ENGLISH LANDSCAPE MOVEMENT

H. T. MOGGRIDGE
Colvin and Moggridge, Filkins, Lechlade, Glos. GL7 3JQ

SUMMARY

Landscape repair differs fundamentally from new design in being based on a work already invented. Discovery of the nature of a park of the English Landscape Movement demands some special techniques. Definition of the original layout must be based primarily upon site observation against which historic maps, sketches and written descriptions need to be checked; the earliest accurate maps are first edition Ordnance Surveys. Much design appears to have been worked out on site without documentation, though later sketches and written descriptions are often revealing. Unlike formal layouts, the layout plans of naturalistic landscapes do not reveal the geometry controlling their perception. Until this is understood it is impossible to prepare a useful restoration plan which correctly conserves the open spaces upon which the design depends. Interpretation of the character of original vegetation is often not revealed merely by establishing its position correctly. Lancelot Brown's handling of vegetation was masterly but his documentation negligible. In such a case a restoration plan needs to follow the evidence as exactly as possible. Some other designs have left a conflict between aesthetic intentions and the way trees mature; then judgements have to be made about how literally to put back the original plants. Analysis of eighteenth century British parks reveals the skill of their creators. As they are now senile, a programme of conservation has become urgent.

INTRODUCTION

This paper describes techniques used in preparing recommendations for the repair of three great naturalistic parks created during the English Landscape Movement. They are Rousham (6 ha) in Oxfordshire by William Kent (1738), Blenheim (1000 ha) in Oxfordshire by Vanburgh and Wise (1705–15) followed by Capability Brown (from 1760) and later by the ninth Duke of Marlborough personally (1890–1920), and Boarn Hill Cottage (3 ha) at Cadland in Hampshire by Brown (1775). It concentrates on those aspects of the working process in which repair differs from original design, in particular the struggle to understand the mind of the original designer and absorb an outlook from another era. As the remnants of many naturalistic parks still survive, these techniques may be applicable elsewhere.

Artificial lakes and rivers and the original land form, often remodelled extensively when the parks were made, appear to have survived intact in many late eighteenth century parks. Ornamental structures have fared less well. For instance, at Painshill, the lakes and land form have lain unaltered beneath encroaching woodland as it smothered the pattern of open spaces and views. But of some thirteen exquisite structures and bridges which originally ornamented these grounds, ruins of only seven have survived. The rest have disappeared so completely that even the exact position of two is uncertain.

Though restoration of architectural features is important, without their setting they are only diminished details from a total composition of solid and void created by tree masses, paths, land forms and water. Primary repair must therefore concern the restoration of plantations, open spaces over land and water, and the paths which guide how the compositions are to be perceived. Once these strategic features are re-established embellishing detail can again be given importance, whether it be architectural or horticultural detail.

DESCRIPTION OF THE ORIGINAL LAYOUT

The first task is to describe as exactly as possible the park which was made in the first place. Often this will consist of elements overlaid from different eras of origin. This is a task different in kind from solving a new problem.

Design *ab initio* cannot lead to a correct solution in the scientific sense, demonstrable as a right interpretation of the facts; rather the imagination of the designer interacts with perception of the site and client's requirements to produce a concept for a landscape which may be one of several satisfactory solutions. Discovery of the design of an historic park is more exact. In theory, it is possible to define the original layout precisely as it was formerly conceived.

The problems attending description of the original layout are those of interpretation of evidence. The most important evidence is the place itself, against which three main types of documentary evidence, maps, sketches and written descriptions, have to be checked. The key map is likely to be an early edition of the Ordnance Survey at 1:2500, a document which usually has the advantage of accuracy. The first edition of the large-scale survey, a complete copy of which is kept at the Map Room of the British Museum, usually records major trees. In the 1870s and 80s, when most first editions were published, many parks had had little added to the eighteenth century layout, so that the bones of the original design may be defined from this map better than from any other source. Earlier plans, both design drawings and surveys of parks soon after completion, tend to be inaccurate due to contemporary vagueness of surveying and drafting techniques. Where design drawings survive they may be no more than indications of what might be done; for instance at Blenheim the Vitruvius Britannicus plans of the classical layout make no distinction between actuality and proposal and L. Brown's partial design drawings are remote from the reality which he created on the ground.

None the less, L. Brown's design plan for Boarn Hill Cottage and W. Kent's design plan of Rousham were both key documents. In both cases it was possible by modern photographic techniques to obtain an O.S. plan and a copy of the original design plans at the same scale on transparent film, so that they could be compared by direct overlaying. The drawn scale on both these plans is, however, certainly wrong and neither was completely accurate. The modifications between the design plan of Rousham and the layout as carried out, reveal a subtle development towards more pictorial design during execution of the work. For instance, on the Rousham plan, the axis of the Elm Walk aligns with the centre point of the terrace in front of Praeneste Arcade; as constructed, this axis is aligned with the south end of the Arcade so that, at the point where the rising path bends out of the wood past an urn, the long view to Apollo in the woods strikes the eye, providing another in the 'series of beautiful scenes with grouped trees, glades and winding paths, interpersed with statues and temples, all inspired by Italian landscape painting' (Sherwood & Pevsner 1974).

Observation of exact ground formations may be the most valuable confirmation of the exact alignment of original features. English parks often involved huge-scale ground modelling by hand; this has usually not been tampered with subsequently, as tree cover and cost have tended to lead to inertia, preventing change (Fig. 26.1). Modern machinery, however, now

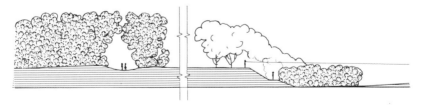

Fig. 26.1. Cross-section of the garden at Cadland. Dense woodland clothed the site before restoration. Tell-tale features of the original design perceptible in the undergrowth were the gentle banking on either side of the inland path (left) and the path half-way down the seashore bank (right); both these features corresponded with Brown's plan in the estate archive.

makes it cheap to remodel ground, a great danger to both evidence of original intentions and to some of the most subtle design effects. For instance at Blenheim the carriage drive north-west of Grand Bridge is raised about 0·5 m above natural ground levels, enough to lift the eye above the level of the bank crest to the south which would otherwise intercept a sudden long sweeping view down the lake.

During the early nineteenth century, professional sketches of views in great parks were frequently prepared both as records for the owner and for sale. These are invaluable records of the kind of pictorial effects created by the landscape, though not many are accurate dimensionally. The Bodleian Library in Oxford has a huge collection of prints and most local museums also have collected prints of their region. The families who first commissioned the park usually will have collected such documents, which can sometimes be spotted hanging unmentioned on their walls by restoration consultants with sharp eyes. The layout of part of Blenheim Park was reconstructed from the numerous prints of views of the Palace from different positions. Firstly, those all too frequent later editions which were merely copied from earlier prints, rather than from the site, were eliminated by spotting tell-tale repeated errors. Then plans were evolved from the prints by reverse alignment through constructed features. Most usefully, the existence of certain unlikely features, such as a group of Lombardy poplars outside the south-east corner of the enclosed courtyard, was confirmed by cross-checking one drawing with another (Figs 26.2 & 26.3).

Contemporary written descriptions are also invaluable. These describe what can be seen from certain places and often mention the species of plant used. For instance, the gardener who supervised the construction of Rousham, wrote a letter to his employers a decade later reminding them of its beauties with details of many features, such as 'a Large handsom Garden Seat, where you set down and have a pretty view of the Arch in Aston Field' and 'one of the noblist Green Serpentine Walks that was ever seen . . . (where) you see the deferant sorts of Flowers, peeping through the deferant sorts of Evergreens, here you think the Laurel produces a Rose, the Holly a Syringa . . .' etc. (MacClarey 1750).

GEOMETRIC IDEAS UNDERLYING THE DESIGN

Having thus built up a picture of what was originally created, a key intellectual task needs to be undertaken before any restoration proposals can be prepared. The restorer must enter the mind of the original designer to work out the geometry underlying the layout.

The geometry of formal layouts is easily grasped; it is exposed by their

FIG. 26.2. A topographical print of 1825 from the Bodleian Library Percy Manning Collection GA Oxon a79. This print is one of several showing Lombardy poplars at the south-east corner of Blenheim Palace, visible in the centre of the picture. It also shows that the planting between the Palace and the Lake was the same in 1825 as it is today, 'a subject worthy of the sublimest pencil' (Mavor).

plans. Even at Chantilly, where le Notre has offset the chateau from the main axis and introduced bent and slanting cross axes, the plan defines the way in which the landscape is perceived. The pictorial landscapes of the English Landscape Movement are not laid out in this way. The plan defines an irregular seemingly casual layout, which an ill-informed commentator might believe could be shifted about freely without change of essential character. Perhaps some second rate parks were thus sloppily laid out, producing thereby a few random pictorial effects. But the great masters, such as L. Brown, seem to have carried in their head a three-dimensional grasp of the topography and the pictures to be created from it, and remodelled the ground and placed the tree masses exactly to achieve their effects. The restoration consultant needs to define on a control drawing the viewlines which create effects which may previously only have been recorded in words. For instance of Blenheim, Mavor wrote: 'The Water, the Palace, the Gardens, the Grand Bridge, the Pillar, Woodstock, and other near and remote objects, open and shut upon the eye like enchantment; and at one point, every change of a few paces furnishes a new scene, each of which would form a subject worthy of the sublimest pencil' (Mavor 1789a). The surroundings of the Lake at Blenheim are indeed a supreme example of precision of design in the creation of such effects. Unseen behind apparently casual naturalness is a complex pattern of straight viewing lines criss-crossing in space. The same blocks of trees around the Lake and

FIG. 26.3. (a) A recent photograph of Bladon Bridge over the artificial river at the southern end of Blenheim Park. (b) A lithograph of 1842 by C. W. Radclyffe from the same spot. The lithograph reveals the intended composition which has subsequently disappeared. In this case, it is not proposed to reopen the view of Bladon Church due to the fine quality of the early twentieth century planting which obscures it; however, the foreground composition of trees framing the bridge can be re-established though this area is no longer part of the gardens.

Grand Bridge provide middle distance composition controllers for at least three different series of views. On entering the Park through the Triumphal Gate, Woodstock, the Lake, Grand Bridge, the westernmost tower of the Palace and the distant wooded ridge of High Park are revealed in a single dramatic instant, each feature perceived beyond a framework of trees; then as the Palace is approached from this position its seven towers and long views into the distance of the park beyond the lake appear in turn, in a series of picturesque views. Perpendicular to these there is another huge encompassing view outwards from the north side of the Palace. Gaps divide the groups of trees above the lakeside into an interesting variety of size, also letting the eye pass through to distant compositions in the space of Great Park. These same gaps become peepholes through which glimpses of water or a single Palace tower are seen from the circular carriage drive within the Park, as it returns to the Grand Bridge from the north-west; from this approach the sequence of scale of scenes is the reverse of that experienced from the Triumphal Gate— little pictures of a single tower or distant lake shore occasionally glimpsed at first, grow in frequency and scale as the Palace is approached, until from the Grand Bridge itself the eye can rove in all directions. At this point the trees are cut away diagonally from the axis so that the whole facade of the Palace is seen exactly framed by trees; a formal axis would of course have reduced this view to the central portico only.

The restorer must prepare a plan which defines such invisible geometry, for it is the real control of perception. Where the viewline passes there must be open space; one of the functions of the ornamental lake is to create, apparently naturally, a space across which the eye can gaze in a complex criss-cross of different angles (Fig. 26.4).

INTERPRETATION OF PLANTING

This research will by now have provided two key plans of the Park—the master plan of the layout as originally carried out (possibly an amalgam of several stages expressed on different plans) and a plan displaying the controlling geometry. The latter is in effect a plan of uninterrupted space which must be left open to preserve the pictorial design. Vegetation will be expressed in the form of masses defined in outline on the plan.

Interpretation of the nature of this vegetation is amongst the most intractable of all the problems of repair. There are very few records of historic planting schemes or how plants were used. In recent years a little information has been collected but hardly enough to give firm guidance (Galpine 1782; Taylor & Hill 1983).

Lancelot Brown left no significant record of his planting instructions. The

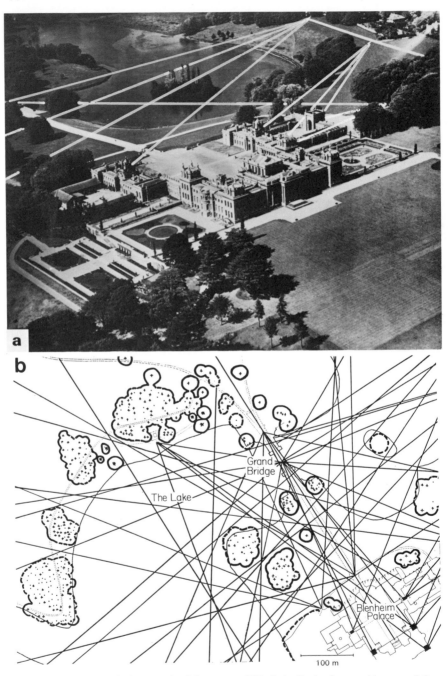

FIG. 26.4. (a) An aerial photograph of the centre of Blenheim Park of 1929 with some of the
view lines between groups of trees overlaid (reproduced with permission of Aerofilms). (b)
Part of the plan of the centre of Blenheim Park on which the invisible geometry of viewlines is
defined.

specification on the Boarn Hill Cottage plan 'Shrubs and Plants that will Grow' is hardly adequate guidance! This appears to have resulted in the belief that he was not interested in plants and their character; yet such evidence as there is does not confirm this prejudice. He was a gardener by origin. His plans almost all include provision for horticultural displays, though they are usually drawn on as blank spaces with a key of land uses: he was preparing a master plan only, like the modern land-use diagram. He seems to have worked out the enrichment of his designs entirely by oral site instructions, often given no doubt to illiterate foremen. For instance, two plans of Lowther Park provide for the following facilities on the key in 1763 'The House, The Offices, The Courts to ditto, The Porter's Lodge, The Approach to the House, The River, The Bridges, The Entrance to the Grounds, A Grotesque Building, A proper place for the Kitchen Garden'—added to which in 1771 are 'A Greenhouse, A Place for the Flower Garden, Menagerie, Sunk Fence, Flues, Hot Walls, Espaliers, Melon Ground, Stove' (Brown 1763, 1771). Mavor's description of Blenheim indicates rich planting in 1789: for instance, 'a deep belt of various trees, evergreens, and deciduous shrubs, whose mingled foliage exhibit the different gradations of tints from the most faint to the most obfuscated green' (Mavor 1789b).

Sketches show carefully sited Lombardy poplar (first introduced to England in 1758), and cedar of Lebanon, already mature by 1815, only 50 years after Brown was consulted. His bold use of simple framework species, beech at Blenheim, evergreen oak and scots pine at Cadland, shows a designer's grasp of scale without necessarily implying lack of horticultural enrichment. None the less, the vagueness of documentary evidence leaves the restorer with little guidance about the correct use of shrubs and ornamental trees.

However, Brown's landscapes show a marvellous grasp of the element of time in landscape design. He seems to have envisaged from conception the process of growth. In their old age his original unchanged plantations still display a lovely subtlety of texture and foliage colour and exemplary beauty of proportion between solid and void so that 'a thousand beauties, originating from design, appear fortuitous to the eye' (Mavor 1789c). Yet the same plantations were already effective in pictures drawn between 1800 and 1825. Thus, in the case of Brown exact replacement both in species and position of any original plantations which survive must be the appropriate policy. Within this framework enrichment has to depend on subjective guesswork.

In contrast, it can be established how the vegetation of William Kent's garden at Rousham was intended to appear. The letter, mentioned above (MacClarey 1750), records many details of shrub species. Kent's design sketches, supported by contemporary drawings of trees in London and sketches of trees from Italy, all combine to give a clear idea of the character of forest tree planting envisaged: tall bare stems with slight heads above about 4 m from ground level (Fig. 26.5). The few surviving forest trees old enough to be original confirm this for they have clean stems (Fig. 26.6). The design plan defines the planting in visual terms 'tall Forrist Trees Standing in Grass: Underwood: Evergreens Standing in Grass' (Fig. 26.7).

However, Rousham's planting contains all the technical errors typical of someone who is not a plantsman at heart; Kent was of course a painter and architect by background. Those skinny young trees have now grown into massive forest specimens (Fig. 26.8). The exact and subtle open spaces have shrunk as the trees have grown great with age. Meanwhile, the dense underplanting of solid yew has matured into magnificent dark groves of tall trees, no longer hiding one part of the garden from another and beneath which no seedling can find light to live. Recent replacement planting has been sited on the outer fringes of this expanding forest canopy; when mature these new saplings will even further encroach upon the shrunk open spaces. In short, the original planting was incapable of sustaining the original design effects.

Not even the most ruthless restorer could wish to start again at Rousham by clear felling all the vegetation. Nor would new plants be any more able to

FIG. 26.5. A pencil and chalk drawing of Villa Borghese by Richard Wilson. This drawing of about 1750 selects bare stemmed trees to depict, characteristic of drawings and paintings of Italy reaching England at the time. The figures are also diminutive in scale.

FIG. 26.6. A beech tree, over 200 years old, in the grove between the paddock and bowling green at Rousham. Probably an original tree, it is typical of the old deciduous trees in the garden in having a stem cleaned of side branches, confirming the evidence of drawings.

grow in the acute-angled triangular plan shapes of the original plan. Therefore, a third plan becomes part of the picture—the vegetation as it is today. At Rousham this reveals continuous conflict with the original plan and the geometric controlling lines. Problems at Rousham are further confounded because the views from the garden, originally towards carefully placed eye-catching elements, are onto land not part of the estate (indeed never so). Therefore, the remoter parts of the layout are all offsite on land owned by Oxford Colleges, thriving farmers and the Property Services Agency, all of whom see the land primarily as a source of money and not part of a great park. Their building programmes make it difficult to sustain the rural idyll of 'the prettiest view in the whole World' (MacClarey 1750) without substantially modifying the historic planting. Therefore, in this case planting has to be reinterpreted in order to perpetuate the original character.

DISCUSSION

Analysis leads on to a restoration master plan, which is the programme for future action. Exploration of the historic aesthetic has to be balanced with modern land uses and management techniques, which at Blenheim was achieved by the involvement of Cobham Resource Consultants, landscape management consultants. In preparing the master plan, judgements have to be

FIG. 26.7. Kent's original design plan of Rousham, including Venus Vale centre right. The plan distinguishes 'Underwood' by dots within a dotted line, 'tall Forrist Trees Standing in Grass' and 'Evergreens Standing in Grass' by round-headed and furry-topped trees on stalks. MacClary's letter (at foot of p. 445) gives the following details of the riverside grove at the centre of the plan and of Venus Vale:

made about problems such as how literal the restoration should be, how to cater for visitors and their cars, whether the number of visitors needs to be limited to conserve the landscape being visited and the time scale of repair. Many landscape projects are faced with comparable problems.

Analysis of landscapes of the eighteenth century English Landscape Movement as a preliminary to restoration, reveals the skill with which they have been designed. Unlike earlier formal parks, they are at present receiving little attention from the artistic establishment in London, perhaps partly because of the problems in grasping the basis of their design which are touched on in this paper. Yet in many parts of the country, such as Oxfordshire, NW Wales or Northumberland, the parks are so close together that they coalesce into a total landscape picture across the whole countryside. The remnants of many of them still exist, having successfully survived changes of land use or ownership and minimal maintenance for long periods. However, surviving original trees are now senile so that massive replanting has become essential to the survival of these great parks into the future. As naturalistic landscapes are one of Britain's principal contributions to European culture, their conservation is as important as it is urgent.

ACKNOWLEDGMENTS

At Blenheim: His Grace the Duke of Marlborough, his agent, Paul Hutton FRICS of Smiths Gore, Bob Smith; estate forester, Ralph Cobham of Cobham Resource Consultants, documents in the Blenheim archive (Figs 26.3b, 26.4), the Bodleian Library (Fig. 26.2).

At Rousham: Charles Cottrell-Dormer, Esq, the Historic Buildings Council for England, documents in the Rousham archive (Figs 26.7, 26.8).

At Cadland: Mr and Mrs Maldwyn Drummond, documents in the Cadland archive.

FIG. 26.7 *continued*

'Through a fine open Grove, of Oaks, Elms, Beach, Alder, Plains, and Horsechestnuts, all in Flower now, and Sixty feet high, this Grove is a Hundred feet Broad, and five Hundred feet Long, on one side runs the River, the other is Backt with all sorts of Evergreens, and Flowering shrubs entermixt,

when you come to the end of this Grove, you comes to two Garden seats, where you set down, but sure no Tongue can express the Beautyfull view that presents itself to your eye, (up Venus Vale)

you see a Fountain four Inches Diameter, playing up fifty feet High, in the middle of a Clump of Old Oaks, and backt with a Cascade, where the Water comes tumbling down from under three Arches, through Ruff Stones,

from hence you carry your Eye on, you see on each side natural Hilloks planted with Large trees of differant sort, and' etc.

Fig. 26.8. (a) Kent's design sketch of Venus Vale. This drawing shows the view up the Vale from a point at the centre of the plan and corresponding with MacClarey's description. The scale of the space and structure is falsified by the figures which are drawn in half height. The character of 'tall Forrist Trees Standing in Grass' is clearly indicated—skinny with stems cleaned of side branches. (b) Venus Vale 15 years ago. The structures and ground form survive. The framing trees and fountains have disappeared.

REFERENCES

Brown, L. (1763, 1771). *Plans of Lowther Park*. Cumbria Record Office, The Castle, Carlisle.

Galpine, J.K. Introduced by **Harvey, J. (1782).** *The Georgian Garden*. An eighteenth century nurseryman's catalogue. Dovecote Press,

MacClarey, J. (1750). (Letter of 1750 or 1760) The Rousham Archive; subsequently published in *Garden History*. Autumn 1983. W.S. Maney and Son, Leeds.

Mavor, W. (1789). *New Description of Blenheim* (1st edn). (a) p. 97 in the 10th edn; (b) p. 126; (c) p. 114. A copy is available in the Ashmolean Library, Ref. OAHS.

Sherwood, J. & Pevsner, N. (1974). *The Buildings of England: Oxfordshire*. p. 402. Penguin, Harmondsworth.

Taylor, M. & Hill, C. (1983). *Hardy Plants Introduced to Britain by 1799* (2nd edn). Cranborne Garden Centre, Cranborne, Dorset.

AUTHOR INDEX

Page numbers shown in italics refer to the lists of references.

SUBJECT INDEX